Representation Theory and Flag Varieties

CONTEMPORARY MATHEMATICS

837

Representation Theory and Flag Varieties

AMS Special Session
Representation Theory and Flag Varieties
September 9–10, 2023
University at Buffalo, Buffalo, New York

Yiqiang Li
Changlong Zhong
Editors

AMS AMERICAN MATHEMATICAL SOCIETY

2020 *Mathematics Subject Classification.* Primary 16G20, 17B10, 17B37, 14M15, 55N34, 14D24, 13D15.

Library of Congress Cataloging-in-Publication Data

Names: AMS Special Session on Representation Theory and Flag Varieties (2023 : Buffalo, N.Y.) author | Li, Yiqiang editor | Zhong, Changlong, 1983- editor

Title: Representation theory and flag varieties : AMS Special Session on Representation Theory and Flag Varieties, September 9–10, 2023, University at Buffalo, Buffalo, NY, USA / Yiqiang Li, Changlong Zhong, editors.

Description: Providence, Rhode Island : American Mathematical Society, [2026] | Series: Contemporary mathematics, 0271-4132 ; volume 837 | Includes bibliographical references.

Identifiers: LCCN 2026004996 | ISBN 9781470477271 paperback | ISBN 9781470485467 ebook

Subjects: LCSH: Representations of groups | Flag manifolds | AMS: Associative rings and algebras – Representation theory of rings and algebras – Representations of quivers and partially ordered sets | Nonassociative rings and algebras – Lie algebras and Lie superalgebras – Representations, algebraic theory (weights) | Nonassociative rings and algebras – Lie algebras and Lie superalgebras – Quantum groups (quantized enveloping algebras) and related deformations | Algebraic geometry – Special varieties – Grassmannians, Schubert varieties, flag manifolds | LCGFT: Conference papers and proceedings | Conference papers and proceedings

Classification: LCC QA174 .A674 2026

LC record available at https://lccn.loc.gov/2026004996

∞ The paper used in this book is acid-free and falls within the guidelines
established to ensure permanence and durability.
Visit the AMS home page at `https://www.ams.org/`

10 9 8 7 6 5 4 3 2 1   31 30 29 28 27 26

Contents

Preface

Representation theory explores the symmetries that arise in mathematics and the natural sciences. Its roots can be traced back to the late 19th century, with Frobenius's pioneering work on the characters of group representations. Since then, the field has evolved into a vast area of study, yielding significant results such as the Springer theory.

One of the key approaches in representation theory is geometric, which leverages the geometry and topology of the varieties involved. Among the most important classes of varieties in this context are flag varieties and Nakajima quiver varieties. This geometric perspective has elevated the theory, creating deep connections with a wide range of mathematical disciplines, including enumerative geometry, integrable systems, algebraic geometry, number theory, combinatorics, and mathematical physics.

Through these explorations, representation theory has not only advanced our understanding of symmetries but has also uncovered hidden structures within the varieties themselves, making it an essential tool for studying these objects.

As the field continues to grow rapidly, this volume aims to make a contribution to the mathematical community by providing an overview of recent developments in representation theory and the geometry of flag and quiver varieties, though it is by no means exhaustive. The contributions in this volume present fresh insights, novel methodologies, and new results, enhancing our understanding of both the theory and its geometric applications.

In addition to original research papers, the volume includes expository articles designed to offer a comprehensive introduction to the topics at hand. This material is particularly valuable for graduate students and early-career researchers, helping them engage with the current state of the field. The volume's accessible nature makes it an ideal resource for those looking to enter these dynamic areas of study and stay informed about the latest advances.

This volume is inspired by the AMS Special Session on Representation Theory and Flag Varieties, held at the University at Buffalo in September 2023. Following the conclusion of the session, it became clear that there was a strong interest in compiling the presented work into a cohesive volume. In addition to curating the contributions, the editors have also contributed their own work. We are pleased to include a paper by Mr. Drew Meyer, a participant of the session, whose contribution aligns well with the theme of the volume.

Contemporary Mathematics
Volume **837**, 2026
https://doi.org/10.1090/conm/837/16788

The Grothendieck ring of the periplectic Lie supergroup and supersymmetric functions

Mee Seong Im, Shifra Reif, and Vera Serganova

Abstract. We show that the Grothendieck ring of finite-dimensional representations of the periplectic Lie supergroup $P(n)$ is isomorphic to the ring of symmetric polynomials in $x_1^{\pm 1}, \ldots, x_n^{\pm 1}$ whose evaluation $x_1 = x_2^{-1} = t$ is independent of t.

Nous démontrons que l'anneau de Grothendieck des représentations de dimension finie du super groupe de Lie périplectique $P(n)$ est isomorphe à l'anneau des polynômes symétriques en $x_1^{\pm 1}, \ldots, x_n^{\pm 1}$ pour lesquels l'évaluation $x_1 = x_2^{-1} = t$ est indépendante de t.

Contents

1. Introduction

Supersymmetric polynomials are polynomials in $x_1, \ldots, x_m, y_1, \ldots, y_n$ which are doubly symmetric and satisfy the additional property that the evaluation $x_1 = y_1 = t$ is independent of t. Many properties of symmetric polynomials generalize to the supersymmetric case: It follows from [**Ste85**] that the ring of supersymmetric polynomials is spanned by the supercharacters of $\mathfrak{gl}(m|n)$-modules which lie in the tensor algebra $T(V)$ of the natural representation V. The super-Schur polynomials were shown in [**BR87, Ser84**] to be supercharacters of simple $\mathfrak{gl}(m|n)$-modules that can be realized using Young tableaux lying in the $(m|n)$-hook.

For semisimple Lie algebras, the ring of characters is isomorphic to the Grothendieck ring, that is, the free abelian group on the objects of the category, modulo the relation $[B] = [A] + [C]$ for every exact sequence $0 \to A \to B \to C \to 0$.

2020 *Mathematics Subject Classification.* Primary: 17B10, 17B05, 17B22, 05E05.

Key words and phrases. Duflo–Serganova functor, Grothendieck ring, periplectic Lie superalgebra, supercharacters, thin Kac modules, translation functors, parabolic induction.

This project was partially supported by ISF Grant No. 1221/17, NSF grant 1701532, and United States Military Academy's Faculty Research Fund.

The ring structure is given by tensor products in the category. For the general linear group, the Grothendieck ring $K[GL(n)]$ of the category of finite-dimensional representations is isomorphic to the ring of symmetric Laurent polynomials

$$K[GL(n)] \cong \mathbb{Z}\left[x_1^{\pm 1}, \dots, x_n^{\pm 1}\right]^{S_n}$$

(see for example, [**FH91**, Sec. 23–24]). Moreover, Schur polynomials are images of irreducible representations under this isomorphism.

This description generalizes to all semisimple complex Lie algebras. In this case, the category admits complete reducibility and the characters of irreducible representations are given explicitly by the Weyl character formula. The Grothendieck ring is then isomorphic to the ring $\mathbb{Z}[P]^W$ of W-invariants in the integral group ring $\mathbb{Z}[P]$, where P is the corresponding weight lattice and W is the Weyl group. The isomorphism is given by the character map.

For modules over Lie superalgebras, one has a parity shift functor Π which does not change the action of the Lie superalgebra. Hence, it is natural to consider one of the two quotients of the ring, either by the relation $[M] = [\Pi M]$ or $[M] = -[\Pi M]$. These quotients are isomorphic to the ring of characters and the ring of supercharacters, respectively.

Describing the ring of characters is harder in the super case. For the general linear Lie superalgebra, not all finite-dimensional modules lie in the tensor algebra $T(V)$. In general, the category of finite-dimensional modules over a simple Lie superalgebra is more subtle: it is not semisimple, characters (and supercharacters) of simple modules are not realized using Young tableaux and are hard to compute. Nevertheless, it was shown by Sergeev and Veselov in [**SV11**], that for basic classical Lie superalgebras, the ring of supercharacters can be realized using supersymmetric functions.

The theorem of A.N. Sergeev and A.P. Veselov states that the ring of supercharacters is equal to the subring of $\mathbb{Z}[P]^W$, admitting an extra condition which corresponds to the isotropic roots of the Lie superalgebra. In the case of the general linear Lie supergroup $GL(m|n)$, the ring of supercharacters is isomorphic to the ring of supersymmetric Laurent polynomials, namely, functions f in $\mathbb{Z}\left[x_1^{\pm 1}, \dots, x_m^{\pm 1}, y_1^{\pm 1}, \dots, y_n^{\pm 1}\right]^{S_m \times S_n}$ for which $f|_{x_1 = y_1 = t}$ is independent of t.

The periplectic Lie superalgebra $\mathfrak{p}(n)$ imposes further difficulties than basic classical Lie superalgebras due to the lack of an invariant bilinear form. It was only recently that its representations were understood and translation functors were computed in [**BDEA$^+$19**] and [**BDEA$^+$20**].

In this paper, we describe the ring of supercharacters for the periplectic Lie supergroup $P(n)$. We show that it is isomorphic to the ring of supersymmetric functions with a suitable supersymmetry condition, that is

THEOREM 1.1. *The ring of supercharacters of $P(n)$ is isomorphic to*

$$J_n := \{f \in \mathbb{Z}[x_1^{\pm 1}, \dots, x_n^{\pm 1}]^{S_n} : f|_{x_1 = x_2^{-1} = t} \text{ is independent of } t\}.$$

The inclusion from left to right for Theorem 1.1 is obtained by restriction to rank-one subalgebras as done in [**SV11**, Prop. 4.3]. The other inclusion is much more involved. A key tool is the ring homomorphism $ds_n : J(P(n)) \to J(P(n-2))$ induced from the Duflo–Serganova functor. We use the realization of ds_n as the evaluation map $f \mapsto f|_{x_n = x_{n-1}^{-1} = t}$, proven in [**HR18**] as well as the description of its kernel. The main step is to prove that ds_n is surjective in order to apply an inductive argument. We construct preimages of ds_n using Euler characteristics of parabolic inductions given in [**GS10**] and translation functors given in [**BDEA$^+$19**].

2. The periplectic Lie superalgebra and its representations

2.1. Lie superalgebras. Given a $\mathbb{Z}_2$-graded vector superspace $V = V_{\overline{0}} \oplus V_{\overline{1}}$, the parity of a homogeneous (even) vector $v \in V_{\overline{0}}$ is defined as $\overline{v} = \overline{0} \in \mathbb{Z}_2 = \{\overline{0}, \overline{1}\}$ while the parity of an odd vector $v \in V_{\overline{1}}$ is defined as $\overline{v} = \overline{1}$. If the parity of a vector v is $\overline{0}$ or $\overline{1}$, we say that v has degree 0 or 1, respectively. We always assume that v is homogeneous whenever the notation $\overline{v}$ appears in expressions. By Π we denote the switch of parity functor.

The Lie superalgebra $\mathfrak{g} = \mathfrak{gl}(n|n)$ is defined to be the endomorphism algebra $\mathsf{End}(V_{\overline{0}} \oplus V_{\overline{1}})$, where $\mathsf{dim} V_{\overline{0}} = \mathsf{dim} V_{\overline{1}} = n$. Then $\mathfrak{g} = \mathfrak{g}_{\overline{0}} \oplus \mathfrak{g}_{\overline{1}}$, where

$$\mathfrak{g}_{\overline{0}} = \mathsf{End}(V_{\overline{0}}) \oplus \mathsf{End}(V_{\overline{1}}) \qquad \text{and} \qquad \mathfrak{g}_{\overline{1}} = \mathsf{Hom}(V_{\overline{0}}, V_{\overline{1}}) \oplus \mathsf{Hom}(V_{\overline{1}}, V_{\overline{0}}).$$

Let $[x, y] = xy - (-1)^{\overline{x}\,\overline{y}} yx$, where x and y are homogeneous elements of $\mathfrak{g}$, and extend $[\ ,\ ]$ linearly to all of $\mathfrak{g}$. By fixing a basis of $V_{\overline{0}}$ and $V_{\overline{1}}$, the superalgebra $\mathfrak{g}$ can be realized as the set of $2n \times 2n$ matrices, where

$$\mathfrak{g}_{\overline{0}} = \left\{ \begin{pmatrix} A & 0 \\ 0 & D \end{pmatrix} : A, D \in M_{n,n}(\mathbb{C}) \right\} \text{ and } \mathfrak{g}_{\overline{1}} = \left\{ \begin{pmatrix} 0 & B \\ C & 0 \end{pmatrix} : B, C \in M_{n,n}(\mathbb{C}) \right\},$$

and $M_{n,n}(\mathbb{C})$ are $n \times n$ complex matrices. Recall that the supertrace is defined by

$$\mathfrak{str} \begin{pmatrix} A & B \\ C & D \end{pmatrix} = \mathsf{tr}(A) - \mathsf{tr}(D).$$

2.1.1. *Periplectic Lie superalgebra.* Let V be an $(n|n)$-dimensional vector superspace equipped with a nondegenerate odd symmetric form

$$\beta : V \otimes V \to \mathbb{C}, \quad \beta(v, w) = \beta(w, v), \quad \text{and} \quad \beta(v, w) = 0 \quad \text{if} \quad \overline{v} = \overline{w}. \tag{2.1}$$

Then $\mathsf{End}_{\mathbb{C}}(V)$ inherits the structure of a vector superspace from V. Let $\mathfrak{p}(n)$ be the Lie superalgebra of all $X \in \mathsf{End}_{\mathbb{C}}(V)$ preserving β, i.e., β satisfies the condition

$$\beta(Xv, w) + (-1)^{\overline{X}\overline{v}} \beta(v, Xw) = 0.$$

With respect to a fixed bases for V, the matrix of $X \in \mathfrak{p}(n)$ has the form $\left(\begin{smallmatrix} A & B \\ C & -A^t \end{smallmatrix}\right)$, where A is an $n \times n$ matrix and B, C are $n \times n$ matrices such that B is symmetric and C is antisymmetric.

For the remainder of this manuscript, we will write $\mathfrak{g} := \mathfrak{p}(n)$. Note that $\mathfrak{str} : \mathfrak{g} \to \mathbb{C}$ is a one-dimensional representation of $\mathfrak{g}$. We will also use the $\mathbb{Z}$-grading $\mathfrak{g} = \mathfrak{g}_{-1} \oplus \mathfrak{g}_0 \oplus \mathfrak{g}_1$ where $\mathfrak{g}_0 = \mathfrak{g}_{\overline{0}} \simeq \mathfrak{gl}(n)$, $\mathfrak{g}_{\pm 1}$ is the annihilator of $V_{\overline{0}}$ (respectively, $V_{\overline{1}}$). By G we denote the algebraic supergroup $P(n)$.

2.2. Root systems. For the periplectic Lie superalgebra $\mathfrak{g}$, fix the standard Cartan subalgebra $\mathfrak{h}$ of diagonal matrices in $\mathfrak{g}_0$ with its standard dual basis $\{\varepsilon_1, \ldots, \varepsilon_n\}$. So we have a root space decomposition $\mathfrak{g} = \mathfrak{h} \oplus \left(\bigoplus_{\alpha \in \Delta} \mathfrak{g}_\alpha\right)$, where $\Delta = \Delta(\mathfrak{g}_{-1}) \cup \Delta(\mathfrak{g}_0) \cup \Delta(\mathfrak{g}_1)$, and

$$\Delta(\mathfrak{g}_0) = \{\varepsilon_i - \varepsilon_j : 1 \le i \ne j \le n\},$$

$$\Delta(\mathfrak{g}_1) = \{\varepsilon_i + \varepsilon_j : 1 \le i \le j \le n\}, \quad \text{and} \quad \Delta(\mathfrak{g}_{-1}) = \{-(\varepsilon_i + \varepsilon_j) : 1 \le i < j \le n\}.$$

The set of simple roots is chosen to be

$$\Pi = \{-2\varepsilon_1, \varepsilon_1 - \varepsilon_2, \ldots, \varepsilon_{n-1} - \varepsilon_n\}.$$

This implies that $\Delta^+(\mathfrak{g}_0) = \{\varepsilon_i - \varepsilon_j : 1 \le i < j \le n\}$. Our Borel subalgebra is then $\mathfrak{b}_0 \oplus \mathfrak{g}_{-1}$, where $\mathfrak{b}_0 = \bigoplus_{\alpha\in\Delta^+(\mathfrak{g}_0)} \mathfrak{g}_\alpha$ and $\mathfrak{g}_{-1} = \bigoplus_{\alpha\in\Delta(\mathfrak{g}_{-1})} \mathfrak{g}_\alpha$.

Let

$$\mathcal{R}_0 = \prod_{\alpha\in\Delta^+(\mathfrak{g}_0)} (1-e^{-\alpha}), \quad \text{and} \quad \mathcal{R}_{-1} = \prod_{\alpha\in\Delta(\mathfrak{g}_{-1})} (1-e^{-\alpha}).$$

For $W = S_n$, the Weyl group of the even subalgebra of $\mathfrak{p}(n)$, $\mathcal{R}_{-1}$ is W-invariant and $e^\rho \mathcal{R}_0$ is W-anti-invariant, where $\rho = \sum_{1\le i\le n} (n-i)\varepsilon_i$.

2.3. Weight spaces. Let $\mathcal{C}_n$ be the category of finite-dimensional representations of $\mathfrak{g}$ and $\mathcal{F}_n$ be the category of finite-dimensional representations of G. Both are abelian symmetric rigid tensor categories. The latter category is equivalent to the category of finite-dimensional $\mathfrak{g}$-modules, integrable over the underlying algebraic group $G_0 = GL(n)$, see [**Ser11, MS17, MZ22**].

The Cartan subalgebra $\mathfrak{h}$ is abelian, so it acts locally-finitely on a finite-dimensional $\mathfrak{g}$-module M. This yields a decomposition of M as a direct sum of generalized weight spaces $M = \oplus_{\lambda\in\mathfrak{h}^*} M_\lambda$ where $M_\lambda = \{v \in M : (h-\lambda(h))^m v = 0 \text{ for all } h \in \mathfrak{h}\} \neq \{0\}$ for some sufficiently large m. If M is a G-module, it is semisimple over $\mathfrak{g}_0$ and hence $\mathfrak{h}$ acts diagonally on M.

Suppose that $M = \bigoplus_{\mu\in\mathfrak{h}^*} M_\mu$ is a weight space decomposition of a $\mathfrak{g}$-module M. Define the character of M as

$$\mathsf{ch}(M) := \sum_{\mu\in\mathfrak{h}^*} \mathsf{dim}(M_\mu) e^\mu,$$

while the supercharacter is defined as

$$\mathsf{sch}(M) := \sum_{\mu\in\mathfrak{h}^*} \mathsf{sdim}(M_\mu) e^\mu,$$

where if M_μ can be written as a $\mathbb{Z}_2$-graded vector space, i.e., $M_\mu = M_{\mu,\bar{0}} \oplus M_{\mu,\bar{1}}$, then the superdimension is the difference of usual dimensions of graded components:

$$\mathsf{sdim}(M_\mu) = \mathsf{dim}(M_{\mu,\bar{0}}) - \mathsf{dim}(M_{\mu,\bar{1}}).$$

Weights of modules in the abelian category $\mathcal{F}_n$ of finite-dimensional representations of the periplectic Lie supergroup $P(n)$ are denoted as

$$\lambda = (\lambda_1, \ldots, \lambda_n) = \sum_{1\le i\le n} \lambda_i \varepsilon_i, \qquad \lambda_i \in \mathbb{Z}.$$

Define the parity of λ as $p(\lambda) = \frac{1}{2}\sum_{1\le i\le n} \lambda_i \pmod 2$ if $\sum_{1\le i\le n} \lambda_i$ is even and $p(\lambda) = \frac{1}{2}(\sum_{1\le i\le n} \lambda_i + 1) \pmod 2$ if $\sum_{1\le i\le n} \lambda_i$ is odd. Note that the standard ordering of the weights for our choice of positive roots is $\varepsilon_i > \varepsilon_j$ when $i < j$ and $\varepsilon_i < 0$ for all i.

A weight λ is dominant if and only if $\lambda_1 \ge \lambda_2 \ge \ldots \ge \lambda_n$. We will denote Λ_n as the set of dominant integral weights. Simple objects in $\mathcal{F}_n$ (up to isomorphism and parity-switch) are parametrized by Λ_n. Denote by $L(\lambda)$ the simple module with highest weight λ with respect to the Borel subalgebra $\mathfrak{b}_0 \oplus \mathfrak{g}_{-1}$, where the parity is taken such that the parity of the highest weight vector is $p(\lambda)$.

2.4. Thin Kac modules. Let $V(\lambda)$ be a simple $\mathfrak{g}_0$-module with highest weight λ with respect to the fixed Borel $\mathfrak{b}_0$ of $\mathfrak{g}_0$. Given a dominant integral weight λ, the thin Kac module corresponding to λ is

$$\nabla(\lambda) = \prod^{n(n-1)/2} \mathsf{Ind}^{\mathfrak{g}}_{\mathfrak{g}_0 \oplus \mathfrak{g}_1} V(\lambda - \gamma) \simeq \mathsf{Coind}^{\mathfrak{g}}_{\mathfrak{g}_0 \oplus \mathfrak{g}_1} V(\lambda) \simeq \Lambda(\mathfrak{g}^*_{-1}) \otimes V(\lambda),$$

where $\gamma = \sum_{\alpha \in \Delta(\mathfrak{g}_{-1})} \alpha = \sum_{i=1}^{n} (1-n)\varepsilon_i$. We will also write $\nabla_{P(n)}(\lambda)$ to specify that the thin Kac module is a representation over the algebraic supergroup $P(n)$.

Recall that $\rho := \sum_{1 \le i \le n} (n-i)\varepsilon_i$. We have:

LEMMA 2.1. *The supercharacter of the thin Kac module $\nabla(\lambda)$ with weight λ is*

$$\mathsf{sch}\,(\nabla(\lambda)) = (-1)^{p(\lambda)} \frac{\mathcal{R}_{-1}}{e^{\rho}\mathcal{R}_0} \sum_{w \in W} (-1)^{\ell(w)} e^{w(\lambda+\rho)}. \tag{2.2}$$

The sign coefficient $(-1)^{p(\lambda)}$ appears in the supercharacter of $\nabla(\lambda)$ since we are working over vector superspaces, and $\mathsf{sch}(\nabla(\lambda))$ incorporates the superdimension, not the dimension, of $\nabla(\lambda)$.

PROOF. Since $\nabla(\lambda) \cong \bigwedge(\mathfrak{g}^*_{-1}) \otimes V(\lambda)$ as $\mathfrak{h}$-modules, the supercharacter of $\bigwedge(\mathfrak{g}^*_{-1})$ and the character of $V(\lambda)$ are

$$\mathsf{sch}\left(\bigwedge(\mathfrak{g}^*_{-1})\right) = \prod_{\beta \in \Delta(\mathfrak{g}_{-1})} (1 - e^{-\beta})$$

and

$$\mathsf{ch}\,(V(\lambda)) = (e^{\rho_0}\mathcal{R}_0)^{-1} \sum_{w \in W} (-1)^{\ell(w)} w(e^{\lambda+\rho_0}),$$

respectively, where $\rho_0 = \frac{1}{2} \sum_{\alpha \in \Delta^+(\mathfrak{g}_0)} \alpha$. Since $\rho - \rho_0 = w\rho - w\rho_0$, we obtain (2.2). $\square$

2.5. Weight diagrams. Let $\{\overline{\lambda}_1, \dots, \overline{\lambda}_n\} \subseteq \mathbb{Z}$ be such that $\lambda + \rho = \sum_{i=1}^{n} \overline{\lambda}_i \varepsilon_i$. The weight diagram d_λ corresponding to a dominant weight λ is the labeling of the line of integers by symbols $\bullet$ and $\circ$, where i has label $\bullet$ if $i \in \{\overline{\lambda}_1, \dots, \overline{\lambda}_n\}$, and $\circ$ if $i \notin \{\overline{\lambda}_1, \dots, \overline{\lambda}_n\}$. For example, d_0 is

...	$\circ$	$\bullet$	$\bullet$	$\bullet$	...	$\bullet$	$\bullet$	$\circ$	$\circ$	...
	-1	0	1	2		$n-2$	$n-1$	n	$n+1$	

since $0 + \rho = (n-1, n-2, \dots, 2, 1, 0)$ and $d_{-3\varepsilon_n - \varepsilon_{n-1}}$ is

...	$\circ$	$\bullet$	$\circ$	$\circ$	$\bullet$	$\circ$	$\bullet$	$\bullet$	...	$\bullet$	$\bullet$
	-4	-3	-2	-1	0	1	2	3		$n-2$	$n-1$

since $-3\varepsilon_n - \varepsilon_{n-1} + \rho = (n-1, n-2, \dots, 3, 2, 0, -3)$.

We say that $\lambda \le \mu$ if and only if $\lambda_i \ge \mu_i$ for each i. In terms of weight diagrams, the i-th black ball in d_λ (counted from left) lies further to the right of the i-th black ball of d_μ.

2.6. The Grothendieck ring. Let Π be the parity switching functor $- \otimes \mathbb{C}^{(0|1)}$, i.e., tensoring with the odd trivial representation $\Pi\mathbb{C}$ on the right. Let $K(P(n))$ be the Grothendieck ring of $\mathcal{F}_n$, and define

$$J(P(n)) = K(P(n))/\langle [M] + [\Pi M] : M \in \mathcal{F}_n \rangle. \tag{2.3}$$

The ring $J(P(n))$ is isomorphic to the reduced Grothendieck ring $K(P(n))/\langle [M] - [\Pi M] : M \in \mathcal{F}_n \rangle$, with the isomorphism given by $[L(\lambda)] \mapsto (-1)^{p(\lambda)}[L(\lambda)]$, where

$L(\lambda)$ is the simple module of highest weight λ. This isomorphism holds since $[M] = [\Pi M]$ in the reduced Grothendieck ring.

One may identify $J(P(n))$ as the ring of supercharacters as follows: let $\Lambda \subseteq \mathfrak{h}^*$ be the abelian group of integral weights of $\mathfrak{g}_0$ and W be the Weyl group of $\mathfrak{g}_0$. The supercharacter function $\mathsf{sch} : J(P(n)) \to \mathsf{Span}_{\mathbb{Z}}\{e^\lambda : \lambda \in \Lambda\}$, sends $[M] \mapsto \mathsf{sch}(M)$. Since isomorphic modules have the same supercharacter and $\mathsf{sch}(M) = -\mathsf{sch}(\Pi M)$, sch is well-defined. Furthermore, sch is injective since two irreducible modules have the same character if and only if they are isomorphic.

Throughout this manuscript, we write $x_i := e^{\varepsilon_i}$ for $i = 1, \dots, n$. We can then identify the ring

$$J_n := \{f \in \mathbb{Z}[x_1^{\pm 1}, \dots, x_n^{\pm 1}]^{S_n} : f|_{x_i = x_j^{-1} = t} \text{ is independent of } t \text{ for } i \neq j\}$$

with a subring of $\mathsf{Span}_{\mathbb{Z}}\{e^\lambda : \lambda \in \Lambda\}$.

The following lemma is proved using restriction to subalgebras of the form $\mathfrak{g}_{-\alpha} \oplus \mathfrak{h} \oplus \mathfrak{g}_\alpha$ where α is an odd root and 2α is not a root.

LEMMA 2.2 ([**SV11**, Prop. 4.3]). *We have $J(P(n)) \subseteq J_n$.*

2.7. Translation functors. Let $\Theta' = - \otimes V : \mathcal{F}_n \to \mathcal{F}_n$ be an endofunctor. Consider the involutive anti-automorphism $\sigma : \mathfrak{gl}(n|n) \to \mathfrak{gl}(n|n)$ defined as

$$\begin{pmatrix} A & B \\ C & D \end{pmatrix}^\sigma := \begin{pmatrix} -D^t & B^t \\ -C^t & -A^t \end{pmatrix}.$$

One can see that $\mathfrak{p}(n) = \{x \in \mathfrak{gl}(n|n) : x^\sigma = x\}$, and we set $\mathfrak{p}(n)^\perp := \{x \in \mathfrak{gl}(n|n) : x^\sigma = -x\}$.

Since $\mathfrak{p}(n)$ and $\mathfrak{p}(n)^\perp$ form maximal isotropic subspaces with respect to the form $\mathfrak{str}XY$ we obtain a nondegenerate bilinear $\mathfrak{p}(n)$-invariant pairing $\langle \cdot\,, \cdot \rangle : \mathfrak{p}(n) \otimes \mathfrak{p}(n)^\perp \to \mathbb{C}$. Choose $\mathbb{Z}$-homogeneous bases $\{X_i\}$ in $\mathfrak{p}(n)$ and $\{X^i\}$ in $\mathfrak{p}(n)^\perp$ such that $\langle X^i, X_j \rangle = \delta_{ij}$. We define the fake Casimir element as

$$\Omega := 2 \sum_{i=1}^{n} X_i \otimes X^i \in \mathfrak{p}(n) \otimes \mathfrak{p}(n)^\perp \subseteq \mathfrak{p}(n) \otimes \mathfrak{gl}(n|n).$$

Given a $\mathfrak{p}(n)$-module M, let $\Omega_M : M \otimes V \to M \otimes V$ be the linear map

$$\Omega_M(m \otimes v) = 2 \sum_{1 \leq i \leq n} (-1)^{\overline{X_i}\,\overline{m}} X_i m \otimes X^i v,$$

where $m \in M$ and $v \in V$ are homogeneous. By [**BDEA$^+$19**, Lemma 4.1.4], we see that Ω_M commutes with the action of $\mathfrak{p}(n)$ on $M \otimes V$ for any $\mathfrak{p}(n)$-module M.

For $k \in \mathbb{C}$, define a functor $\Theta'_k : \mathcal{F}_n \to \mathcal{F}_n$ as $\Theta' = - \otimes V$ followed by the projection onto the generalized k-eigenspace for Ω, i.e.,

$$\Theta'_k(M) := \bigcup_{m>0} \ker(\Omega - k\,\mathrm{Id})^m \Big|_{M \otimes V}.$$

Since $\Theta'_k = 0$ if $k \notin \mathbb{Z}$, we set $\Theta' = \bigoplus_{k \in \mathbb{Z}} \Theta'_k$, and $\Theta_k := \prod^k \Theta'_k$ when $k \in \mathbb{Z}$. The endofunctors Θ_k of $\mathcal{F}_n$ for $k \in \mathbb{Z}$ are exact.

The following is [**BDEA$^+$19**, Prop. 5.2.2].

PROPOSITION 2.3 (Translation of thin Kac modules). *Let $k \in \mathbb{Z}$. Then*

(1) $\Theta'_k\nabla(\lambda) = \nabla(\mu'')$ *if* d_λ *looks as follows at positions* $k-1, k, k+1$, *with* $d_{\mu''}$ *displayed underneath:*

$$\begin{array}{rccc} d_\lambda = & \underset{k-1}{\bullet} & \underset{k}{\bullet} & \underset{k+1}{\circ} \\ d_{\mu''} = & \underset{k-1}{\bullet} & \underset{k}{\circ} & \underset{k+1}{\bullet} \end{array}$$

(2) $\Theta'_k\nabla(\lambda) = \Pi\nabla(\mu')$ *if* d_λ *looks as follows at positions* $k-1, k, k+1$, *with* $d_{\mu'}$ *displayed underneath:*

$$\begin{array}{rccc} d_\lambda = & \underset{k-1}{\circ} & \underset{k}{\bullet} & \underset{k+1}{\bullet} \\ d_{\mu'} = & \underset{k-1}{\bullet} & \underset{k}{\circ} & \underset{k+1}{\bullet} \end{array}$$

(3) *In case* d_λ *looks locally at positions* $k-1, k, k+1$ *as below, there is a short exact sequence*

$$0 \to \nabla(\mu'') \to \Theta'_k\nabla(\lambda) \to \Pi\nabla(\mu') \to 0,$$

where $d_{\mu'}$ *and* $d_{\mu''}$ *are obtained from* d_λ *by moving one black ball away from position* k *(to position* $k-1$, *respectively,* $k+1$*) as follow:*

$$\begin{array}{rccc} d_\lambda = & \underset{k-1}{\circ} & \underset{k}{\bullet} & \underset{k+1}{\circ} \\ d_{\mu'} = & \underset{k-1}{\bullet} & \underset{k}{\circ} & \underset{k+1}{\circ} \\ d_{\mu''} = & \underset{k-1}{\circ} & \underset{k}{\circ} & \underset{k+1}{\bullet} \end{array}$$

(4) $\Theta'_k\nabla(\lambda) = 0$ *in all other cases.*

2.8. Parabolic induction. We consider the standard scalar product on $\mathfrak{h}^*$ such that $(\varepsilon_i, \varepsilon_j) = \delta_{i,j}$. Let γ be some weight. Set

$$\mathfrak{k} := \mathfrak{h} \oplus \bigoplus_{(\alpha,\gamma)=0} \mathfrak{g}_\alpha, \quad \mathfrak{r} := \bigoplus_{(\alpha,\gamma)>0} \mathfrak{g}_\alpha, \quad \mathfrak{q} := \mathfrak{k} \oplus \mathfrak{r}.$$

The subalgebra $\mathfrak{q}$ is called a parabolic subalgebra of $\mathfrak{g}$ with the Levi subalgebra $\mathfrak{k}$ and the nilpotent radical $\mathfrak{r}$. If Q is the corresponding parabolic subgroup of G, then G/Q is a generalized flag supervariety.

Let λ be a weight such that $\lambda(\mathfrak{h} \cap [\mathfrak{k}, \mathfrak{k}]) = 0$. We denote by $\mathcal{O}(-\lambda)$ the line bundle on G/Q induced by the one dimensional representation of Q with weight $-\lambda$. Set

$$\mathcal{E}(\lambda) = \sum_i (-1)^i \mathsf{sch} H^i(G/Q, \mathcal{O}(-\lambda)).$$

By definition $\mathcal{E}(\lambda)$ is in $J(P(n))$. By [**GS10**, Prop. 1],

$$\mathcal{E}(\lambda) = \frac{1}{e^\rho \mathcal{R}_0} \sum_{w\in W} (-1)^w w \left(e^{\lambda+\rho} \prod_{\alpha\in\Delta_1(\mathfrak{r})} (1 - e^{-\alpha}) \right),$$

where $\Delta_1(\mathfrak{r}) := \{\alpha \in \Delta(\mathfrak{g}_1) : (\alpha, \gamma) > 0\}$.

3. Duflo–Serganova homomorphism for $P(n)$

We show that the Duflo–Serganova functor induces a homomorphism between the rings of supercharacters, and discuss the kernel of this homomorphism (cf. [**HR18**]).

3.1. The ds_n homomorphism. Let $x \in \mathfrak{g}_1 \oplus \mathfrak{g}_{-1}$ such that $[x, x] = 0$. Then x^2 is zero in $U(\mathfrak{g})$ and for every $\mathfrak{g}$-module M, we define $M_x = \ker_M x / xM$, and similarly, $\mathfrak{g}_x = \mathfrak{g}^x / [x, \mathfrak{g}]$, where $\mathfrak{g}^x = \{g \in \mathfrak{g} : [x, g] = 0\}$. By [**DS05**, Lemma 6.2], M_x carries a natural $\mathfrak{g}_x$-module structure. Moreover, the Duflo–Serganova functor $DS_x : M \mapsto M_x$ is a symmetric monoidal functor from the category of $\mathfrak{g}$-modules to the category of $\mathfrak{g}_x$-modules. We consider a special case of the DS functor:

$$DS_n : \mathcal{F}_n \to \mathcal{F}_{n-2}, \quad DS_x(M) = M_x,$$

defined as follows. Suppose $x = x_\beta$ for $\beta \in \Delta(\mathfrak{g}_{-1})$. By [**EAS21a**, Lemma 5.1.2], $\mathfrak{g}_x \cong \mathfrak{p}(n-2)$. We embed $\mathfrak{g}_x$ in $\mathfrak{g}$ such that the Cartan subalgebra $\mathfrak{h}_x$ of $\mathfrak{g}_x$ is contained in $\{h \in \mathfrak{h} : \beta(h) = 0\}$. By [**HR18**, Sec. 3], the map

$$ds_x : J(P(n)) \to J(P(n-2)), \quad \text{where} \quad ds_x(\mathsf{sch} M) = \mathsf{sch}(DS_x(M)), \tag{3.1}$$

is well-defined and equal to $(\mathsf{sch} M)|_{\mathfrak{h}_x}$. Since $\beta = -\varepsilon_i - \varepsilon_j$ and $f \in J(P(n))$ satisfies that $f|_{x_i = x_j^{-1} = t}$ is independent of t, we get that $f|_{\mathfrak{h}_x} = f|_{\{h \in \mathfrak{h} : \varepsilon_i(h) = -\varepsilon_j(h)\}} = f|_{x_i = x_j^{-1}}$.

Let $\alpha = -\varepsilon_i - \varepsilon_j \in \Delta(\mathfrak{g}_{-1})$. We can also view

$$ds_n : J(P(n)) \to J(P(n-2)), \quad ds_n = p \circ ds_{x_\alpha}, \tag{3.2}$$

where p is a bijection $p : \{1, \ldots, n\} \setminus \{i, j\} \to \{1, \ldots, n-2\}$. Since the elements in $J(P(n))$ are S_n-invariant, ds_n is independent of i and j. Moreover, the map ds_n extends naturally to J_n by the evaluation $ds_n(f) = f|_{x_{n-1} = x_n^{-1} = t}$ (eventually we show that $J(P(n)) = J_n$ but for now $J_n \subseteq J(P(n))$).

We define

$$ds_n^{(k)} := ds_{n-2k+2} \circ \cdots \circ ds_n \quad \text{for } k \geq 1. \tag{3.3}$$

Define $ds_n^{(0)} := 0$. Note that applying $ds_n^{(k)}$ is the same as applying ds_x for x of higher rank.

3.2. The kernel of ds_n. The following proposition is a straightforward generalization of [**HR18**, Thm. 17].

PROPOSITION 3.1. *The kernel of ds_n is spanned by the supercharacters of thin Kac modules.*

PROOF. Suppose $f \in \ker ds_n$. Then f is divisible by $1 - x_{n-1}x_n$. Since f is W-invariant, f is also divisible by the product $\prod_{i<j}(1 - x_i x_j)$ and hence $f = \prod_{i<j}(1 - x_i x_j) g = \mathcal{R}_{-1} \cdot g$, where $g \in J(P(n))$ must also be W-invariant. By the theory of symmetric functions, write g as a linear combinations of Schur functions

$$g = \sum_{\lambda \in \mathfrak{h}^*}^{\text{finite}} a_\lambda e^{-\rho} \mathcal{R}_0^{-1} \sum_{w \in W} (\mathsf{sgn} w) w(e^{\lambda + \rho}), \quad a_\lambda \in \mathbb{Z},$$

cf., [**Mac15**] and [**HR18**, pages 2177–2178]. Thus f must be of the form $f = \sum_{\lambda\in\mathfrak{h}^*}^{\text{finite}} a_\lambda \nabla(\lambda)$. □

3.3. Translation functors and DS_n. We will need the following statement, see [**EAS21b**, Cor. 3.0.2].

LEMMA 3.2. *The functor DS_n commutes with translations functors Θ'_k.*

COROLLARY 3.3. *If $[M] \in \mathsf{Im}\, ds_n$, then $[\Theta_i(M)] \in \mathsf{Im}\, ds_n$ for every translation functor Θ_i.*

PROOF. Suppose that $DS_n(A) = M$ for some finite-dimensional $P(n)$-module A. By Lemma 3.2, we have $ds_n([\Theta_i(A)]) = [DS_n(\Theta_i(A))] = [\Theta_i(DS_n(A))] = [\Theta_i(M)]$. □

4. Surjectivity of the Duflo–Serganova homomorphism

As explained in Proposition 3.1, the kernel of ds_n is well-understood. To prove Theorem 1.1, we need to understand the image of ds_n as well:

THEOREM 4.1. *The map $ds_n : J(P(n)) \to J(P(n-2))$ is surjective.*

We prove Theorem 4.1 in three steps. We first show that if $[\nabla(0)]$ is in the image of $ds_n^{(k)}$ for some $k \geq 0$ (see (3.3) for the definition), then $[\nabla(\mu)]$ is also in the image of $ds_n^{(k)}$ for every μ. We then show that $[\nabla(0)]$ is in the image of ds_n by explicitly constructing its preimage. Finally, we show that the fact that $[\nabla(\mu)]$ is in the image of $ds_n^{(k)}$ for every μ implies that the map is surjective.

The three steps are given in the following propositions. First we show:

PROPOSITION 4.2. *If $[\nabla(0)] \in \mathsf{Im}\, ds_n^{(k)}$ for some $k \geq 0$, then $[\nabla(\mu)] \in \mathsf{Im}\, ds_n^{(k)}$ for all $\mu \in \Lambda_n$.*

This proposition is proved using the action of translation functors and Corollary 3.3. We make an inductive argument on the distance between the symbols • in the weight diagrams. Thus, we obtain that $[\nabla(\mu)]$ is in the image of $ds_n^{(k)}$ for all $\mu \in \Lambda_n$.

Next we prove that:

PROPOSITION 4.3. *We have $[\nabla(0)] \in \mathsf{Im}\, ds_n^{(k)}$.*

PROOF. The preimage of $[\nabla(0)]$ is constructed using parabolic induction. We choose the parabolic subalgebra $\mathfrak{q}$ associated with $\gamma = -\sum_{l=2k+1}^{n} \varepsilon_l$ and take $\lambda = a(\varepsilon_1 + \ldots + \varepsilon_{2k})$ for $a \in \mathbb{Z}$ (see Section 2.8). Define $\widetilde{ds}_n^{(k)}$ as an evaluation at $x_i = t_i$, $x_{i+k} = t_i^{-1}$ for $i = 1, \ldots, k$, which is the same as $ds_n^{(k)}$ up to a permutation. Let

$$\mathcal{E}(\lambda) = \sum_{w\in S_n} w\left(e^{\lambda} \frac{\prod_{1\leq i\leq n, 2k<j\leq n, i<j} (1 - x_i x_j)}{\prod_{1\leq i<j\leq n} (1 - x_i^{-1} x_j)} \right).$$

We need to show that $\mathcal{E}(\lambda)$ evaluates to $\mathsf{sch}\nabla(0)$ under the map $ds_n^{(k)}$.

If $w \notin S_{2k} \times S_{n-2k}$, then there exists $i \leq k$ such that $w^{-1}(\{i, i+k\}) \not\subset \{1, \ldots, 2k\}$. So for any $w \notin S_{2k} \times S_{n-2k}$, we see that

$$\widetilde{ds}_n^{(k)}\left(w\left(\prod_{1\leq i\leq n, 2k<j\leq n}(1-x_ix_j)\right)\right)=0.$$

So $\mathcal{E}(\lambda)$ under $\widetilde{ds}_n^{(k)}$ is

$$\widetilde{ds}_n^{(k)}(\mathcal{E}(\lambda)) = \sum_{w\in S_{2k}\times S_{n-2k}} \widetilde{ds}_n^{(k)}\left(w\left(e^{\lambda}\frac{\prod_{1\leq i\leq n, 2k<j\leq n, i<j}(1-x_ix_j)}{\prod_{1\leq i<j\leq n}(1-x_i^{-1}x_j)}\right)\right). \tag{4.1}$$

Since λ is $S_{2k}\times S_{n-2k}$-invariant, $\widetilde{ds}_n^{(k)}(e^{\lambda})=1$. Moreover, the expression

$$\frac{\prod_{1\leq i\leq 2k<j\leq n}(1-x_ix_j)}{\prod_{1\leq i\leq 2k<j\leq n}(1-x_i^{-1}x_j)}$$

is also $S_{2k}\times S_{n-2k}$-invariant, so

$$\widetilde{ds}_n^{(k)}\left(\frac{\prod_{1\leq i\leq 2k<j\leq n}(1-x_ix_j)}{\prod_{1\leq i\leq 2k<j\leq n}(1-x_i^{-1}x_j)}\right)=1.$$

We can now simplify (4.1) as

$$\widetilde{ds}_n^{(k)}(\mathcal{E}(\lambda)) = \sum_{w\in S_{2k}\times S_{n-2k}} \widetilde{ds}_n^{(k)}\left(w\left(\frac{\prod_{2k<i<j\leq n}(1-x_ix_j)}{\prod_{1\leq i<j\leq 2k}(1-x_i^{-1}x_j)\prod_{2k<i<j\leq n}(1-x_i^{-1}x_j)}\right)\right),$$

which can be rewritten as

$$\widetilde{ds}_n^{(k)}\left(\left(\sum_{u\in S_{2k}} u\left(\frac{1}{\prod_{1\leq i<j\leq 2k}(1-x_i^{-1}x_j)}\right)\right)\left(\sum_{v\in S_{n-2k}} v\left(\frac{\prod_{2k<i<j\leq n}(1-x_ix_j)}{\prod_{2k<i<j\leq n}(1-x_i^{-1}x_j)}\right)\right)\right).$$

By the denominator identity for $\mathfrak{sl}(2k)$ and $\mathfrak{sl}(n-2k)$, we have

$$\sum_{u\in S_{2k}} u\left(\frac{1}{\prod_{1\leq i<j\leq 2k}(1-x_i^{-1}x_j)}\right) = \sum_{v\in S_{n-2k}} v\left(\frac{1}{\prod_{2k<i<j\leq n}(1-x_i^{-1}x_j)}\right)=1.$$

Therefore, we obtain

$$\widetilde{ds}_n^{(k)}(\mathcal{E}(\lambda)) = \sum_{v\in S_{n-2k}} v\left(\frac{\prod_{2k<i<j\leq n}(1-x_ix_j)}{\prod_{2k<i<j\leq n}(1-x_i^{-1}x_j)}\right) = \prod_{2k<i<j\leq n}(1-x_ix_j),$$

and

$$ds_n^{(k)}(\mathcal{E}(\lambda)) = \prod_{1\leq i<j\leq 2k-i} (1 - x_i x_j) = \mathsf{sch}(\nabla(0)).$$

This concludes the proof. □

Finally, we show that

PROPOSITION 4.4. *If* $\mathsf{Span}\{[\nabla(\lambda)] : \lambda \in \Lambda_{n-2}\} \subseteq \mathsf{Im}\, ds_n$*, then* $J(P(n-2)) \subseteq \mathsf{Im}\, ds_n$*.*

This is proved using an inductive argument on the associated graded algebra of $J(P(n))$ with respect to the filtration

$$0 = \ker ds_n^{(0)} \subseteq \ker ds_n^{(1)} \subseteq \ker ds_n^{(2)} \subseteq \ldots \subseteq \ker ds_n^{(\lfloor \frac{n}{2} \rfloor)} = J(P(n)).$$

Consider the associated graded of $J(P(n))$ with respect to this filtration:

$$\mathsf{Gr}\ J(P(n)) = \bigoplus_{k=1}^{\lfloor \frac{n}{2} \rfloor} \frac{\ker ds_n^{(k)}}{\ker ds_n^{(k-1)}}.$$

Let $\overline{ds}_n : \mathsf{Gr}\ J(P(n)) \to \mathsf{Gr}\ J(P(n-2))$ and $\overline{ds}_{n,k} : \dfrac{\ker ds_n^{(k)}}{\ker ds_n^{(k-1)}} \hookrightarrow \dfrac{\ker ds_{n-2}^{(k-1)}}{\ker ds_{n-2}^{(k-2)}}$ be the corresponding maps.

In order to prove the proposition, it suffices to prove that the associated graded is surjective, by induction on n, i.e., if $\mathsf{Span}\{[\nabla(\lambda)]\}_{\lambda\in\Lambda_{n-2}} \subseteq \mathsf{Im}\,\overline{ds}_n$, then $J(P(n-2)) \subseteq \mathsf{Im}\,\overline{ds}_n$, using the fact that $\ker ds_{n-2k}$ is in the image of ds_{n-2k+2}.

Let $\mathsf{Gr}_k J(P(n)) := \ker ds_n^{(k)} / \ker ds_n^{(k-1)}$. If we consider a specific grade Gr_k of this filtration with $k > 1$, the map $\overline{ds}_{n,k} : \mathsf{Gr}_k J(P(n)) \to \mathsf{Gr}_{k-1} J(P(n-2))$ is injective. Compositing it with the maps $\overline{ds}_{n-2,k-1} : \mathsf{Gr}_{k-1} J(P(n-2)) \to \mathsf{Gr}_{k-2} J(P(n-4))$, etc., one obtains an injective map $\overline{ds}_{n-2k+4,2} \circ \cdots \circ \overline{ds}_{n-2,k-1} \circ \overline{ds}_{n,k} : \mathsf{Gr}_k J(P(n)) \to \mathsf{Gr}_1 J(P(n-2k+2))$. Yet $\mathsf{Gr}_1 J(P(n-2k+2)) = \ker ds_{n-2k+2}$ is in the image of ds_{n-2k+4}, due to Propositions 3.1, 4.2, 4.3, so the map $\overline{ds}_{n-2k+4,2} \circ \cdots \circ \overline{ds}_{n-2,k-1} \circ \overline{ds}_{n,k} : \mathsf{Gr}_k J(P(n)) \to \mathsf{Gr}_1 J(P(n-2k+2))$ is surjective and thus a bijection. This implies that the map $\overline{ds}_{n,k} : \mathsf{Gr}_k J(P(n)) \to \mathsf{Gr}_{k-1} J(P(n-2))$ is also a bijection, as required.

Proof of Theorem 1.1. By Lemma 2.2, we have that $J(P(n)) \subseteq J_n$. Let us show the reverse inclusion. Suppose by induction that $J(P(n-2)) = J_{n-2}$. By Theorem 4.1, the evaluation map

$$ds_n : J_n \to J(P(n-2))$$

given by $ds_n(f) = f|_{x_{n-1}=x_n^{-1}=t}$ is surjective when restricted to $J(P(n))$. Thus, every element of J_n is a sum of elements from $J(P(n))$ and $\ker ds_n$. By Proposition 3.1, $\ker ds_n \subseteq J(P(n))$ and the claim follows. □

References

[BDEA+19] M. Balagovic, Z. Daugherty, I. Entova-Aizenbud, I. Halacheva, J. Hennig, M. S. Im, G. Letzter, E. Norton, V. Serganova, and C. Stroppel, *Translation functors and decomposition numbers for the periplectic Lie superalgebra* $\mathfrak{p}(n)$, Math. Res. Lett. **26** (2019), no. 3, 643–710, DOI 10.4310/MRL.2019.v26.n3.a2. MR4028097

[BDEA+20] M. Balagović, Z. Daugherty, I. Entova-Aizenbud, I. Halacheva, J. Hennig, M. S. Im, G. Letzter, E. Norton, V. Serganova, and C. Stroppel, *The affine VW supercategory*, Selecta Math. (N.S.) **26** (2020), no. 2, Paper No. 20, 42, DOI 10.1007/s00029-020-0541-4. MR4073971

[BR87] A. Berele and A. Regev, *Hook Young diagrams with applications to combinatorics and to representations of Lie superalgebras*, Adv. in Math. **64** (1987), no. 2, 118–175, DOI 10.1016/0001-8708(87)90007-7. MR884183

[DS05] Michel Duflo and Vera Serganova, *On associated variety for Lie superalgebras*, Preprint, `arXiv:math/0507198` (2005), 1–21.

[EAS21a] Inna Entova-Aizenbud and Vera Serganova, *Deligne categories and the periplectic Lie superalgebra*, Mosc. Math. J. **21** (2021), no. 3, 507–565, DOI 10.17323/1609-4514-2021-21-3-507-565. MR4277854

[EAS21b] Inna Entova-Aizenbud and Vera Serganova, *Kac-Wakimoto conjecture for the periplectic Lie superalgebra*, J. Algebra Appl. **20** (2021), no. 1, Paper No. 2140015, 13, DOI 10.1142/S0219498821400156. MR4209969

[FH91] William Fulton and Joe Harris, *Representation theory*, Graduate Texts in Mathematics, vol. 129, Springer-Verlag, New York, 1991. A first course; Readings in Mathematics, DOI 10.1007/978-1-4612-0979-9. MR1153249

[GS10] Caroline Gruson and Vera Serganova, *Cohomology of generalized supergrassmannians and character formulae for basic classical Lie superalgebras*, Proc. Lond. Math. Soc. (3) **101** (2010), no. 3, 852–892, DOI 10.1112/plms/pdq014. MR2734963

[HR18] Crystal Hoyt and Shifra Reif, *Grothendieck rings for Lie superalgebras and the Duflo-Serganova functor*, Algebra Number Theory **12** (2018), no. 9, 2167–2184, DOI 10.2140/ant.2018.12.2167. MR3894432

[Mac15] I. G. Macdonald, *Symmetric functions and Hall polynomials*, 2nd ed., Oxford Classic Texts in the Physical Sciences, The Clarendon Press, Oxford University Press, New York, 2015. With contribution by A. V. Zelevinsky and a foreword by Richard Stanley; Reprint of the 2008 paperback edition [MR1354144]. MR3443860

[MS17] Akira Masuoka and Taiki Shibata, *Algebraic supergroups and Harish-Chandra pairs over a commutative ring*, Trans. Amer. Math. Soc. **369** (2017), no. 5, 3443–3481, DOI 10.1090/tran/6751. MR3605977

[MZ22] A. Masuoka and A. N. Zubkov, *Group superschemes*, J. Algebra **605** (2022), 89–145, DOI 10.1016/j.jalgebra.2022.04.027. MR4418962

[Ser84] A. N. Sergeev, *Tensor algebra of the identity representation as a module over the Lie superalgebras* $\mathrm{Gl}(n, m)$ *and* $Q(n)$ (Russian), Mat. Sb. (N.S.) **123(165)** (1984), no. 3, 422–430. MR735715

[Ser11] Vera Serganova, *Quasireductive supergroups*, New developments in Lie theory and its applications, Contemp. Math., vol. 544, Amer. Math. Soc., Providence, RI, 2011, pp. 141–159, DOI 10.1090/conm/544/10753. MR2849718

[Ste85] John R. Stembridge, *A characterization of supersymmetric polynomials*, J. Algebra **95** (1985), no. 2, 439–444, DOI 10.1016/0021-8693(85)90115-2. MR801279

[SV11] Alexander N. Sergeev and Alexander P. Veselov, *Grothendieck rings of basic classical Lie superalgebras*, Ann. of Math. (2) **173** (2011), no. 2, 663–703, DOI 10.4007/annals.2011.173.2.2. MR2776360

Department of Mathematical Sciences, United States Military Academy, West Point, New York 10996

Current address: Department of Mathematics, Johns Hopkins University, Baltimore, Maryland 21218

Email address: `meeseong@jhu.edu`

Department of Mathematics, Bar-Ilan University, Ramat Gan, Israel

Email address: `shifra.reif@biu.ac.il`

Department of Mathematics, University of California at Berkeley, Berkeley, California 94720

Email address: `serganov@math.berkeley.edu`

Contemporary Mathematics
Volume **837**, 2026
https://doi.org/10.1090/conm/837/16789

Divided difference operators for partial flag varieties

Julianna S. Tymoczko

ABSTRACT. Divided difference operators are well-known degree-reducing operators on the cohomology of flag varieties that are used to compute algebraic invariants of the ring (for instance, structure constants). We identify divided difference operators on the equivariant cohomology of G/P for arbitrary partial flag varieties of arbitrary Lie type, resolving ambiguity in the literature about these operators. We give explicit presentations of the equivariant cohomology of Grassmannians in classical Lie types and show how to use these divided difference operators in the *ordinary* cohomology of G/P as well.

We provide three applications. The first shows that all Schubert classes of partial flag varieties can be generated from a sequence of divided difference operators on the highest-degree Schubert class. The second is a generalization of Billey's formula for the localizations of equivariant Schubert classes of flag varieties to arbitrary partial flag varieties. The third gives a choice of Schubert polynomials for partial flag varieties as well as an explicit formula for each. We focus on the example of maximal Grassmannians, including Grassmannians of k-planes in a complex n-dimensional vector space.

CONTENTS

2020 *Mathematics Subject Classification.* Primary 14M15, 05E15, 57R91.

Key words and phrases. Divided difference operators, Grassmannians, equivariant cohomology, Schubert polynomials.

The author gratefully acknowledges helpful conversations with David Anderson, Anders Buch, Charles Cadman, Sergey Fomin, William Fulton, Mark Shimozono, Frank Sottile, and John Stembridge, as well as invaluable assistance from Rebecca Goldin and Megumi Harada, and useful feedback from the referee. The author was partially supported by NSF grants DMS–0402874, DMS–1800773, and DMS–2349088.

1. Introduction

The divided difference operators of Bernstein-Gelfand-Gelfand and Demazure [**BGG**], [**D**] are an essential tool in the Schubert calculus of flag varieties; a recent sample from diverse contexts includes [**Ma**], [**Pr**], [**LLM**], [**CLL**], [**Bi1**], [**KM**], [**F**]. This paper introduces nontrivial divided difference operators on the cohomology of *partial* flag varieties, including the classical Grassmannian of k-dimensional vector subspaces of $\mathbb{C}^n$. The divided difference operators generate all Schubert classes from the top-degree Schubert class; they can be used to give a direct proof of a positive formula for localizations of equivariant Schubert classes; and the operators suggest a natural choice of Schubert polynomials, along with a closed formula for each polynomial.

The results in this paper apply to all partial flag varieties G/P, where G is a semisimple complex linear algebraic group and P is a parabolic subgroup containing a fixed Borel subgroup B. For the sake of simplicity, this introduction presents the main results in the special case of Grassmannians of k-planes in $\mathbb{C}^n$.

The classical i^{th} divided difference operator ∂_i is a 'discrete derivative' defined on the polynomial ring $\mathbb{Z}[x_1, \ldots, x_n]$ by

$$\partial_i f = \frac{f(\cdots, x_i, x_{i+1}, \cdots) - f(\cdots, x_{i+1}, x_i, \cdots)}{x_i - x_{i+1}}.$$

The kernel of each divided difference operator contains the ideal I generated by nonconstant homogeneous symmetric polynomials. Thus divided difference operators act on the ring $\mathbb{Z}[x_1, \ldots, x_n]/I$, which is isomorphic to the cohomology ring of the full flag variety by a classical result of Borel [**Bo**].

The torus-equivariant cohomology ring of the full flag variety can be presented similarly. Denote the elementary symmetric functions in the variables $x_1, \ldots, x_n$ by $e_1(x), e_2(x), \ldots, e_n(x)$. The equivariant cohomology of the flag variety is isomorphic to $\mathbb{Z}[x_1, \ldots, x_n; y_1, \ldots, y_n]/J$ for the ideal J generated by differences $e_i(x) - e_i(y)$. Divided difference operators exist on the equivariant cohomology ring, both in the x variables and in the y variables; however only the operators in the x variables descend to ordinary cohomology, via the projection setting each y_i to zero.

The problem with extending this theory to Grassmannians is that the cohomology of Grassmannians is a submodule of the cohomology of flag variety. Divided difference operators in the x variables do not preserve this submodule. Divided difference operators in the y variables do, but they disappear in ordinary cohomology.

This note uses a different presentation of the equivariant cohomology ring, first given for G/B by Kostant-Kumar and Arabia [**KK, A**] and often called *the GKM presentation* because of Goresky-Kottwitz-MacPherson's construction of equivariant cohomology for a wide class of varieties [**GKM**].

The GKM presentation describes equivariant classes in terms of their localizations, so an equivariant class is a tuple of polynomials satisfying certain linear compatibility conditions. For example, let $G(k, n)$ be the Grassmannian of k-planes in $\mathbb{C}^n$. The torus T of $n \times n$ diagonal invertible matrices acts on $\mathbb{C}^n$ by multiplication; this induces an action on $G(k, n)$. Let $\mathcal{B}_k$ be the set of n-bit binary strings with exactly k ones and denote elements of $\mathcal{B}_k$ by $\mathbf{b} = b_1 b_2 \cdots b_n$. Then

$$H_T^*(G(k,n);\mathbb{C}) \cong \left\{ p : \mathcal{B}_k \to \mathbb{C}[t_1,\ldots,t_n] : \begin{array}{c} \text{for each } \mathbf{b}, \mathbf{b}' \in \mathcal{B}_k \text{ such that} \\ b_i = b'_i \text{ unless } i \in \{j_1, j_2\} \\ \text{the difference } p(\mathbf{b}) - p(\mathbf{b}') \\ \text{is in the ideal } (t_{j_1} - t_{j_2}) \end{array} \right\}.$$

(The variables t_i are weights of the torus T; the earlier variables x_i and y_i are Chern classes of certain line bundles on the flag variety, and correspond to torus weights if appropriately reindexed.) Section 2 treats GKM theory in detail.

The divided difference operators δ_i were observed for the Grassmannian $G(k,n)$ in Knutson-Tao's work [**KT**], for the flag variety in unpublished work of A. Knutson [**K**], and in greater generality by M. Brion [**B**]. However, these all misidentified δ_i as the Bernstein-Gelfand-Gelfand/Demazure divided difference operators. M. Brion gives a geometric description of the operators δ_i in [**B**] but while proving Properties (1)-(3) quotes proofs from [**D**] which treat the operators ∂_i. In fact, the traditional divided difference operators ∂_i of Bernstein-Gelfand-Gelfand [**BGG**] and Demazure [**D**] were extended to the localized presentation of the equivariant cohomology ring by B. Kostant and S. Kumar [**KK**] and in [**AJS**]. Arabia proved that the Kostant-Kumar and Bernstein-Gelfand-Gelfand operators agree when viewed as algebra homomorphisms [**A**, Theorem 3.3.1]. A more detailed exposition is in [**T2**]. The divided difference operators δ_i were also defined by D. Peterson in unpublished lecture notes [**P**]. Unfortunately, what is available in the literature from Peterson does not contain many proofs. Comments in the text below highlight results which Peterson also proves, using different methods.

We define the divided difference operators δ_i using a natural action of the permutation group S_n on the equivariant cohomology of the Grassmannian obtained by simultaneously permuting both the entries of $\mathcal{B}_k$ and the variables $t_1,\ldots,t_n$. This action is detailed in Section 4 and has been studied in other contexts [**A**, page 131], [**B**], [**T**]. Our construction of the divided difference operators on equivariant cohomology is combinatorial and explicit, and induces operators on the ordinary cohomology of the Grassmannian as well. Our divided difference operators δ_i satisfy three key properties:

(1) *Explicit formula (Proposition* 5.2*).* If $p \in H_T^*(G(k,n);\mathbb{C})$ the map δ_i sends

$$\delta_i p = \frac{p - s_i \cdot p}{t_i - t_{i+1}} \tag{1.1}$$

where $s_i \cdot p$ denotes the action of $s_i = (i, i+1)$ on the equivariant class p;

(2) *Compatibility with Schubert classes (Lemma* 4.4*).* The ring $H_T^*(G(k,n);\mathbb{C})$ has a natural basis of equivariant Schubert classes $p_{\mathbf{b}}$ indexed by the elements $\mathbf{b} \in \mathcal{B}_k$. The divided difference operators are compatible with equivariant Schubert classes, in the sense that

$$\delta_i p_{\mathbf{b}} = \begin{cases} p_{s_i \mathbf{b}} & \text{if } b_i < b_{i+1} \text{ as integers, and} \\ 0 & \text{otherwise;} \end{cases} \tag{1.2}$$

(3) *NilCoxeter relations (Proposition* 5.6*).* The divided difference operators satisfy the nilCoxeter relations: first, if w is a permutation with a minimal-length factorization $w = s_{i_1} s_{i_2} \cdots s_{i_k}$ then the divided difference operator δ_w is well-defined by $\delta_w = \delta_{i_1}\delta_{i_2}\cdots\delta_{i_k}$; second, we have $\delta_i^2 = 0$ for each i.

We suspect the superficial similarity between Formula (1.1) and the formula for the classical divided difference operator ∂_i in the Borel presentation caused much of the confusion in the literature between δ_i and ∂_i in the GKM presentation.

One main purpose of this paper is to provide examples of the moment graphs for orthogonal and Lagrangian Grassmannians, which complements work akin to that of Pragacz et alia, e.g. [**DP**, **Pr**, **PR**], Tamvakis et alia, e.g. [**Tam**, **BKT**], or Ikeda et alia, e.g. [**I**, **IMN**, **IN**]. Section 3.4 has these examples.

In addition, we give three applications of the divided difference operator δ_i. The first, Theorem 6.1, states that all Schubert classes in the (equivariant) cohomology of the partial flag variety can be generated by successive applications of divided difference operators to the top-degree class. (In the case of $G(k,n)$, the top-degree class is $p_{\mathbf{b}}$ for the bit-string $\mathbf{b}$ with $n-k$ zeroes followed by k ones.) This generalizes work of [**BGG**] and [**D**] and gives a new description of the ring $H_T^*(G(k,n);\mathbb{C})$.

The second extends to partial flag varieties the formula for localizations at each torus-fixed point of the equivariant Schubert classes for flag varieties due to Billey and Andersen-Jantzen-Soergel [**Bi2**], [**AJS**]. In the case of $G(k,n)$, this formula gives each polynomial $p_{\mathbf{b}}(\mathbf{b}')$. Our result is stated in Theorem 7.1. The proof given here applies to all partial flag varieties G/P and so extends the beautiful determinantal formula for type A_n given in [**LRS**] and the combinatorial formulas for maximal Grassmannians in several classical types of [**Kr**], [**IN**]. We give a direct proof, unlike the non-constructive geometric proof in [**Ku**, Corollary 11.3.14]. This section also demonstrates explicitly an isomorphism from a subring of the equivariant cohomology of flag varieties to that of Grassmannians.

Finally, Section 8 gives the third application: a collection of Schubert polynomials for all partial flag varieties that satisfy key conditions highlighted by Fomin and Kirillov [**FK**]. In the case of $G(k,n)$ these are:

(1) the Schubert polynomial $\mathfrak{S}_{\mathbf{b}}$ is a homogenous polynomial of degree equal to the minimal length of permutations w satisfying $w(\mathbf{1^k0^{n-k}}) = \mathbf{b}$;
(2) the polynomial $\mathfrak{S}_{\mathbf{b}}$ is identified from the Grassmannian Schubert class $p_{\mathbf{b}}$;
(3) the polynomial $\mathfrak{S}_{\mathbf{b}}$ is positive in the simple roots $t_i - t_{i+1}$;
(4) and the structure constants satisfy $\mathfrak{S}_{\mathbf{b}}\mathfrak{S}_{\mathbf{b}'} = \sum c_{\mathbf{b}\mathbf{b}'}^{\mathbf{c}}\mathfrak{S}_{\mathbf{c}} \mod I$, namely they are the structure constants defined by the Schubert classes in the ordinary cohomology of the Grassmannian up to the ideal I.

(Note that I is independent of k.) More unusually, we give a closed formula for calculating Schubert polynomials for partial flag varieties using Billey's formula. If w is the permutation satisfying Condition (1) for $\mathbf{b}$, the Grassmannian Schubert polynomial $\mathfrak{S}_{\mathbf{b}}$ is defined to be the average of the localizations of the equivariant Schubert class $p_{w^{-1}}$ in the full flag variety (Theorem 8.2). In the flag variety, these Schubert polynomials are the Bernstein-Gelfand-Gelfand Schubert polynomials. Unlike the double Schubert polynomials for Grassmannian permutations defined by Lascoux [**L**], our (single) Schubert polynomials can be computed from a closed formula and demonstrate invariance under the permutation subgroup $\langle s_i : i \neq k\rangle$. Like Lascoux's, no divided difference operators act on our Grassmannian Schubert polynomials, though the δ_i act on the corresponding localized Schubert classes.

1.1. Note. Since this manuscript was posted to the arXiv, considerable work has been done related to and extending the analysis of equivariant cohomology of Grassmannians of classical Lie types contained herein. A small selection of highlights include:

- an equivariant Pieri formula for isotropic Grassmannians of types B_n, C_n, and D_n due to Li and Ravikumar [**LR**];
- constructions of equivariant Giambelli formulas, due to Ikeda and Matsumura [**IM**] building from Tamvakis and Wilson [**TW**];
- other generalizations of similar divided difference operators, e.g. by Mihalcea, Naruse, and Su [**MNS**] or as described by Anderson and Fulton [**AF**, Remark, end of Chapter 16];
- analyses and explicit constructions of moment graphs for other varieties related to Schubert calculus, like the degenerate flag varieties described by Cerulli-Irelli, Feigin, and Reineke [**CIFR**];
- and a beautiful conjecture relating equivariant structure constants for Grassmannians of different classical Lie types, due to Robichaux, Yadav, and Yong [**RYY**].

1.2. Notation. More details about the general theory may be found in classical references, e.g. [**H**] or [**Sp**]. More details on the combinatorics of Weyl groups and roots may be found in [**BB**, Chapter 2].

Let G be a complex semisimple linear algebraic group and P be a parabolic subgroup in standard position, namely containing a fixed Borel B and maximal torus $T \subseteq B$. If $g \in G$ then denote the corresponding flag in G/B by $[g]$. The Weyl group corresponding to G is the quotient $N(T)/T$, where $N(T)$ denotes the normalizer of the torus. Denote the Weyl group by W. Let W_J be the subgroup of W defined by P. This means there is a subset J of the integers $\{1, 2, \ldots, \mathrm{rk}(G)\}$ so that W_J is the subgroup of W generated by the corresponding simple reflections $W_J = \langle s_i : i \in J\rangle$. The quotient W/W_J indexes the Schubert cells of G/P and, more generally, plays the same role for G/P as W does for G/B.

If $G = GL_n(\mathbb{C})$ then the subgroup B may be chosen to be the upper-triangular invertible matrices. In this case T is the subgroup of diagonal matrices, W is the permutation group S_n, and the simple reflection s_i is the transposition $(i, i+1)$.

Let $w \in W$ and choose a minimal-length factorization of w into simple reflections, say $w = s_{i_1} \cdots s_{i_k}$. Then the *length* of w is k, and is denoted $\ell(w)$. We denote the coset corresponding to $w \in W$ by $[w] \in W/W_J$. Each $[w] \in W/W_J$ has a unique minimal-length representative of $[w]$ in W, because P is parabolic. Suppose that w' is this minimal-length representative. The length of $[w]$ in W/W_J is defined to be $\ell(w')$ and is denoted $\ell_P([w])$. The minimal-length representatives of the cosets in W/W_J are often denoted W^J. The Bruhat order on W is defined by $w \geq v$ if and only if there is a reduced word $w = s_{i_1} \cdots s_{i_k}$ such that v can be written as a substring $v = s_{i_{j_1}} \cdots s_{i_{j_l}}$. The Bruhat order on W/W_J is the ordinary Bruhat order on the minimal representatives, so that if $[w], [v] \in W/W_J$ are cosets with minimal-length representatives $w', v' \in W$ then $[w] \geq_P [v]$ if and only if $w' \geq v'$.

Denote the Lie algebras of G, B, P, and T by $\mathfrak{g}$, $\mathfrak{b}$, $\mathfrak{p}$, and $\mathfrak{t}$ respectively. The root system associated to G is Δ. The roots are the eigenfunctions of the adjoint action of T on $\mathfrak{g}$. (If $t \in T$ and $X \in \mathfrak{g}$, then the adjoint action is given by $\mathrm{Ad}_t(X) = tXt^{-1}$.) Given a root α, the root space $\mathfrak{g}_\alpha \subseteq \mathfrak{g}$ is the eigenspace of α. The positive roots Δ^+ are the roots with $\mathfrak{b} = \oplus_{\alpha \in \Delta^+} \mathfrak{g}_\alpha$. The simple roots in Δ^+ are denoted $R = \{\alpha_1, \alpha_2, \ldots, \alpha_n\}$, where n is the rank of G. The roots are partially ordered by the condition $\alpha > \beta$ if and only if $\alpha - \beta$ is a sum of positive roots.

In the case of $GL_n(\mathbb{C})$, the Lie algebra $\mathfrak{g}$ consists of all $n \times n$ matrices. The roots are the formal differences $t_i - t_j$ for $i, j \in \{1, \ldots, n\}$. The positive roots are $t_i - t_j$ for $1 \leq i < j \leq n$ and the simple roots are $t_i - t_{i+1}$ for $i = 1, 2, \ldots, n-1$.

The subalgebra $\mathfrak{p}$ decomposes into a maximal semisimple part $\mathfrak{l}$ and a nilpotent part $\mathfrak{n}$ so that $\mathfrak{p} = \mathfrak{l} \oplus \mathfrak{n}$. If $\mathfrak{g}_\alpha$ is a root subspace then either $\mathfrak{g}_\alpha \subseteq \mathfrak{p}$ or not. Let $\mathfrak{n}^-$ be the span $\mathfrak{n}^- = \oplus_{\mathfrak{g}_\alpha \not\subseteq \mathfrak{p}} \mathfrak{g}_\alpha$. By construction $\mathfrak{g} = \mathfrak{p} \oplus \mathfrak{n}^-$.

If V is a T-stable vector subspace of $\mathfrak{g}$, denote the roots whose root subspaces are contained in V by Δ_V. We will be concerned with $\Delta_{\mathfrak{n}^-}$, which is called $\Delta_{G,P}$ in [**GHZ**]. Observe that $\Delta_{\mathfrak{n}^-} = -\Delta_{\mathfrak{n}}$ since $\Delta_{\mathfrak{l}} = -\Delta_{\mathfrak{l}}$ and $\Delta = \Delta_{\mathfrak{p}} \sqcup \Delta_{\mathfrak{n}^-}$.

The following lemma is included for the sake of completeness.

LEMMA 1.1.

$$\Delta_{\mathfrak{n}^-} = -\{\alpha : \alpha > \alpha_i \textit{ for at least one } i \notin J\}.$$

PROOF. The semisimple part $\mathfrak{l}$ is the Lie algebra generated by the simple root spaces $\{\mathfrak{g}_{\pm\alpha_i} : i \in J\}$ together with $\mathfrak{t}$. The root system for $\mathfrak{l}$ is consequently

$$\Delta_{\mathfrak{l}} = \left\{ \alpha \in \Delta : \alpha = \sum_{i \in J} c_i \alpha_i \text{ for some } c_i \in \mathbb{Z} \right\}.$$

Since $\mathfrak{p} \supseteq \mathfrak{b}$ and $\mathfrak{p} \supseteq \mathfrak{l}$, we know that

$$\Delta_{\mathfrak{p}} = \Delta_{\mathfrak{l}} \cup \Delta^+.$$

Since $\Delta_{\mathfrak{n}^-} = \Delta - \Delta_{\mathfrak{p}}$, we conclude

$$\Delta_{\mathfrak{n}^-} = -\{\alpha : \alpha > \alpha_i \text{ for at least one } i \notin J\}.$$

□

2. GKM theory for G/P

This section is a self-contained introduction to GKM theory, which is a combinatorial construction of equivariant cohomology. (Here, as in all parts of this paper, we use cohomology with complex coefficients.)

Let X be a compact projective algebraic variety with an algebraic action of an n-dimensional torus T. Suppose that T acts on X with finitely-many fixed points and finitely-many one-dimensional orbits. In this case, if O is a one-dimensional T-orbit in X, then the closure $\overline{O}$ is homeomorphic to $\mathbb{C}P^1$ and consists of the union of O with two distinct T-fixed points in X [**GKM**, Section (7.1)].

If in addition X satisfies a condition called *equivariant formality*, then X is called a GKM space. Equivariant formality refers to the degeneration of a certain spectral sequence; see [**GKM**, Section (1.2)] for a precise definition. Every smooth complex algebraic variety is equivariantly formal, as is any variety with no odd-dimensional ordinary cohomology [**GKM**, Theorem (14.1)]. In particular, each partial flag variety G/P is equivariantly formal.

Under these circumstances, the equivariant cohomology of X is given by a combinatorial algorithm. Denote the symmetric algebra in $\mathfrak{t}^*$ by $S(\mathfrak{t}^*)$.

PROPOSITION 2.1. *(Goresky, Kottwitz, and MacPherson) Let $v_1, \ldots, v_r$ be the T-fixed points of X. If O_{ij} is a one-dimensional T-orbit whose closure contains the two fixed points v_i and v_j, write α_{ij} for the weight of the torus action on O_{ij}. Then*

$$H_T^*(X) \cong \left\{ \begin{array}{ll} p : \{v_1, \ldots, v_r\} \to S(\mathfrak{t}^*) : & \textit{for each one-dimensional orbit } O_{ij}, \\ & \textit{the difference } p(v_i) - p(v_j) \in (\alpha_{ij}) \end{array} \right\}.$$

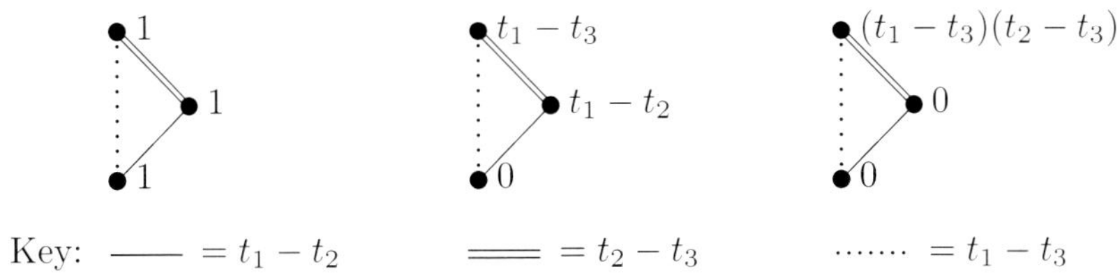

FIGURE 1. Examples of GKM classes for $\mathbb{CP}^2$

It is frequently useful to think of p as a tuple of polynomials, namely an element of $S(\mathfrak{t}^*)^r$. Then the ring structure of $H_T^*(X)$ is defined by coordinate-wise addition and multiplication. When $X = G/B$, this formulation was discovered by [**KK**] and [**A**]. The condition that the difference $p(v_i) - p(v_j) \in (\alpha_{ij})$ is often called *the GKM condition*. We will choose a basis for $\mathfrak{t}^*$ and identify $\mathbb{C}[t_1, \ldots, t_n]$ with $S(\mathfrak{t}^*)$ through the natural isomorphism.

The GKM conditions can be described combinatorially using the *moment graph* of the variety X. The vertices of the moment graph are the T-fixed points of X. There is an edge between two vertices in the graph if and only if there is a one-dimensional T-orbit whose closure contains the corresponding two fixed points. Each edge is labeled by the T-weight on its corresponding one-dimensional orbit. We assume that the graph has been directed by the choice of a suitably generic one-dimensional torus. (This graph is the 1-skeleton of X under the image of the moment map.)

With these conventions, the previous theorem can be rewritten. An equivariant class can be described as a copy of the moment graph whose vertices are labeled by elements of $S(\mathfrak{t}^*)$ such that the labels on each pair of vertices connected by an edge satisfy the GKM condition, namely $p(v_i) - p(v_j) \in (\alpha_{ij})$. Figure 1 gives examples of equivariant classes for $\mathbb{CP}^2$.

The ordinary cohomology ring is the quotient of the equivariant cohomology ring by a maximal ideal in $S(\mathfrak{t}^*)$:

$$H^*(X) = \frac{H_T^*(X)}{(\alpha_1, \ldots, \alpha_n)S(\mathfrak{t}^*)}.$$

2.1. The flow-up basis for $H_T^*(G/P)$. The vertices of the moment graph are partially ordered by the rule that $v \leq w$ if and only a sequence of directed edges leads from w to v in the graph. We write $w \overset{\alpha}{\mapsto} v$ if there is an edge directed from w to v and labeled α.

DEFINITION 2.2. A *flow-up class* $p_w \in H_T^*(X)$ corresponding to the T-fixed point w is a class that satisfies three conditions:

(1) the class p_w is supported exactly on the T-fixed points v such that $v \geq w$;
(2) the class p_w is homogeneous; and
(3) the localization $p_w(w) = \prod_{\substack{\alpha \text{ such that} \\ w \overset{\alpha}{\mapsto} v}} \alpha.$

Each of the classes in Figure 1 is in fact a flow-up class. When flow-up classes exist for each T-fixed point in X, the set $\{p_w : w \in X^T\}$ forms a basis for $H_T^*(X)$, see [**GZ**, Theorem 2.4.4] or [**T**, Proposition 2.13]. This is the case for a broad class of manifolds called Palais-Smale manifolds, which includes G/P. In $H_T^*(G/P)$ the flow-up classes are equivariant Schubert classes, corresponding to the dual to the closure of $[Bw]$ in the sense of Graham [**G**, Proposition 2.1]. (Note that this is not the Poincaré dual.) The class p_w was called ξ^w in [**KK**].

Calculations in $H_T^*(X)$ are much easier if $H_T^*(X)$ has a basis of flow-up classes. In a later section, we compute an explicit formula for the localizations $p_{[w]}([v])$ of flow-up classes in $H_T^*(G/P)$.

3. The moment graph of G/P

In this section, we characterize the moment graph of G/P with respect to the T-action in several ways. The results here generalize Carrell's work [**C**] on generalized flag varieties G/B to all partial flag varieties G/P. The moment graph is described in the following theorem, which is a variation of work of Guillemin-Holm-Zara [**GHZ**, Theorem 2.4].

THEOREM 3.1. *The moment graph of G/P consists of the following data:*

(1) *The vertices are the cosets $[w] \in W/W_J$.*
(2) *There is an edge between two fixed points $[w] \neq [v]$ if and only if $[v] = [s_\alpha w]$ for some reflection $s_\alpha \in W$ with $\alpha > 0$ and $w^{-1}(\alpha) \in \pm\Delta_{\mathfrak{n}^-}$. In this case the edge is labeled α.*
(3) *Suppose there is an edge between two vertices $[w] \neq [s_\alpha w]$ for $\alpha > 0$. If $w^{-1}(\alpha) < 0$ then the edge is directed $[w] \mapsto [s_\alpha w]$. Otherwise $[s_\alpha w] \mapsto [w]$.*

Section 3.1 describes the Schubert cells of G/P in terms of certain subsets of roots and then uses this data to describe the T-fixed points and one-dimensional T-orbits in G/P. In the algebraic context, this simplifies the proof of [**GHZ**, Theorem 2.4]. We include our proof here because it can be used to construct explicitly the unique minimal-length representative for each coset in W/W_J. We do this in Section 3.2. If w is a minimal-length representative for the coset $[w] \in W/W_J$, then we will see that the following are equal:

(1) the length of w in W;
(2) the length of $[w]$ in W/W_J;
(3) the dimension of the Schubert cell $[Bw]$ in G/P;
(4) and the number of edges directed out of $[w]$ in the moment graph of G/P.

Section 3.3 shows that the moment graph of G/P can be realized combinatorially as the quotient of the moment graph of G/B by the action of W_J. Furthermore, the subgraph of covering relations of the moment graph of G/P (namely, the edges between elements of W/W_J that differ by length one) is the subgraph induced from the moment graph of G/B by the minimal-length representatives of W/W_J. This expands on the results of [**S**, Proposition 1.1] that show the quotient W/W_J carries a poset structure induced by the Bruhat order on the minimal-length representatives. The final part of this section gives several examples.

3.1. Cells and roots.

The next lemma is the essential step in this section. It establishes a T-equivariant bijection between a certain subgroup of the unipotent group and the Schubert cell $[Bw]$ in G/P. We will use this to explicitly identify the moment graph of G/P as well as the minimal-length representatives for each $[w] \in W/W_J$.

In what follows, we will frequently treat $w \in W$ as if it were an element in G. There are two facts we will use repeatedly:

(1) The element $[wP] \in G/P$ is well-defined for all $w \in W$ because $T \subseteq P$. Indeed, for all $g \in G$ the element $[gwP]$ is well-defined.
(2) If $\mathfrak{g}_\alpha$ is a root subspace then $w\mathfrak{g}_\alpha w^{-1}$ is well-defined, and equals $\mathfrak{g}_{w(\alpha)}$. In fact let U_α be the subgroup of G with $\mathrm{Lie}(U_\alpha) = \mathfrak{g}_\alpha$. Then $wU_\alpha w^{-1}$ is well defined and equals $U_{w(\alpha)}$. In general if $u \in U_\alpha$ then wuw^{-1} is defined up to conjugation by T. (See Humphreys for more [**H**, Theorem in Section 26.3].)

LEMMA 3.2. *For each $[w] \in W/W_J$, let U_w^P be the subgroup of the unipotent group in B defined by the condition*

$$\mathrm{Lie}(U_w^P) = \mathfrak{b} \cap \left(w\mathfrak{n}^- w^{-1}\right).$$

The map $U_w^P \to [Bw]$ given by $u \mapsto [uw]$ is a T-equivariant continuous bijection, where T acts on U_w^P by conjugation and on G/P by (left) multiplication.

PROOF. First we show that U_w^P is a subgroup. The unipotent subgroup U can be factored into root subgroups $U = \prod_{\alpha\in\Delta^+} U_\alpha$ and the factorization is unique up to (arbitrary) fixed ordering of the roots [**H**, Proposition 28.1]. The definition of U_w^P implies that U_w^P is a product of root subgroups U_α, so we may write

$$U_w^P = \prod_{\substack{\alpha \in \Delta^+ s.t. \\ w^{-1}(\alpha) \in \Delta_{\mathfrak{n}^-}}} U_\alpha. \tag{3.1}$$

(Again, the ordering is fixed but arbitrary.) The Chevalley commutator relation [**Sp**, page 207] states that if $u \in U_\alpha$ and $v \in U_\beta$ then

$$uvu^{-1}v^{-1} \in \prod_{\substack{i,j \geq 1 \text{ s.t.} \\ i\alpha + j\beta \in \Delta}} U_{i\alpha+j\beta}.$$

Recall that $\mathfrak{n}^-$ is the direct sum of root spaces $\mathfrak{g}_\alpha$ such that $\alpha < -\alpha_i$ for at least one $i \notin J$. Hence if $w^{-1}(\alpha), w^{-1}(\beta) \in \Delta_{\mathfrak{n}^-}$ and $\alpha+\beta$ is a root then $w^{-1}(\alpha+\beta) \in \Delta_{\mathfrak{n}^-}$. The Chevalley commutator relation thus implies that U_w^P is a subgroup.

Let U' be the product of root subgroups U_α such that $w^{-1}(\alpha) \in \Delta_{\mathfrak{p}}$. Order the roots $\alpha \in \Delta$ so that the unipotent subgroup factors $U = U_w^P U'$. We will use this factorization to show that the map $U_w^P \to [Uw]$ is bijective. (The Schubert cell $[Bw] = [Uw]$ because $B = UT$ and $w \in W$.)

First we show surjectivity. Let $[uw]$ be any element of $[Uw] \subseteq G/P$ and write $u = u_1u_2$ where $u_1 \in U_w^P$ and $u_2 \in U'$. Then $w^{-1}u_2w \subseteq P$ by definition of U'. Hence $[uw] = [u_1w]$ has a representative in $[U_w^P w]$.

Now we show injectivity. The representative u_1 is unique because if $u_1, u' \in U_w^P$ have $[u_1w] = [u'w]$ then $w^{-1}(u_1^{-1}u')w \subseteq P$. The only element of U_w^P satisfying this condition is $u_1^{-1}u' = id$.

Finally we analyze the map $U_w^P \to [U_w^P w]$. It is T-equivariant by definition of the T-actions, and because W is a quotient of the normalizer of T. Let g_w be any element of G in the coset $w \in W$. Then both the multiplication map $U_w^P \hookrightarrow U_w^P g_w$ and the quotient map $U_w^P g_w \to [U_w^P w]$ are continuous. The map $u \mapsto [uw]$ is their composition, and hence is continuous. □

The next corollary proves that the moment graph of G/P has vertices, edges, and labels on edges described in Theorem 3.1.

COROLLARY 3.3. *The T-fixed points in G/P are the cosets $[w] \in W/W_J$. The T-stable curves in G/P are*

$$\{[U_\alpha w] \cup [s_\alpha w] : \text{ for each } [w] \in W/W_J \text{ and root subgroup } U_\alpha \subseteq U_w^P\}.$$

The weight of the T-action on the curve $[U_\alpha w] \cup [s_\alpha w]$ is α.

PROOF. Let $[w] \in W/W_J$. The only T-fixed point under the adjoint action of T on U_w^P is the identity; it maps to $[w] \in G/P$ by the T-equivariant bijection of Lemma 3.2. Similarly, the one-dimensional T-orbits of U_w^P are exactly the non-identity elements of $U_\alpha \subseteq U_w^P$, so the one-dimensional orbits in the Schubert cell $[Uw]$ are exactly the sets $[U_\alpha w] - [w]$ for each root subgroup $U_\alpha \subseteq U_w^P$. By inspection, the weight of the T-action on the orbit $[U_\alpha w] - [w]$ is α.

The closure of each one-dimensional T-orbit in any GKM space (including G/P) consists of the T-orbit together with two extra points, each of which is fixed by the T-action [**GKM**, Section (7.1)]. The set $[U_\alpha w] \subseteq [Uw]$ already contains one T-fixed point in G/P; we must find the other. The closure of $[U_\alpha w]$ in G/B is $[U_\alpha w] \cup [s_\alpha w]$ by [**C**]. The projection $G/B \to G/P$ is continuous so the closure of $[U_\alpha w]$ in G/P contains $[s_\alpha w]$. If $[s_\alpha w] \neq [w]$ in W/W_J then we conclude that $[s_\alpha w]$ must be the second T-fixed point in the one-dimensional orbit in G/P. By hypothesis the root $w^{-1}(\alpha) \in \Delta_{\mathfrak{n}^-}$ so $U_w^P \neq U_{s_\alpha w}^P$. Using this hypothesis, Lemma 3.5 will show $[w] \neq [s_\alpha w]$. □

3.2. The minimal word for $[w]$. We now describe explicitly the minimal-length word in W for each $[w] \in W/W_J$. Our main result is that the edges directed out of $[w]$ in the moment graph for G/P are labeled exactly by roots whose corresponding root subgroups are contained in U_w^P, which in turn are exactly the "inversions" of the minimal-length representative of $[w]$, as defined below. This implies that the geometric quantity $\dim[Bw]$, the combinatorial quantity $\#\{\text{edges directed out of } [w]\}$, and the algebraic quantity $\ell_P([w])$ are all equal. We also include here several technical combinatorial results about roots and Weyl group elements. We begin with the following definition.

DEFINITION 3.4. The P-inversions at w are the roots in the set

$$\mathcal{I}_w^P = \{\beta \in \Delta^+ : w^{-1}(\beta) \in \Delta_{\mathfrak{n}^-}\}.$$

(A similar set is often called $N(w)$, e.g. [**BB**, page 102].) If $P = B$ or equivalently $\Delta_{\mathfrak{n}^-} = \Delta^-$ we write $\mathcal{I}_w$ instead of $\mathcal{I}_w^B$. Lemma 3.2 showed that $\mathcal{I}_w^P$ consists of the labels on the edges directed out of w in the moment graph for G/P. The next lemma follows from the definitions.

LEMMA 3.5. *If $[w] = [v]$ in G/P then $\mathcal{I}_w^P = \mathcal{I}_v^P$.*

PROOF. Recall that if s_i is a simple reflection then $s_i(\alpha) = \alpha + c_i\alpha_i$ for some $c_i \in \mathbb{Z}$. Hence if $u \in W_J$ then $u^{-1}(\alpha) = \alpha + \sum_{i \in J} c_i\alpha_i$. Lemma 1.1 implies that $u^{-1}w^{-1}(\beta) \in \Delta_{\mathfrak{n}^-}$ if and only if $w^{-1}(\beta) \in \Delta_{\mathfrak{n}^-}$. □

The previous lemma implies that $\mathcal{I}_{[w]}^P$ is well-defined on cosets $[w] \in W/W_J$, by setting $\mathcal{I}_{[w]}^P = \mathcal{I}_w^P$ for any (and hence every) representative $w \in [w]$.

LEMMA 3.6. *For each $w \in W$, there is a unique element $v \in W$ with $\mathcal{I}_v = \mathcal{I}_w^P$.*

PROOF. We use Kostant's result [**Ko**] that the set $\{\mathcal{I}_v : v \in W\}$ is in bijection with the collection of subsets S of Δ^+ that satisfy the two conditions

(1) if $\alpha, \beta \in S$ and $\alpha + \beta \in \Delta^+$ then $\alpha + \beta \in S$; and
(2) if $\alpha, \beta \in \Delta^+ - S$ and $\alpha + \beta \in \Delta^+$ then $\alpha + \beta \notin S$.

We also use the characterization of $\Delta_{\mathfrak{n}^-}$ as the set of roots α with $\alpha < -\alpha_i$ for at least one $i \notin J$, as in Lemma 1.1.

Observe that if $\alpha, \beta \in \mathcal{I}_w^P$ and $\alpha + \beta \in \Delta$ then $w^{-1}(\alpha) + w^{-1}(\beta) < -\alpha_i$ for at least one $i \notin J$, for instance the i such that $w^{-1}(\alpha) < -\alpha_i$. Thus $\alpha + \beta \in \mathcal{I}_w^P$.

Suppose $\alpha, \beta \in \Delta^+ - \mathcal{I}_w^P$ and $\alpha + \beta \in \Delta$. Expand in terms of simple roots as $w^{-1}(\alpha) = \sum c_i \alpha_i$ and $w^{-1}(\beta) = \sum d_i \alpha_i$. We describe the c_i (respectively d_i):

- If $\alpha \in \mathcal{I}_w - \mathcal{I}_w^P$ then the c_i are nonpositive, and nonzero only when $i \in J$.
- If $\alpha \in \Delta^+ - \mathcal{I}_w$ then the c_i are positive.

Consequently the coefficients in the sum $\sum (c_i + d_i)\alpha_i$ can be negative only for $i \in J$. Thus $w^{-1}(\alpha + \beta) \in \Delta_{\mathfrak{p}}$ and so $\alpha + \beta \notin \mathcal{I}_w^P$.

The set $\mathcal{I}_w^P$ satisfies conditions (1) and (2) so there is a unique Weyl group element v with $\mathcal{I}_v = \mathcal{I}_w^P$. □

COROLLARY 3.7. *For each $w \in W$, the Weyl group element v defined by*

$$\mathcal{I}_v = \mathcal{I}_w^P$$

is the unique minimal-length element of the coset $[w] \in W/W_J$.

PROOF. It is well-known that for each $w \in W$ the length $\ell(w) = \#\mathcal{I}_w$ (see e.g. [**BB**, Proposition 4.4.4]). If $u \in [w]$ then $\mathcal{I}_u \supseteq \mathcal{I}_u^P$ with equality if and only if $u = v$. This proves the claim. □

It follows that the length $\ell_P([w])$ can also be characterized as the number of edges directed out of $[w]$ in the moment graph of G/P. If $[w]$ and $[s_\alpha w]$ are two fixed points joined by an edge then $\ell_P([w]) \neq \ell_P([s_\alpha w])$. (These are combinatorial implications of the geometric statement that G/P is Palais-Smale.)

LEMMA 3.8. *The left-multiplication action of the Weyl group W on W/W_J induces a collection of graph automorphisms on the moment graph of G/P.*

PROOF. Left multiplication by elements of the group W permutes the cosets of W/W_J, namely the vertices of the moment graph. We need only confirm that the edges of the moment graph are preserved by this action. Inside the Weyl group

$$ws_\alpha v = (ws_\alpha w^{-1})wv = s_{w(\alpha)}wv.$$

Note that $[s_{w(\alpha)}wv] = [wv]$ if and only if $v^{-1}s_\alpha v \in W_J$, equivalently $[s_\alpha v] = [v]$. This means that there is an edge between two distinct cosets $[s_\alpha v]$ and $[v]$ in the moment graph for G/P if and only if there is an edge between $[ws_\alpha v] = [s_{w(\alpha)}wv]$ and $[wv]$ in the moment graph for G/P. □

The next lemma characterizes how $[s_i w]$ and $[w]$ differ. It is a partial converse to a result of Deodhar [**D**]. Unlike in G/B, it is possible for $[s_i w] = [w]$ in G/P.

LEMMA 3.9. *Let $w \in W$ and let s_i be a simple reflection.*

(1) *If $[s_i w] = [w]$ then $s_i \mathcal{I}_w^P = \mathcal{I}_w^P$.*
(2) *If $[s_i w] >_P [w]$ then $s_i \mathcal{I}_{s_i w}^P = \mathcal{I}_w^P \cup \{-\alpha_i\}$.*

PROOF. From Lemma 3.8, we know that multiplication by each simple reflection s_i induces a graph automorphism of the moment graph of G/P. Suppose $[v] \mapsto [s_\alpha v]$ is an edge with $\alpha \neq \alpha_i$. Then this automorphism sends the edge $[v] \mapsto [s_\alpha v]$ to the edge $[s_i v] \mapsto [s_i s_\alpha v] = [s_{s_i(\alpha)}(s_i v)]$ because

- $v^{-1}(\alpha)$ and $(s_i v)^{-1}(s_i(\alpha))$ have the same sign, and
- $s_i(\alpha)$ is positive as long as $\alpha \neq \alpha_i$.

If $[s_i w] = [w]$ then $\mathcal{I}^P_{s_i w} = \mathcal{I}^P_w$ by Lemma 3.5. Hence s_i permutes the edges directed out of $[w]$, namely $s_i \mathcal{I}^P_w = \mathcal{I}^P_w$.

If $[s_i w] >_P [w]$ then α_i labels an edge directed from $[s_i w]$ to $[w]$ in the moment graph of G/P. Hence $\alpha_i \in \mathcal{I}^P_{s_i w}$ and $\alpha_i \notin \mathcal{I}^P_w$. Let $\beta \neq \alpha_i$ be a positive root. Note that $s_i(\beta) \in \mathcal{I}^P_w$ if and only if $s_i(\beta)$ is positive and $\beta \in \mathcal{I}^P_{s_i w}$. The simple reflection s_i permutes the positive roots not equal to α_i. We conclude $\mathcal{I}^P_{s_i w} = s_i \mathcal{I}^P_w \cup \{\alpha_i\}$. □

COROLLARY 3.10. *If s_i is a simple reflection in W and $[w]$ is a coset in W/W_J then $|\ell_P([s_i w]) - \ell_P([w])| \leq 1$.*

COROLLARY 3.11. *If $w \in W^J$ and s_i is an arbitrary simple reflection then either $[s_i w] = [w]$ in W/W_J or $s_i w \in W^J$.*

PROOF. We know $\mathcal{I}^P_w = \mathcal{I}_w$ since w is the minimal-length representative of $[w]$.

If $[s_i w] >_P [w]$ then $\alpha_i \notin \mathcal{I}^P_w$ and $\alpha_i \in \mathcal{I}^P_{s_i w}$. Hence $\alpha_i \in \mathcal{I}_{s_i w}$ so $s_i w > w$. It follows that $\ell(s_i w) = \ell(w) + 1$. Since $\ell_P([s_i w]) = \ell_P([w]) + 1$ and $\ell_P([w]) = \ell(w)$, we conclude $\ell(s_i w) = \ell_P([s_i w])$. This means that $\mathcal{I}_{s_i w} = \mathcal{I}^P_{s_i w}$ and $s_i w \in W^J$.

Now suppose $[s_i w] <_P [w]$. Let u be the minimal-length representative of $[s_i w]$. Since $[s_i u] = [w]$, the previous case proved $s_i u$ is a minimal-length representative of $[w]$. Each coset of W/W_J has a unique minimal-length element so $s_i u = w$. □

3.3. The moment graph of G/P as a quotient graph and as a subgraph. The Bruhat order is traditionally described by the graph of its covering relations, namely the set of Weyl group elements together with edges between v and w if $v > w$ and v and w differ by length one. The moment graph of G/B contains the Bruhat order as a subgraph but also includes additional edges corresponding to reflections that change length by more than one. The next result generalizes this statement to G/P. It is paired with a dual result. The net outcome is that the moment graph of G/P can be constructed combinatorially from that of G/B either as a subgraph (with additional edges) or as a quotient graph. The quotient W/W_J is endowed with a poset structure inherited from the Bruhat order on the minimal-length representatives of W/W_J. This is proven in [**S**, Proposition 1.1]; we include a proof here because of its brevity.

Recall that if (V, E) is a graph and $V_1, V_2, \ldots, V_k$ partitions V into k disjoint parts, then the *quotient graph* has vertex set $\{1, 2, \ldots, k\}$ and has edge (i, j) if and only if for some $u \in V_i$ and $v \in V_j$ there is an edge $(u, v) \in E$ [**MS**, 59.4.2].

THEOREM 3.12.

(1) *The moment graph of G/P is the quotient graph of the moment graph of G/B using the partition of W into cosets W/W_J.*
(2) *The subgraph of the moment graph of G/B induced by the Weyl group elements W^J is a subgraph of the moment graph of G/P that contains all the covering relations in the moment graph of G/P.*

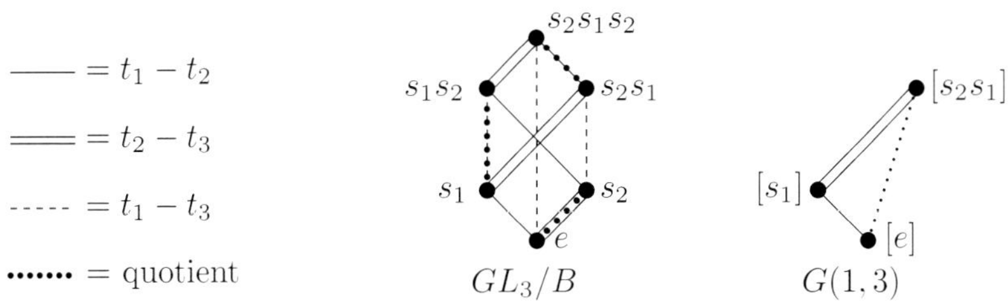

FIGURE 2. The moment graph for GL_3/B and for $G(1,3)$

PROOF. By Theorem 3.1, the vertices of the moment graph of G/P are the cosets W/W_J. There is an edge between two vertices $[w], [v] \in W/W_J$ if and only if $[v] = [s_\alpha w]$ for some α; equivalently, if and only if there is an edge between two representatives of the cosets in the moment graph of G/B. By definition, the moment graph of G/P is the quotient of the moment graph of G/B by W_J.

We now prove the second part. Suppose w and $s_\alpha w$ are both in W^J and that there is an edge between w and $s_\alpha w$ in the moment graph for G/B. Then $[w] \neq [s_\alpha w]$ because W^J is a set of distinct coset representatives for W/W_J, so there is an edge between $[w]$ and $[s_\alpha w]$ in the moment graph of G/P.

Conversely, suppose $w, v \in W^J$ differ in length by one, and that there is an edge from $[w]$ to $[v]$ in the moment graph of G/P, say labeled α. This means $\alpha \in \mathcal{I}_w^P$. Since $w \in W^J$ we know that $\mathcal{I}_w = \mathcal{I}_w^P$ and so $s_\alpha w < w$ in the Bruhat order. Thus $\ell(s_\alpha w) \leq \ell(v)$. It follows that $s_\alpha w = v$ since $[s_\alpha w] = [v]$ and v has minimum length in the coset $[v]$. Hence the edge from w to v is in the moment graph of G/B. □

REMARK 3.13. The covering relations determine the entire moment graph of G/P. In fact, any edge $[w] \overset{\alpha}{\mapsto} [s_\alpha w]$ is determined by the edges $[v] \mapsto [s_i v]$ for simple reflections s_i together with transitivity of left multiplication. To see this, denote an undirected edge in the moment graph by $[w] \leftrightarrow [v]$, including the self-edges like $[s_i w] = [w]$. If s_α is a reflection and $s_\alpha = s_{i_1} s_{i_2} \cdots$ is any minimal-length factorization, then the (undirected) path $[w] \leftrightarrow [s_{i_1} w] \leftrightarrow [s_{i_2} s_{i_1} w] \leftrightarrow \cdots \leftrightarrow [s_\alpha w]$ determines the edge $[w] \leftrightarrow [s_\alpha w]$ in the moment graph of G/P.

For instance, Figure 2 shows the moment graph for the flag variety GL_3/B, namely full flags in $\mathbb{C}^3$, together with the moment graph for the Grassmannian of lines in $\mathbb{C}^3$, namely $\mathbb{P}^1$. If the graph for $G(1,3)$ is constructed as a quotient graph, then the edges marked by large dots are collapsed in the moment graph of GL_3/B. Note that the edges between cosets are well-defined. If the graph for $G(1,3)$ is constructed as a subgraph, then it is the graph induced by the vertices e, s_1, and $s_2 s_1$ from the moment graph of GL_3/B. Any vertex that is not on an edge $t_i - t_{i+1}$ can be considered to have a self-edge labeled $t_i - t_{i+1}$. We can then reconstruct the rest of the edges by transitivity.

3.4. Examples. We describe moment graphs for the Grassmannian of k-planes in $\mathbb{C}^n$ as well as the isotropic Grassmannians of classical types.

3.4.1. *Type A_n maximal Grassmannians.* It is common to denote the vertices of the moment graph for $G(k,n)$ as n-bit binary strings with exactly k ones [**KT**]. The permutation $w \in S_n$ acts on the n-bit binary string $b_1 \cdots b_n$ by the formula $w(b_1 \cdots b_n) = b_{w(1)} \cdots b_{w(n)}$. If $\mathbf{b}$ and $\mathbf{c}$ are distinct bit strings with $(ij)\mathbf{b} = \mathbf{c}$ then the vertices $\mathbf{b}$ and $\mathbf{c}$ are joined by an edge labeled $t_i - t_j$. If $j > i$ the edge is directed from the bit string whose j^{th} bit is one to the bit string whose j^{th} bit is zero, for instance $1010 \mapsto 1100$ in $G(2,4)$. To recover the permutation corresponding to a bit string $\mathbf{b}$, choose a minimal-length directed path from $\mathbf{b}$ to 1^k0^{n-k} whose edges are labeled by simple roots $t_i - t_{i+1}$, for instance

$$\mathbf{b} \overset{t_{i_m}-t_{i_m+1}}{\mapsto} (i_m, i_m+1)\mathbf{b} \overset{t_{i_{m-1}}-t_{i_{m-1}+1}}{\mapsto} \cdots \overset{t_{i_1}-t_{i_1+1}}{\mapsto} 1^k0^{n-k}.$$

The minimal-length permutation for $\mathbf{b}$ is $(i_m, i_m+1)(i_{m-1}, i_{m-1}+1)\cdots(i_1, i_1+1)$. Figure 2 gave the moment graph of $G(1,3) = \mathbb{P}^1$. The moment graph for $G(2,4)$ is an octahedron and can be found in [**KT**, Figure 7] or in [**T3**, Figure 11].

3.4.2. *Type C_n: isotropy with respect to a skew-symmetric form.* The general (maximal) Grassmannian is the collection of k-planes in $\mathbb{C}^N$ that are isotropic with respect to some linear form, namely planes V that satisfy the condition $\langle v_1, v_2\rangle = 0$ for all $v_1, v_2 \in V$. Their moment graphs are similar to that of the Grassmannian $G(k,n)$. We describe each Lie type in turn.

First we describe the isotropic k-planes in $\mathbb{C}^{2n}$ that satisfy a skew-symmetric form, namely maximal Grassmannians of type C_n. Define a skew-symmetric form on the basis vectors e_i by

- $\langle e_i, e_{2n+1-i}\rangle = 1$ if $i < 2n+1-i$ and $\langle e_i, e_{2n+1-i}\rangle = -1$ if $i > 2n+1-i$;
- $\langle e_i, e_j\rangle = 0$ otherwise.

The vertices of the moment graph for the corresponding isotropic Grassmannian are the $2n$-bit binary strings $b_1\cdots b_{2n}$ with exactly k ones such that $b_ib_{2n+1-i} = 0$ for all i. The whole permutation group S_n does not act on this set of bit strings, but the subgroup generated by the following permutations does:

Type I: $(ij)(2n+1-i, 2n+1-j)$ for all $i < j \leq n$,
Type II: $(i, 2n+1-j)(j, 2n+1-i)$ for all $i < j \leq n$, and
Type III: $(i, 2n+1-i)$ for all $i \leq n$.

Let s be one of the permutations listed above. If $\mathbf{b}$ and $\mathbf{c}$ are two different bit strings such that $s\mathbf{b} = \mathbf{c}$ then $\mathbf{b}$ and $\mathbf{c}$ are joined by an edge. This edge is labeled $t_i - t_j$ if s is Type I, $t_i + t_j$ if s is Type II, and $2t_i$ if s is Type III. The directions are induced from the directions in the moment graph of the ordinary Grassmannian, in the following sense. If s is one of the permutations listed above then $s\mathbf{b} \geq b$ if the relevant condition below holds.

s **is Type I:** $b_i \leq b_j$ and $b_{2n+1-j} \leq b_{2n+1-i}$;
s **is Type II:** $b_i \leq b_{2n+1-j}$ and $b_j \leq b_{2n+1-i}$; and
s **is Type III:** $b_i \leq b_{2n+1-i}$.

If $s\mathbf{b} \neq \mathbf{b}$ then at least one of the inequalities in this list will be a strict equality, and the corresponding edge will be directed $s\mathbf{b} \mapsto \mathbf{b}$. The simple reflections for type C_n are Type I reflections of the form

$$s_i = (i, i+1)(2n-i, 2n+1-i) \text{ for } i < n$$

and the Type III reflection $s_n = (n, n+1)$. The minimum-length permutation associated to the fixed point $\mathbf{b}$ is found in the same way as for $G(k,n)$, namely if

$$\mathbf{b} \mapsto s_{i_m}\mathbf{b} \mapsto s_{i_{m-1}}s_{i_m}\mathbf{b} \mapsto \cdots \mapsto s_{i_1}\cdots s_{i_{m-1}}s_{i_m}\mathbf{b} = 1^k0^{n-k}$$

is a minimal-length path from $\mathbf{b}$ to 1^k0^{n-k} labeled by simple roots then a minimum-length permutation for $\mathbf{b}$ is $s_{i_m}s_{i_{m-1}}\cdots s_{i_1}$. Figure 3 shows the subgraph of the moment graph of $IG(2,6)$ induced from the edges labeled $t_1 - t_2$, $t_2 - t_3$, and $2t_3$, namely those corresponding to simple reflections. The other edges have been

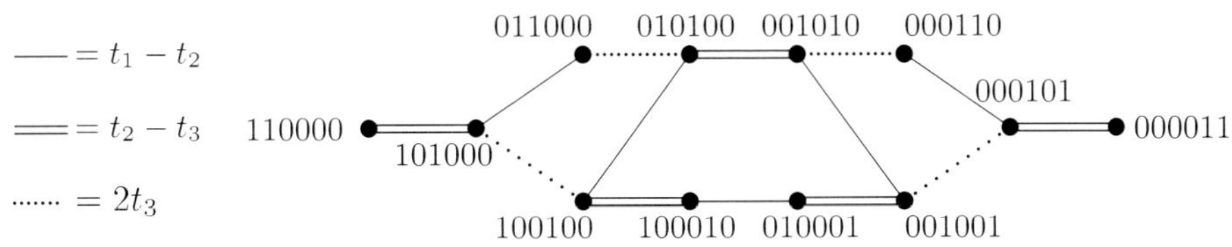

FIGURE 3. The moment graph of $IG(2, 6)$ of type C_3

omitted to make the Figure more readable. They can be reconstructed by the procedure described in Remark 3.13; for instance the full moment graph has an edge $010001 \mapsto 010100$ labeled $t_1 - t_3$. (The figure has also been rotated so that edges are directed from right to left rather than up to down.)

3.4.3. *Type B_n and D_n: isotropy with respect to a symmetric form.* The moment graphs for Grassmannians of isotropic k-planes in $\mathbb{C}^N$ that satisfy a symmetric bilinear form, namely maximal Grassmannians of type B_n and D_n, are similar. Both use the symmetric bilinear form defined on the basis vectors e_i by $\langle e_i, e_{N+1-i}\rangle = 1$ and $\langle e_i, e_j\rangle = 0$ otherwise. In each case, we will describe the fixed points, the simple reflections, and the directions of the edges corresponding to simple reflections in the moment graph. The edges corresponding to simple reflections will determine the rest of the edges as described in Remark 3.13. (The simple reflections generate all the reflections inductively according to the rule that if s_α and s_β are two reflections then $s_\beta s_\alpha s_\beta$ is also a reflection, corresponding to the root $s_\beta(\alpha)$.) In the next two sections, we will describe how simple reflections act on roots; this action extends naturally to all reflections. The minimal-length Weyl group element corresponding to each fixed point can be reconstructed analogously to types A_n and C_n.

3.4.4. *Type B_n maximal Grassmannians.* In type B_n, we consider the Grassmannian of isotropic k-planes in $\mathbb{C}^{2n+1}$. The fixed points are indexed by bit strings of length $2n+1$ with exactly k ones such that $b_i b_{2n+2-i} = 0$ for each i. (In particular, this implies that $b_{n+1} = 0$.) The group that acts on these fixed points is generated by the simple reflections

- $s_i = (i, i+1)(2n+1-i, 2n+2-i)$ for all $i = 1, \dots, n-1$ and
- $s_n = (n, n+2)$.

When $i < n$ the edge between $\mathbf{b}$ and $s_i\mathbf{b}$ is labeled $t_i - t_{i+1}$. When $i = n$ the edge between $\mathbf{b}$ and $s_i\mathbf{b}$ is labeled t_n. The edge is directed $s_i\mathbf{b} \mapsto \mathbf{b}$ if

- $b_i \leq b_{i+1}$ and $b_{2n+1-i} \leq b_{2n+2-i}$ for $i = 1, \dots, n-1$ (with at least one strict inequality), or
- $b_n < b_{n+2}$ for $i = n$.

If $i \neq n$ the simple reflection s_i exchanges the variables t_i and t_{i+1} and fixes the other variables; the simple reflection s_n sends t_n to $-t_n$ and fixes the other variables. An arbitrary reflection s_α factors into a product of simple reflections, extending this action from simple reflections to all reflections; Remark 3.13 now determines the moment graph.

3.4.5. *Type D_n maximal Grassmannians.* In type D_n, we consider the Grassmannian of isotropic k-planes in $\mathbb{C}^{2n}$. The fixed points are indexed by bit strings of length $2n$ with exactly k ones for $k = 1, \ldots, n-2$ or $k = n$, that also satisfy $b_i b_{2n+1-i} = 0$ (as in type C_n). The group that acts on these fixed points is generated by the simple reflections

- $s_i = (i, i+1)(2n-i, 2n+1-i)$ for $i = 1, \ldots, n-1$ and
- $s_n = (n, n+2)(n-1, n+1)$.

When $i < n$ the edge between $\mathbf{b}$ and $s_i\mathbf{b}$ is labeled $t_i - t_{i+1}$. When $i = n$ the edge between $\mathbf{b}$ and $s_i\mathbf{b}$ is labeled $t_{n-1} + t_n$. The edge is directed $s_i\mathbf{b} \mapsto \mathbf{b}$ if

- $b_i \leq b_{i+1}$ and $b_{2n-i} \leq b_{2n+1-i}$ for $i = 1, \ldots, n-1$ (with at least one strict inequality), and
- $b_n \leq b_{n+2}$ and $b_{n-1} \leq b_{n+1}$ for $i = n$ (with at least one strict inequality).

If $i \neq n$ the simple reflection s_i exchanges the variables t_i and t_{i+1} and fixes the other variables; the simple reflection s_n sends t_n to $-t_n$ and t_{n-1} to $-t_{n-1}$, and fixes the other variables. Remark 3.13 now determines the moment graph.

4. A W-action on G/P

In this section we define an action of the Weyl group W on $H_T^*(G/P)$ and give an explicit formula for the action of simple reflections on the basis of equivariant Schubert classes.

Recall that W acts on $\mathfrak{t}^*$ by reflections, with the action of $w \in W$ on the root α denoted $w(\alpha)$. This action extends naturally to a W-action on $S(\mathfrak{t}^*)$; each $w \in W$ can be considered as a $\mathbb{C}$-algebra homomorphism of $S(\mathfrak{t}^*)$.

LEMMA 4.1. *For each $w \in W$ and $p : W/W_J \to S(\alpha_1, \ldots, \alpha_n)$, the element $w \cdot p : W/W_J \to S(\alpha_1, \ldots, \alpha_n)$ is defined by the condition*

$$(w \cdot p)([v]) = wp(w^{-1}[v]).$$

This makes each $w \in W$ a well-defined twisted $S(\mathfrak{t}^*)$-module homomorphism on $H_T^*(G/P)$, namely if $c \in S(\mathfrak{t}^*)$ and $p \in H_T^*(G/P)$ then $w \cdot (cp) = (w(c))(w \cdot p)$.

PROOF. The group action of W on functions $W/W_J \to S(\alpha_1, \ldots, \alpha_n)$ is well-defined because

$$(w_1 \cdot (w_2 \cdot p))([v]) = w_1(w_2 \cdot p)(w_1^{-1}[v]) = w_1 w_2 p(w_2^{-1} w_1^{-1}[v]) = ((w_1 w_2) \cdot p)([v]).$$

We confirm that W acts on $H_T^*(G/P)$, namely that if p is a class in $H_T^*(G/P)$ then $w \cdot p$ is also in $H_T^*(G/P)$. To do this, we check that

$$(w \cdot p)([s_\alpha v]) - (w \cdot p)([v]) \in \big(\alpha\big)$$

for all vertices $[v]$ and edges $[s_\alpha v] \mapsto [v]$. The identity

$$w^{-1}[s_\alpha v] = [(w^{-1} s_\alpha w) w^{-1} v] = [s_{w^{-1}(\alpha)} w^{-1} v]$$

implies that

$$(w \cdot p)([s_\alpha v]) - (w \cdot p)([v]) = wp([s_{w^{-1}(\alpha)} w^{-1} v]) - wp([w^{-1} v]).$$

Lemma 3.8 proved that each w is a graph automorphism on the moment graph of G/P. Since $[s_\alpha v] \mapsto [v]$ is an edge in the moment graph of G/P, we conclude that $[s_{w^{-1}(\alpha)} w^{-1} v] \mapsto [w^{-1} v]$ is also an edge, and is labeled $w^{-1}(\alpha)$. Thus

$$wp([s_{w^{-1}(\alpha)} w^{-1} v]) - wp([w^{-1} v]) \in \big(\alpha\big),$$

namely $w \cdot p \in H_T^*(G/P)$.

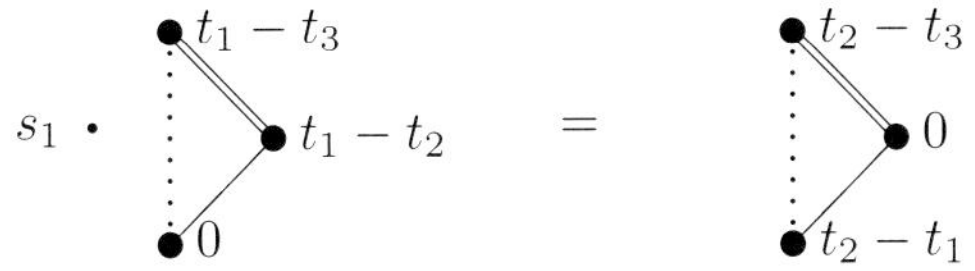

FIGURE 4. The action of s_1 on a Schubert class in $H_T^*(\mathbb{P}^2)$

Finally, we show each $w \in W$ acts as a twisted $S(\mathfrak{t}^*)$-module homomorphism. Addition in $H_T^*(G/P)$ is defined coordinate-wise, so for each $w \in W$ and cohomology classes $p_1, p_2 \in H_T^*(G/P)$ we have $w \cdot (p_1 + p_2) = w \cdot p_1 + w \cdot p_2$. If $p \in H_T^*(G/P)$ and $c \in S(\mathfrak{t}^*)$ then for each $[v] \in W/W_J$ we have

$$(w \cdot (cp))[v] = w(cp)([w^{-1}v]) = w(c)((w \cdot p)[v]).$$

So the action of $w \in W$ twists the $S(\mathfrak{t}^*)$-action on $H_T^*(G/P)$. □

This action can be described explicitly for simple reflections s_i. If we consider $p \in H_T^*(G/P)$ as a tuple of polynomials, one polynomial for each vertex of the moment graph, then the action of s_i permutes the indices in each polynomial by s_i and exchanges polynomials on either side of an edge labeled α_i. Figure 4 demonstrates this, using the same notation as Figure 1.

4.1. A formula for the action of simple reflections. We give an explicit formula for the action of a simple reflection on the equivariant Schubert basis of $H_T^*(G/P)$. This formula is the basis for our divided difference operators.

We begin with a short corollary of the results on P-inversions, included here for completeness. (If we use specific representatives $w \in W^J$, this corollary follows from results on G/B [**T**, Proposition 4.7].) Recall that Lemma 3.5 implies that for each $[w] \in W/W_J$, the set $\mathcal{I}_{[w]}^P$ is well-defined by the rule $\mathcal{I}_{[w]}^P = \mathcal{I}_w^P$.

COROLLARY 4.2. *Let $[w] \in W/W_J$. Suppose $p_{[w]}$ is a corresponding flow-up class in $H_T^*(G/P)$ and s_i is a simple reflection such that $[s_i w] \mapsto [w]$ then*

$$p_{[w]}([s_i w]) = \prod_{\alpha \in \mathcal{I}_{[w]}^P} s_i(\alpha) = s_i p_{[w]}([w]).$$

PROOF. Let

$$q = \prod_{\alpha \in \mathcal{I}_{[s_i w]}^P, \alpha \neq \alpha_i} \alpha.$$

Since $\ell_P([s_i w]) = \ell_P([w]) + 1$ and $p_{[w]}$ is a flow-up class, every vertex $[v] \neq [w]$ with an edge $[s_i w] \mapsto [v]$ satisfies $p_{[w]}([v]) = 0$. By comparing degrees, we conclude that $p_{[w]}([s_i w])$ is a scalar multiple cq for some $c \in \mathbb{C}$.

Lemma 3.9 shows that $s_i q = p_{[w]}([w])$ and that neither q nor $p_{[w]}([w])$ is divisible by α_i. It follows that $q - p_{[w]}([w]) \in (\alpha_i)$ and that $c = 1$ is the the only scalar satisfying $cq - p_{[w]}([w]) \in (\alpha_i)$. The GKM conditions imply $p_{[w]}([s_i w]) = q$. □

Next we generalize to G/P a classical result for G/B [**BGG**, Lemma 2.4].

LEMMA 4.3. *Suppose that $[s_i w] >_P [w]$. If there is an edge $[s_\alpha w] \mapsto [w]$ with $\ell_P([s_\alpha w]) = \ell_P([w]) + 1$ in the moment graph of G/P then the configuration of Figure 5 is in the moment graph.*

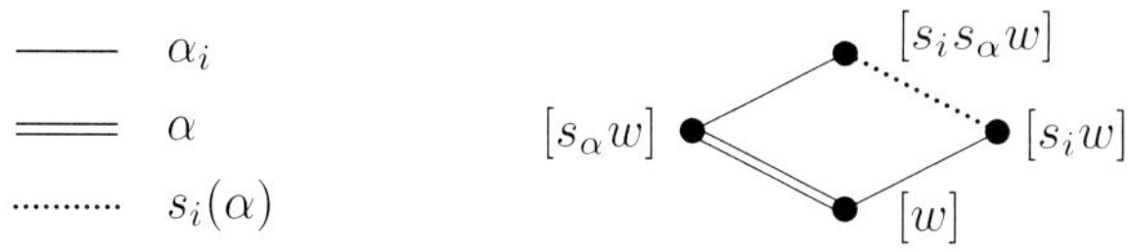

FIGURE 5. Schematic of diamond shape (angles not to scale)

PROOF. There is an edge $[s_\alpha w] \mapsto [w]$ so the root α is in $\mathcal{I}^P_{[s_\alpha w]}$. Lemma 3.9 shows that $s_i\left(\mathcal{I}^P_{[s_\alpha w]} - \{\alpha_i\}\right) \subseteq \mathcal{I}^P_{[s_i s_\alpha w]}$ regardless of whether $[s_i s_\alpha w] = [s_\alpha w]$ or, if not, the direction of the edge between them. In particular $s_i(\alpha) \in \mathcal{I}^P_{[s_i s_\alpha w]}$. The reflection corresponding to $s_i(\alpha)$ is $s_i s_\alpha s_i$. Since $(s_i s_\alpha s_i)[s_i s_\alpha w] = [s_i w]$, we conclude $[s_i s_\alpha w] \mapsto [s_i w]$. In particular $\ell_P([s_i s_\alpha w]) \geq \ell_P([s_i w]) + 1$. Lemma 3.10 says $\ell_P([s_i s_\alpha w]) \leq \ell_P([s_\alpha w]) + 1$. Since $\ell_P([s_\alpha w]) = \ell_P([s_i w]) = \ell_P([w]) + 1$, we see that $\ell_P([s_i s_\alpha w]) = \ell_P([w]) + 2$ with the moment graph from Figure 5. □

We will now prove the main result of this section, an explicit formula for $s_i \cdot p_{[w]}$.

LEMMA 4.4. *Let $p_{[w]}$ be a flow-up class for $[w]$ in $H^*_T(G/P)$ and let s_i be a simple reflection. Then*

$$s_i \cdot p_{[w]} = \begin{cases} p_{[w]} & \textit{if } [s_i w] \geq [w] \textit{ and} \\ p_{[w]} - \alpha_i p_{[s_i w]} & \textit{if } [s_i w] <_P [w]. \end{cases}$$

PROOF. We decompose $s_i \cdot p_{[w]} = \sum c_{[v]} p_{[v]}$ in terms of the basis of flow-up classes. If $\ell_P([v]) > \ell_P([w])$ then $c_{[v]} = 0$ because $s_i \cdot p_{[w]}$ is a homogeneous class of degree $\ell_P([w])$ and the coefficients $c_{[v]}$ have degree at least zero.

Let $[u]$ be a minimal element of W/W_J with $c_{[u]} \neq 0$. We will evaluate the expression $s_i \cdot p_{[w]} = \sum c_{[v]} p_{[v]}$ at $[u]$. Each flow-up class $p_{[v]}$ is supported only on $[u]$ with $[u] \geq_P [v]$, so

$$(s_i \cdot p_{[w]})([u]) = \sum c_{[v]} p_{[v]}([u]) = c_{[u]} p_{[u]}([u]) \tag{4.1}$$

by the minimality hypothesis on $[u]$. The reflection s_i changes the length of a fixed point by at most one. If $\ell_P([v]) < \ell_P([w])$ then $\ell_P([s_i v]) \leq \ell_P([w])$ and in particular $[s_i v] \not\geq [w]$ unless $[s_i v] = [w]$. Hence $(s_i \cdot p_{[w]})([u]) = s_i p_{[w]}([s_i u]) = 0$ unless $u = s_i w$. We conclude that $c_{[v]} = 0$ if $\ell_P([v]) < \ell_P([w])$ except perhaps when both $[u] = [s_i w]$ and $[u] < [w]$.

Thus the only possible nonzero $c_{[v]}$ are $[v]$ with $\ell_P([v]) = \ell_P([w])$ and, in the case $[s_i w] < [w]$, the coefficient $c_{[s_i w]}$.

Consider the case when $[s_i w] \geq [w]$ first. Let $[u]$ be a minimal element of W/W_J with $c_{[u]} \neq 0$. Equation (4.1) implies that $(s_i \cdot p_{[w]})([u]) \neq 0$, and in particular the moment graph has an edge $[s_i u] \overset{\alpha}{\mapsto} [w]$ for some α. If $[u] \neq [w]$ then $\alpha \neq \alpha_i$. Then $\alpha \in \mathcal{I}^P_{[s_i u]}$ implies that $s_i \alpha \in \mathcal{I}^P_{[u]}$ by Lemma 3.9. The reflection $s_{s_i(\alpha)} = s_i s_\alpha s_i$ so

$$[s_{s_i(\alpha)} u] = [s_i s_\alpha s_i u] = [s_i w].$$

Hence the moment graph has an edge $[u] \mapsto [s_i w]$ and so $\ell_P([u]) > \ell_P([s_i w])$, which by hypothesis is at least $\ell_P([w])$. This contradiction shows that $[u] = [w]$ is the only minimal element of W/W_J with $c_{[u]} \neq 0$, and so also the only element of length $\ell_P([w])$ with $c_{[u]} \neq 0$. Moreover Equation (4.1) implies that

$$c_{[w]} = \frac{(s_i \cdot p_{[w]})([w])}{p_{[w]}([w])} = \frac{s_i p_{[w]}([s_i w])}{p_{[w]}([w])}.$$

If $[s_i w] \mapsto [w]$ then Corollary 4.2 shows $c_{[w]} = 1$. If $[s_i w] = [w]$ then Lemma 3.9 shows $c_{[w]} = 1$.

Now suppose $[s_i w] <_P [w]$. All fixed points $[v]$ with $[v] < [s_i w]$ have $c_{[v]} = 0$ so Equation (4.1) applies to $u = s_i w$. Together with Corollary 4.2, it implies that

$$c_{[s_i w]} = \frac{s_i p_{[w]}([w])}{p_{[s_i w]}([s_i w])} = -\alpha_i.$$

We will now identify $c_{[u]}$ for fixed points $[u]$ of length $\ell_P([w])$ by evaluating

$$s_i \cdot p_{[w]} = -\alpha_i p_{[s_i w]} + \sum c_{[v]} p_{[v]}$$

at various fixed points. Since the sum is taken over $[v]$ with $\ell_P([v]) = \ell_P([w])$, the equation obtained by evaluation at $[u]$ is

$$(s_i \cdot p_{[w]})([u]) = -\alpha_i p_{[s_i w]}([u]) + c_{[u]} p_{[u]}([u]). \tag{4.2}$$

We begin with $[u] = [w]$. Corollary 4.2 says $-\alpha_i p_{[s_i w]}([w]) = -\alpha_i s_i p_{[s_i w]}([s_i w])$, which is $-p_{[w]}([w])$ by Lemma 3.9. Since $(s_i \cdot p_{[w]})([w]) = 0$ we obtain $c_{[w]} = 1$.

Next let $[u] \neq [w]$ be a fixed point with $\ell_P([u]) = \ell_P([w])$. Equation (4.2) shows that $c_{[u]} = 0$ unless $p_{[s_i w]}([u]) \neq 0$ or $p_{[w]}([s_i u]) \neq 0$. The former implies that there is an edge $[u] \mapsto [s_i w]$ while the latter implies that there is an edge $[s_i u] \mapsto [w]$ by length considerations and the definition of flow-up classes. Lemma 4.3 shows that if $[u] \mapsto [s_i w]$ then there is an edge $[s_i u] \mapsto [w]$. Conversely, if $[s_i u] \mapsto [w]$ is an edge, say labeled α, then $[u] = [(s_i s_\alpha s_i) s_i w]$ and so $[u] \mapsto [s_i w]$ is an edge labeled $s_i(\alpha)$. Hence $c_{[u]} = 0$ unless the following picture is in the moment graph:

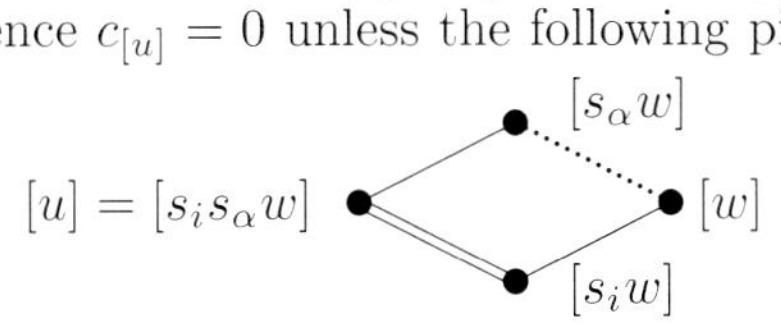

Lemma 3.9 showed that if the downward pointing edges at $[s_i w]$ and $[u]$ are labeled by $\mathcal{I}^P_{[s_i w]}$ and $\mathcal{I}^P_{[u]}$ respectively, then the downward pointing edges at $[w]$ and $[s_i u]$ are labeled $s_i \mathcal{I}^P_{[s_i w]} \cup \{\alpha_i\}$ and $s_i \mathcal{I}^P_{[u]} \cup \{\alpha_i\}$, respectively. Let

$$q = \prod_{\substack{\alpha \in \mathcal{I}^P_{[u]} s.t. \\ [s_\alpha u] \neq [s_i w]}} \alpha.$$

Each edge $[u] \mapsto [s_\alpha u] \neq [s_i w]$ points to a vertex with $p_{[s_i w]}([s_\alpha u]) = 0$ by definition of flow-up classes. Thus the localization $p_{[s_i w]}([u])$ is in the ideal (q). A similar argument shows that $p_{[w]}([s_i u]) \in \big((\alpha_i)(s_i q)\big)$.

Comparing degrees, we see that $p_{[s_i w]}([v])$ is in fact a scalar multiple of q, say cq for $c \in \mathbb{C}$. The scalar c is determined by the GKM condition

$$cq - p_{[s_i w]}([s_i w]) \in \big(s_i(\alpha)\big)$$

because $s_i(\alpha)$ is relatively prime to q. (Indeed, the edges at a fixed vertex in the moment graph of a GKM space are labeled by positive roots, none of which are multiples or repetitions of each other.) The same scalar c satisfies the equation $c\alpha_i(s_i q) - \alpha_i s_i p_{[s_i w]}([s_i w]) \in (\alpha)$, which is the GKM condition for the directed edge $[s_i u] \mapsto [w]$ because $p_{[w]}([w]) = \alpha_i s_i p_{[s_i w]}([s_i w])$. No other scalar c satisfies the equation because α is relatively prime to $\alpha_i(s_i q)$. So $p_{[w]}([s_i u]) = \alpha_i s_i p_{[s_i w]}([u])$. In other words $(s_i \cdot p_{[w]})([u]) = -\alpha_i p_{[s_i w]}([u])$. Equation (4.2) implies that $c_{[u]} = 0$. $\square$

We reformulate the previous result for later use.

COROLLARY 4.5. *Let $p_{[w]} \in H_T^*(G/P)$ be the flow-up class corresponding to $[w] \in W/W_J$ and let s_i be a simple reflection. Then for each $[v] \in W/W_J$*

$$p_{[w]}([v]) = \begin{cases} s_i(p_{[w]}([s_i v])) & \text{if } [s_i w] \geq [w] \text{ and} \\ s_i(p_{[w]}([s_i v])) + \alpha_i p_{[s_i w]}([v]) & \text{if } [s_i w] < [w]. \end{cases}$$

5. A divided difference operator

Bernstein-Gelfand-Gelfand and Demazure simultaneously defined divided difference operators in order to determine the relationship between two presentations of the cohomology ring of the flag variety: the Borel presentation (as a quotient of a polynomial ring by an ideal of symmetric polynomials), and the geometric presentation (in terms of the basis of Schubert classes) [**BGG**], [**De**]. We refer to their operators as BGG divided difference operators. In this section we define a new divided difference operator on both $H_T^*(G/P)$ and on $H^*(G/P)$, for arbitrary complex semisimple linear algebraic groups G and arbitrary parabolics P.

We begin with a review of the key aspects of BGG divided difference operators, from our perspective. The BGG divided difference operator is a degree-lowering rational operator on $S(\mathfrak{t}^*)$. The formula for the action of the ith BGG divided difference operator ∂_i on $f \in S(\mathfrak{t}^*)$ is

$$\partial_i f = \frac{f - s_i f}{\alpha_i}.$$

We will use two properties of the BGG divided difference operator. First:

- The BGG divided difference operator satisfies the two nil-Coxeter relations. First, if $s_{i_1} \cdots s_{i_k}$ and $s_{j_1} \cdots s_{j_k}$ are two reduced factorizations for the same word, then $\partial_{i_1} \cdots \partial_{i_k} = \partial_{j_1} \cdots \partial_{j_k}$. Second, $\partial_i^2 = 0$ for each i.

The second property describes how divided difference operators act on Schubert polynomials. A Schubert polynomial corresponding to $w \in W$ is a representative $\mathfrak{S}_w \in S(\mathfrak{t}^*)$ of the Schubert class in $H^*(G/P)$ corresponding to w. There is an extensive literature on how best to choose the representative, e.g. [**LS**], [**FK**]. For the purposes of this paper, any convention can be chosen as long as the following are satisfied:

- If ∂_i is a BGG divided difference operator and $\mathfrak{S}_w$ is a Schubert polynomial corresponding to $w \in W$ then
$$\partial_i(\mathfrak{S}_w) = \begin{cases} \mathfrak{S}_{ws_i} & \text{if } ws_i < w \text{ and} \\ 0 & \text{otherwise.} \end{cases}$$
- The Schubert polynomial $\mathfrak{S}_e$ is a positive integer.

The Schubert polynomials defined in [**BGG**] satisfy these properties, as do others.

BGG operators satisfy a Leibniz formula, proven for type A_n in several places [**M**, Equation 2.2] and generalized in the following lemma.

LEMMA 5.1. *For each $f, g \in S(\mathfrak{t}^*)$ and BGG divided difference operator ∂_i*

$$\partial_i(fg) = \partial_i(f)g + (s_i f)\partial_i(g).$$

PROOF. Expand:

$$\begin{aligned}\partial_i(fg) &= \frac{fg-(s_if)(s_ig)}{\alpha_i}\\ &= \frac{fg-(s_if)g}{\alpha_i}+\frac{(s_if)g-(s_if)(s_ig)}{\alpha_i} = \partial_i(f)g+(s_if)\partial_i(g). \quad \square\end{aligned}$$

We now define a second collection of divided difference operators δ_i on the GKM presentation of the $S(\mathfrak{t}^*)$-module $H_T^*(G/P)$. We use δ_i to distinguish the new divided difference operators from the BGG divided difference operators ∂_i, which will also be used in what follows. At the end of this section, we discuss results for ordinary cohomology, as well as how ∂_i and δ_i are related. Mark Shimozono helpfully pointed out that the following result was also given by D. Peterson [**P**].

PROPOSITION 5.2. *For each simple reflection $s_i \in W$ define a divided difference operator δ_i that acts on the class $p \in H_T^*(G/P)$ as a $\mathbb{C}$-module homomorphism by*

$$\delta_i p = \frac{p - s_i \cdot p}{\alpha_i}$$

Each δ_i is well-defined and $\delta_i^2 = 0$. It satisfies a "Leibniz formula":

$$\delta_i(cp_{[v]}) = (\partial_i c)p_{[v]} + (s_ic)\delta_i p_{[v]} \tag{5.1}$$

where ∂_i is the BGG divided difference operator on $S(\mathfrak{t}^*)$. The action of δ_i on the basis of flow-up classes is given by

$$\delta_i p_{[v]} = \begin{cases} p_{[s_iv]} & \text{if } [s_iv] <_P [v] \text{ and} \\ 0 & \text{otherwise.}\end{cases} \tag{5.2}$$

PROOF. The formula for δ_i is $\mathbb{C}$-linear because the W-action is. Write p in terms of the basis of flow-up classes as $p = \sum c_{[v]}p_{[v]}$. We obtain

$$\delta_i p = \frac{\sum_{[v]}(c_{[v]}p_{[v]} - s_i \cdot (c_{[v]}p_{[v]}))}{\alpha_i} = \sum_{[v]} \delta_i(c_{[v]}p_{[v]}).$$

To show $\delta_i p \in H_T^*(G/P)$ it suffices to show $\delta_i(cp_{[v]}) \in H_T^*(G/P)$ for each $c \in S(\mathfrak{t}^*)$ and flow-up class $p_{[v]} \in H_T^*(G/P)$. Equation (5.1) follows from

$$\delta_i(cp_{[v]}) = \frac{cp_{[v]} - (s_ic)p_{[v]}}{\alpha_i} + \frac{(s_ic)p_{[v]} - (s_ic)(s_i \cdot p_{[v]})}{\alpha_i}$$

For every $c \in S(\mathfrak{t}^*)$ the BGG divided difference satisfies $\partial_i c \in S(\mathfrak{t}^*)$, and also $s_ic \in H_T^*(G/P)$. By Equation (5.1) it suffices to show $\delta_i p_{[v]} \in H_T^*(G/P)$. But $\delta_i p_{[v]} \in H_T^*(G/P)$ by Equation (5.2), which is a direct restatement of Lemma 4.4. Thus δ_i is well-defined on $H_T^*(G/P)$.

The relation $\delta_i^2 = 0$ is a result of the expression

$$\frac{p - s_i \cdot p}{\alpha_i} - s_i \cdot \left(\frac{p - s_i \cdot p}{\alpha_i}\right) = \frac{p - s_i \cdot p}{\alpha_i} - \frac{s_i \cdot p - s_i^2 \cdot p}{-\alpha_i}. \qquad \square$$

Figure 6 gives all nonzero examples of the divided difference operator δ_i in the equivariant cohomology of the projective plane.

The next proposition proves that these δ_i satisfy the braid relations. The main step uses Equation (5.1) to reduce the problem to a calculation on the BGG divided difference operators on $H^*(G/B)$. This key calculation is performed in Lemma 5.7 following the proof of the proposition.

We use the following three definitions.

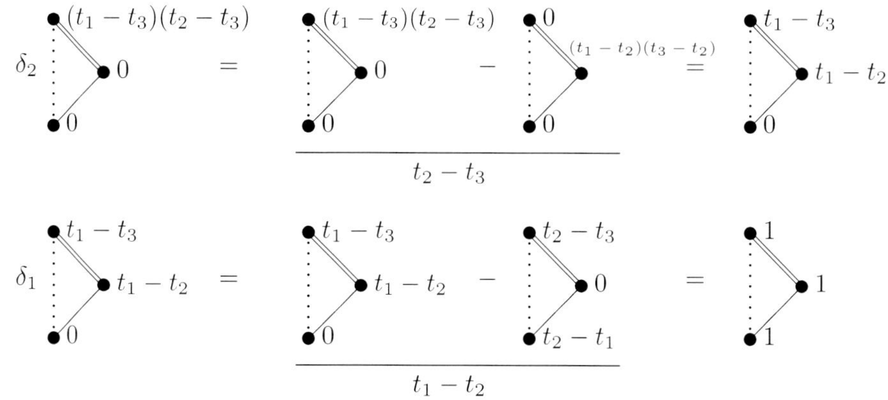

FIGURE 6. Divided difference operators on Schubert classes in $H_T^*(\mathbb{P}^2)$

DEFINITION 5.3. For each $w \in W$, the set $R(w)$ consists of all reduced words for w in the simple reflections $\{s_1, s_2, \ldots, s_n\}$.

DEFINITION 5.4. Given a fixed reduced factorization $w = w_1 w_2 \cdots w_k \in W$ and a reduced word $u \leq w$, define $\mathcal{S}_u^w$ to be the set of ordered integer sequences $1 \leq b_1 < b_2 < \cdots < b_{\ell(u)} \leq k$ such that $w_{b_1} w_{b_2} \cdots w_{b_{\ell(u)}} \in R(u)$.

DEFINITION 5.5. Given a reduced factorization $w = w_1 w_2 \cdots w_k \in W$ and an integer sequence $\mathbf{b} = b_1 b_2 \cdots b_l$ satisfying $1 \leq b_1 < b_2 < \cdots < b_l \leq k$, define $\iota_w(\mathbf{b})$ to be the word in $\{s_1, s_2, \ldots, s_n, \partial_1, \partial_2, \ldots, \partial_n\}$ obtained from w by switching the letters that are *not* in positions $b_1, b_2, \ldots, b_l$ from s_i to ∂_i.

For instance, in type A_n if $w = s_1 s_2 s_1$ then $R(w) = \{s_1 s_2 s_1, s_2 s_1 s_2\}$. The set $\mathcal{S}_{s_1}^{s_1 s_2 s_1} = \{1, 3\}$ which gives the words $\iota_{s_1 s_2 s_1}(1) = s_1 \partial_2 \partial_1$ and $\iota_{s_1 s_2 s_1}(3) = \partial_1 \partial_2 s_1$.

PROPOSITION 5.6. *The divided difference operators $\{\delta_i\}$ satisfy the braid relations.*

PROOF. Suppose $(s_i s_j)^m = (s_j s_i)^m$ is a braid relation defining W for $i \neq j$. We will compare $(\delta_i \delta_j)^m (c p_{[v]})$ to $(\delta_i \delta_j)^m (c p_{[v]})$ for an arbitrary $c \in S(\mathfrak{t}^*)$ and flow-up class $p_{[v]} \in H_T^*(G/P)$. The flow-up classes form an $S(\mathfrak{t}^*)$-module basis of $H_T^*(G/P)$, so additivity of the divided difference operators δ_i implies the claim.

Note that if $u < (s_i s_j)^m$ then u has a unique minimal-length factorization in W. Thus δ_u can be defined as the sequence of δ_i and δ_j corresponding to u. Repeated application of Equation (5.1) shows that

$$\begin{aligned}(\delta_i \delta_j)^m (c p_{[v]}) &= (s_i s_j)^m (c) \left((\delta_i \delta_j)^m p_{[v]} \right) \\ &\quad + \sum_{\substack{u < (s_i s_j)^m \text{ s.t.} \\ \ell_P([uv]) = \\ \ell_P([v]) - \ell(u)}} (\delta_u p_{[v]}) \left(\sum_{\mathbf{b} \in \mathcal{S}_u^{(s_i s_j)^m}} \iota_{(s_i s_j)^m}(\mathbf{b}) c \right)\end{aligned}$$

and similarly for $(\delta_j \delta_i)^m (c p_{[v]})$. The coefficient of $\delta_u p_{[v]}$ is a sum of a sequence of BGG divided difference operators applied to elements of $S(\mathfrak{t}^*)$. Lemma 5.7 proves

$$\sum_{\mathbf{b} \in \mathcal{S}_u^{(s_i s_j)^m}} \iota_{(s_i s_j)^m}(\mathbf{b}) c = \sum_{\mathbf{b} \in \mathcal{S}_u^{(s_j s_i)^m}} \iota_{(s_j s_i)^m}(\mathbf{b}) c$$

for each $u \leq (s_i s_j)^m$. Since

$$\{u \in W : u < (s_i s_j)^m\} = \{u \in W : u < (s_j s_i)^m\}$$

we have

$$(\delta_i \delta_j)^m (c p_{[v]}) = (\delta_j \delta_i)^m (c p_{[v]})$$

for all $c \in S(\mathbf{t}^*)$ if and only if

$$(\delta_i \delta_j)^m p_{[v]} = (\delta_j \delta_i)^m p_{[v]}.$$

If $\ell_P([(s_i s_j)^m v]) = \ell_P([v]) - 2m$ then repeated application of Equation (5.2) shows

$$(\delta_i \delta_j)^m p_{[v]} = p_{[(s_i s_j)^m v]} = (\delta_j \delta_i)^m p_{[v]}.$$

If not, the same Equation gives $(\delta_i \delta_j)^m p_{[v]} = 0 = (\delta_j \delta_i)^m p_{[v]}$. The claim follows. □

LEMMA 5.7. *Assume $i \neq j$ and that $(s_i s_j)^m = (s_j s_i)^m$ is a braid relation in W. Let $u \in W$ be* any *reduced word with $u \leq (s_i s_j)^m$. Then*

$$\sum_{\mathbf{b} \in \mathcal{S}_u^{(s_i s_j)^m}} \iota_{(s_i s_j)^m}(\mathbf{b}) = \sum_{\mathbf{b} \in \mathcal{S}_u^{(s_j s_i)^m}} \iota_{(s_j s_i)^m}(\mathbf{b}). \tag{5.3}$$

PROOF. The proof is by induction on the length of u. If u has length zero then $\iota_{(s_i s_j)^m}(\emptyset) = (\partial_i \partial_j)^m$ and $\iota_{(s_j s_i)^m}(\emptyset) = (\partial_j \partial_i)^m$. BGG divided difference operators satisfy the braid relations so $\iota_{(s_i s_j)^m}(\emptyset) = \iota_{(s_j s_i)^m}(\emptyset)$. Similarly, when we have the factorization $u = (s_i s_j)^m = (s_j s_i)^m$ the sets $\mathcal{S}_{(s_i s_j)^m}^{(s_i s_j)^m}$ and $\mathcal{S}_{(s_j s_i)^m}^{(s_j s_i)^m}$ both equal $\{\mathbf{b} = (1, 2, \ldots, 2m)\}$ and so $\iota_{(s_i s_j)^m}(\mathbf{b}) = (\partial_i \partial_j)^m = \iota_{(s_j s_i)^m}(\mathbf{b})$.

Assume that if $\ell(u) \leq k$ then Equation (5.3) holds for u. For every pair $f, g \in S(\mathbf{t}^*)$, repeated application of Lemma 5.1 shows that

$$(\partial_i \partial_j)^m (fg) = \sum_{u \leq (s_i s_j)^m} (\partial_u g) \left(\sum_{\mathbf{b} \in \mathcal{S}_u^{(s_i s_j)^m}} \iota_{(s_i s_j)^m}(\mathbf{b}) f \right)$$

and similarly for $(\partial_j \partial_i)^m (fg)$. Note that $u \leq (s_i s_j)^m$ if and only if $u \leq (s_j s_i)^m$.

If $u < (s_i s_j)^m$ then u has a unique factorization into simple reflections. Choose $u_0 < (s_i s_j)^m$ of length $k+1 \neq m$ and let g be any Schubert polynomial $\mathfrak{S}_{u_0}$ satisfying the conditions set out at the beginning of this Section. In particular $\partial_i \mathfrak{S}_u = 0$ whenever $u s_i \geq u$. It follows that if $\ell(u) \geq k+1$ and $u \neq u_0$ then $\partial_u \mathfrak{S}_{u_0} = 0$. The inductive hypothesis proved that if $\ell(u) \leq k$ then

$$(\partial_u \mathfrak{S}_{u_0}) \left(\sum_{\mathbf{b} \in \mathcal{S}_u^{(s_i s_j)^m}} \iota_{(s_i s_j)^m}(\mathbf{b}) f \right) = (\partial_u \mathfrak{S}_{u_0}) \left(\sum_{\mathbf{b} \in \mathcal{S}_u^{(s_j s_i)^m}} \iota_{(s_j s_i)^m}(\mathbf{b}) f \right).$$

It follows that

$$\begin{aligned} &(\partial_i \partial_j)^m (f \mathfrak{S}_{u_0}) - (\partial_j \partial_i)^m (f \mathfrak{S}_{u_0}) = \\ &(\partial_{u_0} \mathfrak{S}_{u_0}) \left(\textstyle\sum_{\mathbf{b} \in \mathcal{S}_{u_0}^{(s_i s_j)^m}} \iota_{(s_i s_j)^m}(\mathbf{b}) f - \sum_{\mathbf{b} \in \mathcal{S}_{u_0}^{(s_j s_i)^m}} \iota_{(s_j s_i)^m}(\mathbf{b}) f \right). \end{aligned} \tag{5.4}$$

Recall that $\partial_{u_0} \mathfrak{S}_{u_0}$ was assumed to be a nonzero integer. The nil-Coxeter relations for the BGG divided difference operators imply that $(\partial_i \partial_j)^m = (\partial_j \partial_i)^m$ and so Equation (5.4) is zero. Since f was arbitrary Equation (5.3) holds for u_0. By induction it is true for all $u \leq (s_i s_j)^m$. □

Together, these results show δ_w is well-defined for each $w \in W$. By Proposition 5.6, we have $\delta_{i_1} \cdots \delta_{i_k} p_{[u]} = p_{[s_{i_1} \cdots s_{i_k} u]}$ for each flow-up class $p_{[u]} \in H_T^*(G/P)$ with $\ell_P([s_{i_1} \cdots s_{i_k} u] = \ell_P([u]) - k$. By Proposition 5.2, we have $\delta_{i_1} \cdots \delta_{i_k} p_{[u]} = 0$ otherwise. Hence $\delta_w = \delta_{i_1} \cdots \delta_{i_k}$ for any reduced factorization $w = s_{i_1} \cdots s_{i_k}$.

REMARK 5.8. Divided difference operators are $\mathbb{C}$-module homomorphisms defined on $H_T^*(G/P)$. As a $\mathbb{C}$-module, the ordinary cohomology $H^*(G/P)$ is the $\mathbb{C}$-span of the flow-up classes. (The ring structure of $H^*(G/P)$ is obtained from the ring isomorphism

$$H^*(G/P) \cong \frac{H_T^*(G/P)}{(t_1, t_2, \ldots, t_n) H_T^*(G/P)}$$

[**GKM**, Equation (1.2.4)].) This means the divided difference operators δ_i on $H_T^*(G/P)$ also apply to ordinary cohomology, viewed as a submodule of $H_T^*(G/P)$.

REMARK 5.9. Kostant-Kumar defined a different divided difference operator on $H_T^*(G/B)$ using the GKM presentation [**KK**]. We denote this divided difference operator ∂_i^{KK} because Arabia proved that it induces the BGG divided difference operator on ordinary cohomology [**A**]. If $p \in H_T^*(G/B)$ is a class in the GKM presentation then for each $v \in W$

$$(\partial_i^{KK} p)(v) = \frac{p(v) - p(vs_i)}{-v(\alpha_i)}.$$

We remark that Kostant-Kumar's divided difference operator is not well-defined on $H_T^*(G/P)$. For a more detailed treatment see [**T2**].

6. Generating Schubert classes

A classical result for flag varieties states that successive application of BGG divided difference operators on the Schubert class p_{w_0} corresponding to the longest element $w_0 \in W$ generates all Schubert classes [**BGG**]. We prove this for Grassmannians with the divided difference operators δ_i.

THEOREM 6.1. *Each Schubert class $p_{[w]}$ in $H^*(G/P)$ and in $H_T^*(G/P)$ can be obtained as*

$$\delta_v p_{[w_0]} = p_{[w]}$$

for some $v \in W$. To find v explicitly, choose any minimal-length path

$$[w_0] \overset{\alpha_{i_k}}{\mapsto} s_{i_k}[w_0] \overset{\alpha_{i_{k-1}}}{\mapsto} s_{i_{k-1}} s_{i_k}[w_0] \overset{\alpha_{i_{k-2}}}{\mapsto} \cdots \overset{\alpha_{i_1}}{\mapsto} s_{i_1} \cdots s_{i_k}[w_0] = [w]$$

from $[w_0]$ to $[w]$ in the moment graph of G/P and set $v = s_{i_1} s_{i_2} \cdots s_{i_k}$.

PROOF. Recall that there is an order-reversing bijection $\varphi : W/W_J \to W/W_J$ that satisfies $[w] \leq_P [u]$ if and only if $\varphi([u]) \leq_P \varphi([w])$. If w_0 is the longest element of W then φ can be defined by $\varphi([u]) = [w_0 u]$. (Among others, Björner-Brenti prove this [**BB**, Proposition 2.5.4].)

The shortest element of W/W_J is the identity $[e]$ and is sent to the longest element $\varphi([e]) = [w_0]$. The edge $[u] \mapsto [s_\alpha u]$ maps to the edge $[w_0 s_\alpha u] \mapsto [w_0 u]$, which is labeled $w_0(\alpha)$. If s_i is a simple reflection then $w_0 s_i w_0^{-1}$ is also simple.

Each element $[w_0^{-1} u] \in W/W_J$ has a reduced factorization into simple reflections, say $[w_0^{-1} u] = s_{i_1'} \cdots s_{i_k'}[e]$ with $\ell_P([w_0^{-1} u]) = k$. Choosing this factorization corresponds to choosing a minimal-length path from $[w_0^{-1} u]$ to $[e]$ in the moment graph. Suppose that $w_0 s_{i_j'} w_0^{-1} = s_{i_j}$ for each j. Applying φ shows that

$[u] = s_{i_1} \cdots s_{i_k}[w_0]$ and $\ell_P([w_0]) = \ell_P([u]) + k$, giving a minimal-length path from $[w_0]$ to $[u]$ in the moment graph. Equation (5.2) proves $\delta_{i_1} \cdots \delta_{i_k} p_{[w_0]} = p_{[u]}$. □

7. A generalized Billey formula for flow-up classes

Billey and Andersen-Jantzen-Soergel independently gave a formula for the localizations of equivariant Schubert classes in flag varieties [**AJS**], [**Bi2**]. We use Corollary 4.5 to generalize Billey's formula to arbitrary partial flag varieties. The new formula is similar to the original: to compute the localization $p_{[w]}([v])$, use minimal-length representatives $w \in [w]$ and $v \in [v]$ and compute $p_w(v)$ in the equivariant cohomology of the full flag variety. In fact, we will show that v may be chosen arbitrarily from the coset $[v]$, though w must be minimal-length in $[w]$. Billey's proof used the divided difference formula of Kostant-Kumar [**KK**], which is not well-defined on $H_T^*(G/P)$, so we modify as follows.

THEOREM 7.1. *Let $[v], [w]$ be two elements of W/W_J and assume that v, w are minimal in $[v], [w]$. Fix a reduced word for v, say $v = b_1 \cdots b_l$. Let $R(w)$ denote the set of reduced words for w and for each i let $r_b(i) = b_1 b_2 \cdots b_{i-1}(\alpha_i)$. Then*

$$p_{[w]}([v]) = p_w(v) = \sum_{b_{i_1} \cdots b_{i_k} \in R(w)} r_b(i_1) r_b(i_2) \cdots r_b(i_k)$$

where the sum is over $1 \leq i_1 < i_2 < \cdots < i_k \leq l$ with $b_{i_1} \cdots b_{i_k} \in R(w)$.

PROOF. The proof is by induction on the length of w. If $[w]$ is the maximal element in W/W_J then $p_{[w]}([v]) = 0$ unless $[v] = [w]$, in which case

$$p_{[w]}([w]) = \prod_{\alpha \in \mathcal{I}^P_{[w]}} \alpha = r_b(1) r_b(2) r_b(3) \cdots r_b(l)$$

by definition of flow-up classes and the exchange condition [**BB**, Corollary 1.4.4].

Assume by induction that the claim holds for all elements of W/W_J with length at least $k+1$. Let $[w] \in W/W_J$ have $\ell_P([w]) = k$ and take w to be in W^J, namely w is the minimal-length representative of the coset $[w]$. Since $[w]$ is not the longest element of W/W_J, there is a simple reflection s_i with $[s_i w] >_P [w]$. In particular, no reduced word for w begins with s_i.

We now compute $p_{[s_i w]}([v])$ for each $[v] \in W/W_J$. We will then use the formula from Corollary 4.5 to deduce $p_{[w]}([v])$. The proof is divided into two cases: when $[s_i v] <_P [v]$ and when $[s_i v] \geq_P [v]$. Assume that v has been chosen in W^J, namely that v is minimal-length in the coset $[v]$.

If $[v] >_P [s_i v]$ then $s_i v$ is the minimal-length element of $[s_i v]$ by Lemma 3.11. In this case we may take $b_1 = s_i$ in the reduced word for v. By induction we know

$$p_{[s_i w]}([v]) = \sum_{b_{i_0} \cdots b_{i_k} \in R(s_i w)} r_b(i_0) \cdots r_b(i_k)$$

which decomposes into the sum

$$p_{[s_i w]}([v]) = \alpha_i \sum_{b_{i_1} \cdots b_{i_k} \in R(w)} r_b(i_1) \cdots r_b(i_k) + \sum_{b_{i_0} \cdots b_{i_k} \in R(s_i w), i_0 \neq 1} r_b(i_0) \cdots r_b(i_k).$$

Similarly we conclude

$$p_{[s_iw]}([s_iv]) = s_i \left(\sum_{b_{i_0}\cdots b_{i_k} \in R(s_iw), i_0 \neq 1} r_b(i_0)\cdots r_b(i_k) \right).$$

Consequently

$$p_{[s_iw]}([v]) - s_i p_{[s_iw]}([s_iv]) = \alpha_i \sum_{b_{i_1}\cdots b_{i_k} \in R(w)} r_b(i_1)\cdots r_b(i_k).$$

Corollary 4.5 states that this quantity equals $\alpha_i p_{[w]}([v])$, which proves the claim.

If $[v] \leq_P [s_iv]$ then the element $s_iv > v$ again by Lemma 3.11. Choose a reduced factorization $b_0b_1\cdots b_{\ell(v)}$ for s_iv so that both $b_0 = s_i$ and $b = b_1\cdots b_{\ell(v)}$ is a reduced factorization for v. The inductive hypothesis shows that

$$\begin{aligned} p_{[s_iw]}([s_iv]) &= \sum_{b_{i_0}\cdots b_{i_k} \in R(s_iw)} r_{b_0b}(i_0)\cdots r_{b_0b}(i_k) \\ &= \alpha_i s_i \left(\sum_{b_{i_1}\cdots b_{i_k} \in R(w), i_1 \geq 1} r_b(i_1)\cdots r_b(i_k) \right) \\ &\quad + s_i \left(\sum_{b_{i_0}\cdots b_{i_k} \in R(s_iw), i_0 \geq 1} r_b(i_0)\cdots r_b(i_k) \right) \\ &= \alpha_i s_i \left(\sum_{b_{i_1}\cdots b_{i_k} \in R(w), i_1 \geq 1} r_b(i_1)\cdots r_b(i_k) \right) + s_i p_{[s_iw]}([v]). \end{aligned}$$

Again we conclude

$$\frac{p_{[s_iw]}([v]) - s_i p_{[s_iw]}([s_iv])}{\alpha_i} = \sum_{b_{i_1}\cdots b_{i_k} \in R(w)} r_b(i_1)\cdots r_b(i_k) = p_{[w]}([v]). \qquad \square$$

Billey proved that this expression is independent of the word chosen for v [**Bi2**]. It is not independent of the word chosen for $[w]$: for instance, both $s_1s_2s_1$ and s_1s_2 represent the class $[s_1s_2]$ in $G(1,3)$, but $p_{s_1s_2s_1}$ is a class of degree three while $p_{s_1s_2}$ has degree two. While $w \in [w]$ must be minimal-length, we will show that the localizations of p_w in $H_T^*(G/B)$ agree on all the different representatives in $[v]$.

To prove this, we use Kostant-Kumar's divided difference operator on the GKM presentation of $H_T^*(G/B)$ (see [**KK**] or [**Bi2**]). Kostant-Kumar showed that for each $w \in W$ the image $\partial_i^{KK} p_w = 0$ if $ws_i > w$. The next lemma follows immediately.

LEMMA 7.2. *For each $w \in W$ and flow-up class $p_w \in H_T^*(G/B)$, if s_i is a simple reflection such that $ws_i > w$ then $p_w(v) = p_w(vs_i)$ for all $v \in W$.*

PROOF. If $ws_i > w$ then $\partial_i^{KK} p_w = 0$, so $p_w(v) = p_w(vs_i)$ for all $v \in W$. $\square$

COROLLARY 7.3. *For each $w \in W^J$, flow-up class $p_w \in H_T^*(G/B)$, $v \in W$, and $u \in W_J$,*

$$p_w(v) = p_w(vu).$$

PROOF. If $w \in W^J$ then $ws_i > w$ for each $s_i \in W_J$. (This is a classical result proven by, e.g., Björner-Brenti [**BB**, Lemma 2.4.3].) The previous lemma showed $p_w(v) = p_w(vs_i)$ for all $v \in W$ and $s_i \in W_J$. The subgroup W_J is generated by $\{s_i : s_i \in W_J\}$, so $p_w(v) = p_w(vu)$ for all $u \in W_J$. $\square$

minimal $w \in [w]$	e	s_2	s_3s_2	s_1s_2	$s_1s_3s_2$	$s_2s_1s_3s_2$
$p_{[s_2]}([w])$	0	$t_2 - t_3$	$t_2 - t_4$	$t_1 - t_3$	$t_1 - t_4$	$t_1 - t_4 + t_2 - t_3$

FIGURE 7

Figure 7 gives the localizations of the class $p_{[s_2]}$ in $H_T^*(G(2,4))$, the Grassmannian of 2-planes in $\mathbb{C}^4$. The roots α_i are written as $t_i - t_{i+1}$.

Billey's formula also implies the following.

COROLLARY 7.4. *The map $\varphi : H_T^*(G/B) \to H_T^*(G/P)$ that sends the class $p = (p(v))_{v\in W}$ to $\varphi(p) = (p(v))_{v\in W^J}$ restricts to an $S(\mathfrak{t}^*)$-algebra isomorphism on the $S(\mathfrak{t}^*)$-span of $\{p_w : w \in W^J\}$.*

PROOF. Billey's formula proves that for each $w \in W^J$, the class $\varphi(p_w)$ satisfies all the edge relations specified by the GKM conditions for G/P. It also shows that $\varphi(p_w) = p_{[w]}$ is the flow-up class corresponding to $[w] \in W/W_J$. This confirms the image of the map $\varphi : \langle p_w : w \in W^J\rangle \to H_T^*(G/P)$. The algebra structure in the GKM presentation for both $H_T^*(G/B)$ and $H_T^*(G/P)$ is defined coordinate-wise, so φ is a $S(\mathfrak{t}^*)$-algebra isomorphism by definition. □

This isomorphism is inverse to the geometric map $H_T^*(G/P) \hookrightarrow H_T^*(G/B)$ induced from the projection $G/B \to G/P$ (e.g. [**Ku**, Corollary 11.3.14]). We could use this fact to give a different, geometric proof of the main theorem of this section. Indeed, the map $\pi : G/B \to G/P$ is a smooth fiber bundle so we have

$$\pi^{-1}(Y[w]) = Y(w)$$

where $Y[w]$ is the (opposite) Schubert variety $\overline{B^- wP/P}$ and similarly for $Y(w)$ in the flag variety. It follows that $\pi^*(p_[w]) = p_w$. Analyzing the fixed points in each case gives the claim.

8. Schubert polynomials for Grassmannians

Most of this paper has focused on a geometric construction of the (equivariant) cohomology of Grassmannians. This section will focus on an algebraic construction of the cohomology ring. In it, we propose an explicit collection of Schubert polynomials for the (ordinary) cohomology of Grassmannians. These Schubert polynomials are defined using the intrinsic geometry of Schubert varieties, as the average of localizations of a corresponding equivariant Schubert class. Lascoux also defined Grassmannian double Schubert polynomials by restricting the ordinary double Schubert polynomials to Grassmannian permutations [**L**], and see also [**IMN**], [**Tam**]. Our approach 1) clearly demonstrates the W_J-invariance of the Grassmannian Schubert polynomials, and 2) gives an explicit closed formula for the Grassmannian Schubert polynomials.

First we review the algebraic construction of the cohomology of the flag variety. Recall that the Weyl group W acts on the symmetric algebra $S(\mathfrak{t}^*)$, and let I be the ideal in $S(\mathfrak{t}^*)$ generated by W-invariant elements $f \in S(\mathfrak{t}^*)$ that satisfy $f(0) = 0$. Borel and Atiyah-Hirzebruch proved that there is a W-equivariant ring isomorphism $\Theta : S(\mathfrak{t}^*)/I \to H^*(G/B)$ [**Bo**], [**AH**]. Seminal work of Bernstein-Gelfand-Gelfand and Demazure identified explicitly a family of so-called Schubert polynomials $\mathfrak{S}_w \in S(\mathfrak{t}^*)$ that represent the Schubert classes [**BGG**], [**D**].

The choice of polynomial representatives of Schubert classes in $S(\mathfrak{t}^*)/I$ is not unique. Combinatorists have discussed passionately which properties of Schubert polynomials are essential. We use a modified version of Fomin-Kirillov's [**FK**]:

(1) the polynomial $\mathfrak{S}_w$ is homogeneous of degree $\ell(w)$;
(2) the polynomial $\mathfrak{S}_w$ is positive in the α_i, namely each coefficient is a non-negative integer;
(3) the $\mathfrak{S}_w$ respect BGG divided difference operators in the sense that

$$\partial_i \mathfrak{S}_w = \begin{cases} \mathfrak{S}_{ws_i} & \text{if } ws_i < w \text{ and} \\ 0 & \text{otherwise;} \end{cases}$$

(4) and the product of Schubert polynomials gives the structure constants c_{uv}^w in $H^*(G/B)$ up to the ideal I via the expression

$$\mathfrak{S}_u \mathfrak{S}_v = \sum c_{uv}^w \mathfrak{S}_w \mod I.$$

(In fact Fomin-Kirillov require positivity in a different set of variables.)

Bernstein-Gelfand-Gelfand and Demazure both defined Schubert polynomials by choosing a top-degree Schubert polynomial and then using divided difference operators (and Property (3) in the previous list) to define the others [**BGG**], [**D**]. For brevity, we refer to these polynomials as the BGG Schubert polynomials. The BGG Schubert polynomials satisfy Properties (1)–(4).

We will directly construct Schubert polynomials for $H^*(G/P)$. We will use another classical result of Bernstein-Gelfand-Gelfand: that $H^*(G/P)$ is isomorphic to the W_J-invariant subring of $S(\mathfrak{t}^*)/I$ [**BGG**, Theorem 5.5]. Our Schubert polynomials will also satisfy Properties (1)–(4). Note that the ideal I that appears in Property (4) is determined by G and is independent of the parabolic P.

Our first proposition, due to C. Cadman, gives a formula for the BGG Schubert polynomials in terms of the flow-up classes.

PROPOSITION 8.1. *(Cadman) The map $\kappa : H_T^*(G/B) \to S(\mathfrak{t}^*)/I$ defined by*

$$\kappa(p) = \frac{\sum_{w \in W} p(w)}{|W|}$$

is a degree-preserving $S(\mathfrak{t}^*)$-module homomorphism that restricts to a $\mathbb{C}$-module isomorphism on the submodule $H^*(G/B)$. The map sends $\kappa(p_w) = \mathfrak{S}_{w^{-1}}$ for each $w \in W$. The map κ is W-equivariant, namely $\kappa(w \cdot p) = w(\kappa(p))$ for all choices of $w \in W$ and $p \in H_T^*(G/B)$. Moreover, the map κ commutes with divided difference operators in the sense that $\kappa \circ \delta_i = \partial_i \circ \kappa$ for each i.

PROOF. The map κ is an $S(\mathfrak{t}^*)$-module homomorphism because for each element $w \in W$, the localization map $H_T^*(G/B) \to H_T^*(pt)$ given by $p \mapsto p(w)$ is an $S(\mathfrak{t}^*)$-module homomorphism. (It preserves degree by definition.) The submodule $H^*(G/B)$ is the $\mathbb{C}$-span in $H_T^*(G/B)$ of the equivariant Schubert classes $\{p_w : w \in W\}$, so κ restricts to a $\mathbb{C}$-module homomorphism on $H^*(G/B)$. For each $p \in H_T^*(G/B)$ and $w \in W$ we have

$$\frac{\sum_{v \in W} wp(w^{-1}v)}{|W|} = w\left(\frac{\sum_{v \in W} p(w^{-1}v)}{|W|}\right) = w\left(\frac{\sum_{v \in W} p(v)}{|W|}\right)$$

which proves that $\kappa(w \cdot p) = w(\kappa(p))$. In addition, for each $p \in H_T^*(G/B)$

$$\frac{\sum_{v \in W} (\delta_i p)(v)}{|W|} = \frac{\sum_{v \in W} (p(v) - s_i p(s_i v))}{|W| \alpha_i} = \partial_i \left(\frac{\sum_{v \in W} p(v)}{|W|}\right).$$

We prove that $\kappa(p_w) = \mathfrak{S}_{w^{-1}}$ by induction on the length of w. The BGG Schubert polynomial for the longest permutation is defined to be $\mathfrak{S}_{w_0} = \frac{\prod_{\alpha \in \Delta^+} \alpha}{|W|}$. The Schubert class p_{w_0} has localizations $p_{w_0}(w_0) = \prod_{\alpha \in \Delta^+} \alpha$ and $p_{w_0}(v) = 0$ if $v \neq w_0$. So the claim holds for the longest permutation.

Assume the claim holds for w of length k. If s_i is a simple reflection with $s_i w < w$ then $\delta_i p_w = p_{s_i w}$. Since κ commutes with δ_i, we conclude that we have $\kappa(p_{s_i w}) = \partial_i \mathfrak{S}_{w^{-1}} = \mathfrak{S}_{w^{-1}s_i}$. The claim holds by induction.

The $\mathbb{C}$-module homomorphism $\kappa : H^*(G/B) \to S(\mathfrak{t}^*)/I$ bijectively maps a free basis in the domain to a free basis in the range, so it is an isomorphism. □

Bernstein-Gelfand-Gelfand explicitly identified the images of their Schubert polynomials under the map of Borel and Atiyah-Hirzebruch; with our conventions, they found [**BGG**, Section I, Theorem (ii)]

$$\Theta\left(\mathfrak{S}_w I\right) = p_w. \tag{8.1}$$

The maps κ and Θ both use the same W-action on $S(\mathfrak{t}^*)/I$, but *different* W-actions on $H_T^*(G/B)$. The W-action on $H^*(G/B)$ that is used in [**Bo**], [**AH**], [**BGG**] is very complicated, making it difficult to identify W_J-invariant polynomials. (This W-action is discussed in the GKM setting in [**KK**]. In fact the graded $\mathbb{C}$-module $H^*(G/B)$ with the Kostant-Kumar W-action is the regular representation [**T2**].)

The W-action on $H_T^*(G/B)$ that we use in this paper is much simpler. Bernstein-Gelfand-Gelfand showed that $H^*(G/P)$ is isomorphic to the W_J-stable elements of $S(\mathfrak{t}^*)/I$ [**BGG**, Theorem 5.5]. We now construct Schubert polynomials for Grassmannians by restricting the map κ to W_J-stable classes in $H_T^*(G/B)$. To do that, we use minimal representatives of *left* W_J cosets.

THEOREM 8.2. *The ring $H^*(G/P)$ is isomorphic to the subring of $S(\mathfrak{t}^*)/I$ spanned by $\{\mathfrak{S}_w : w \in W^J\}$, where each $\mathfrak{S}_w$ is the image $\kappa(p_{w^{-1}})$. The Grassmannian Schubert polynomial $\mathfrak{S}_w$ is*

(1) *a homogeneous polynomial of degree $\ell_P([w])$;*
(2) *positive in the simple roots α_i;*
(3) *and gives the structure constants c_{uv}^w in $H^*(G/P)$ up to the ideal I via the expression*

$$\mathfrak{S}_u \mathfrak{S}_v = \sum c_{uv}^w \mathfrak{S}_w \mod I.$$

PROOF. The map κ is W-equivariant, so it suffices to identify W_J-invariant elements of $H_T^*(G/B)$. The equivariant Schubert class $p_w \in H_T^*(G/B)$ is W_J-invariant if and only if $s_i \cdot p_w = p_w$ for each $i \in J$, since the set $\{s_i : i \in J\}$ generates W_J. By Lemma 4.4, this is equivalent to the condition $s_i w > w$ for each $i \in J$, namely w is a minimal representative of its *left* W_J coset. In other words, the set $\{p_{w^{-1}} : w \in W^J\}$ is precisely the collection of W_J-invariant Schubert classes.

Applying κ, we see that $\{\mathfrak{S}_w : w \in W^J\}$ is a W_J-invariant set of polynomials. Each polynomial is positive in the simple roots α_i and homogeneous of the appropriate degree because the flow-up classes p_w are. BGG Schubert polynomials are linearly independent in $S(\mathfrak{t}^*)/I$ since they form a basis for $H^*(G/B)$, so $\{\mathfrak{S}_w : w \in W^J\}$ is linearly independent in the W_J-invariant submodule of $S(\mathfrak{t}^*)/I$. The map κ preserves degree so each $\mathfrak{S}_w$ is a homogeneous polynomial of degree $\ell(w) = \ell_P([w])$. Comparing degrees, we conclude that $\{\mathfrak{S}_w : w \in W^J\}$ is a basis for the W_J-invariant submodule of $S(\mathfrak{t}^*)/I$.

To see that these BGG Schubert polynomials satisfy the condition that

$$\mathfrak{S}_u \mathfrak{S}_v = \sum c_{uv}^w \mathfrak{S}_w \mod I$$

where c_{uv}^w are the structure constants of the Schubert classes $p_{[u]}$, $p_{[v]}$, and $p_{[w]}$ in $H^*(G/P)$, consider the composition $\Theta \circ \kappa$:

$$\begin{aligned} \Theta\left(\kappa\left(\langle p_{w^{-1}} : w \in W^J \rangle\right)\right) &= \Theta\left(\langle \mathfrak{S}_w : w \in W^J \rangle\right) \\ &= \langle p_w : w \in W^J \rangle. \end{aligned}$$

The map Θ is a $\mathbb{C}$-algebra isomorphism so the structure constants in the expansion $\mathfrak{S}_u \mathfrak{S}_v = \sum c_{uv}^w \mathfrak{S}_w \mod I$ agree with those in the expression $p_u p_v = \sum c_{uv}^w p_w$ for the full flag variety. By Theorem 7.1, if $w \in W^J$ then restricting the class $p_w \in H_T^*(G/B)$ to the fixed points in W^J gives the class $p_{[w]} \in H_T^*(G/P)$. Hence the coefficient $c_{[u],[v]}^{[w]}$ in the expression $p_{[u]} p_{[v]} = \sum c_{[u],[v]}^{[w]} p_{[w]}$ for the partial flag variety G/P satisfies $c_{[u],[v]}^{[w]} = c_{u,v}^w$ for $u, v, w \in W^J$. □

We stress that these Grassmannian Schubert polynomials are completely general, and in particular depend neither on the Lie type of the group nor on particular properties of the parabolic P.

Note that these Grassmannian Schubert polynomials have no divided difference operators, since ∂_i disrupts right cosets in W/W_J. Kostant-Kumar's divided difference operator [**KK**] is well-defined on the classes $\{p_{w^{-1}} : w \in W^J\}$, but it does not commute with the map κ.

8.1. Calculating Schubert polynomials. The most serious objection to Schubert polynomials without divided difference operators is: how could one compute them? Billey's formula in fact gives a closed formula for the Schubert polynomials we propose. (In fact, Billey's formula gives the BGG Schubert polynomials as well, though we only give examples for Grassmannians.) Lemma 4.4 shows that if $w \in W^J$ then $\kappa(p_w)$ can be computed either by averaging the entire class in G/B or by averaging the class in G/P and then summing over its W_J-orbit, namely

$$\kappa(p_w) = \frac{\sum_{v \in W_J} v\left(\sum_{u \in W^J} p_w(v)\right)}{|W|}.$$

(This formula does not reduce to scalar multiplication, since W_J can act nontrivially on the polynomial in the parentheses.) The Schubert polynomials in $\mathbb{P}^2$ are 1, $\frac{1}{3}(2t_1 - t_2 - t_3)$, and $\frac{1}{3}(t_1 - t_2)(t_1 - t_3)$. The reader may confirm that these polynomials are invariant under the subgroup generated by s_2. Note that they are positive not in the t_i but in the simple roots $\alpha_1 = t_1 - t_2$ and $\alpha_2 = t_2 - t_3$. Table 8 shows the Schubert polynomials for $G(2, 4)$. The Billey formula is particularly useful for maximal Grassmannians of classical type, where the moment graph has explicit combinatorial descriptions (such as in Section 3.4).

$w \in W^J$	$\mathfrak{S}_w$
$s_2s_1s_3s_2$	$\frac{1}{6}\alpha_2(\alpha_1+\alpha_2)(\alpha_2+\alpha_3)(\alpha_1+\alpha_2+\alpha_3)$
$s_1s_3s_2$	$\frac{\alpha_2(\alpha_1+\alpha_2)(\alpha_1+\alpha_2+\alpha_3)+\alpha_2(\alpha_2+\alpha_3)(\alpha_1+\alpha_2+\alpha_3)}{24}$ $+\frac{(\alpha_1+\alpha_2)(\alpha_2+\alpha_3)(\alpha_1+\alpha_2+\alpha_3)+\alpha_2(\alpha_1+\alpha_2)(\alpha_2+\alpha_3)}{24}$
s_1s_2	$\frac{1}{12}\left(\alpha_2(\alpha_1+\alpha_2)+(\alpha_2+\alpha_3)(\alpha_1+\alpha_2+\alpha_3)+2\alpha_3^2\right)$
s_3s_2	$\frac{1}{12}\left(\alpha_2(\alpha_2+\alpha_3)+(\alpha_1+\alpha_2)(\alpha_1+\alpha_2+\alpha_3)+2\alpha_1^2\right)$
s_2	$\frac{1}{2}(\alpha_1+2\alpha_2+\alpha_3)$
e	1

FIGURE 8. Schubert polynomials for $G(2,4)$

References

[A] Alberto Arabia, *Cohomologie T-équivariante de la variété de drapeaux d'un groupe de Kac-Moody* (French, with English summary), Bull. Soc. Math. France **117** (1989), no. 2, 129–165. MR1015806

[AH] M. F. Atiyah and F. Hirzebruch, *Vector bundles and homogeneous spaces*, Proc. Sympos. Pure Math., Vol. III, Amer. Math. Soc., Providence, RI, 1961, pp. 7–38. MR139181

[AJS] H. H. Andersen, J. C. Jantzen, and W. Soergel, *Representations of quantum groups at a pth root of unity and of semisimple groups in characteristic p: independence of p* (English, with English and French summaries), Astérisque **220** (1994), 321. MR1272539

[AF] David Anderson and William Fulton, *Equivariant cohomology in algebraic geometry*, Cambridge Studies in Advanced Mathematics, vol. 210, Cambridge University Press, Cambridge, 2024. MR4655919

[BGG] I. N. Bernšteĭn, I. M. Gel′fand, and S. I. Gel′fand, *Schubert cells, and the cohomology of the spaces G/P* (Russian), Uspehi Mat. Nauk **28** (1973), no. 3(171), 3–26. MR429933

[Bi1] Sara Billey, *Transition equations for isotropic flag manifolds*, Discrete Math. **193** (1998), no. 1-3, 69–84, DOI 10.1016/S0012-365X(98)00135-6. Selected papers in honor of Adriano Garsia (Taormina, 1994). MR1661363

[Bi2] Sara C. Billey, *Kostant polynomials and the cohomology ring for G/B*, Duke Math. J. **96** (1999), no. 1, 205–224, DOI 10.1215/S0012-7094-99-09606-0. MR1663931

[BB] Anders Björner and Francesco Brenti, *Combinatorics of Coxeter groups*, Graduate Texts in Mathematics, vol. 231, Springer, New York, 2005. MR2133266

[Bou] N. Bourbaki, *Groupes et Algèbres de Lie, Chp. IV-VI*, Masson, Paris, 1981.

[Bo] Armand Borel, *Sur la cohomologie des espaces fibrés principaux et des espaces homogènes de groupes de Lie compacts* (French), Ann. of Math. (2) **57** (1953), 115–207, DOI 10.2307/1969728. MR51508

[B] M. Brion, *Equivariant Chow groups for torus actions*, Transform. Groups **2** (1997), no. 3, 225–267, DOI 10.1007/BF01234659. MR1466694

[BKT] Anders Skovsted Buch, Andrew Kresch, and Harry Tamvakis, *A Giambelli formula for even orthogonal Grassmannians*, J. Reine Angew. Math. **708** (2015), 17–48, DOI 10.1515/crelle-2013-0071. MR3420327

[C] James B. Carrell, *The Bruhat graph of a Coxeter group, a conjecture of Deodhar, and rational smoothness of Schubert varieties*, Algebraic groups and their generalizations: classical methods (University Park, PA, 1991), Proc. Sympos. Pure Math., vol. 56, Amer. Math. Soc., Providence, RI, 1994, pp. 53–61, DOI 10.1090/pspum/056.1/1278700. MR1278700

[CLL] William Y. C. Chen, Bingqing Li, and J. D. Louck, *The flagged double Schur function*, J. Algebraic Combin. **15** (2002), no. 1, 7–26, DOI 10.1023/A:1013217015135. MR1879657

[CIFR] Giovanni Cerulli Irelli, Evgeny Feigin, and Markus Reineke, *Degenerate flag varieties: moment graphs and Schröder numbers*, J. Algebraic Combin. **38** (2013), no. 1, 159–189, DOI 10.1007/s10801-012-0397-6. MR3070124

[De] Michel Demazure, *Invariants symétriques entiers des groupes de Weyl et torsion* (French), Invent. Math. **21** (1973), 287–301, DOI 10.1007/BF01418790. MR342522

[D] Vinay V. Deodhar, *Some characterizations of Bruhat ordering on a Coxeter group and determination of the relative Möbius function*, Invent. Math. **39** (1977), no. 2, 187–198, DOI 10.1007/BF01390109. MR435249

[DP] Haibao Duan and Piotr Pragacz, *Divided differences of type D and the Grassmannian of complex structures*, Algebraic combinatorics and quantum groups, World Sci. Publ., River Edge, NJ, 2003, pp. 31–60, DOI 10.1142/9789812775405_0003. MR2035129

[FK] Sergey Fomin and Anatol N. Kirillov, *Combinatorial B_n-analogues of Schubert polynomials*, Trans. Amer. Math. Soc. **348** (1996), no. 9, 3591–3620, DOI 10.1090/S0002-9947-96-01558-9. MR1340174

[F] William Fulton, *Flags, Schubert polynomials, degeneracy loci, and determinantal formulas*, Duke Math. J. **65** (1992), no. 3, 381–420, DOI 10.1215/S0012-7094-92-06516-1. MR1154177

[GKM] Mark Goresky, Robert Kottwitz, and Robert MacPherson, *Equivariant cohomology, Koszul duality, and the localization theorem*, Invent. Math. **131** (1998), no. 1, 25–83, DOI 10.1007/s002220050197. MR1489894

[G] William Graham, *Positivity in equivariant Schubert calculus*, Duke Math. J. **109** (2001), no. 3, 599–614, DOI 10.1215/S0012-7094-01-10935-6. MR1853356

[GHZ] V. Guillemin, T. Holm, and C. Zara, *A GKM description of the equivariant cohomology ring of a homogeneous space*, J. Algebraic Combin. **23** (2006), no. 1, 21–41, DOI 10.1007/s10801-006-6027-4. MR2218848

[GZ] V. Guillemin and C. Zara, *1-skeleta, Betti numbers, and equivariant cohomology*, Duke Math. J. **107** (2001), no. 2, 283–349, DOI 10.1215/S0012-7094-01-10724-2. MR1823050

[H] James E. Humphreys, *Linear algebraic groups*, Graduate Texts in Mathematics, No. 21, Springer-Verlag, New York-Heidelberg, 1975. MR396773

[I] Takeshi Ikeda, *Schubert classes in the equivariant cohomology of the Lagrangian Grassmannian*, Adv. Math. **215** (2007), no. 1, 1–23, DOI 10.1016/j.aim.2007.04.008. MR2354984

[IM] Takeshi Ikeda and Tomoo Matsumura, *Pfaffian sum formula for the symplectic Grassmannians*, Math. Z. **280** (2015), no. 1-2, 269–306, DOI 10.1007/s00209-015-1423-x. MR3343907

[IMN] Takeshi Ikeda, Leonardo C. Mihalcea, and Hiroshi Naruse, *Double Schubert polynomials for the classical groups*, Adv. Math. **226** (2011), no. 1, 840–886, DOI 10.1016/j.aim.2010.07.008. MR2735777

[IN] Takeshi Ikeda and Hiroshi Naruse, *Excited Young diagrams and equivariant Schubert calculus*, Trans. Amer. Math. Soc. **361** (2009), no. 10, 5193–5221, DOI 10.1090/S0002-9947-09-04879-X. MR2515809

[K] A. Knutson, *A Schubert calculus recurrence from the noncomplex W-action on G/B*, `arXiv:math.CO/0306304`, 2003.

[KT] Allen Knutson and Terence Tao, *Puzzles and (equivariant) cohomology of Grassmannians*, Duke Math. J. **119** (2003), no. 2, 221–260, DOI 10.1215/S0012-7094-03-11922-5. MR1997946

[KM] Allen Knutson and Ezra Miller, *Gröbner geometry of Schubert polynomials*, Ann. of Math. (2) **161** (2005), no. 3, 1245–1318, DOI 10.4007/annals.2005.161.1245. MR2180402

[Ko] Bertram Kostant, *Lie algebra cohomology and the generalized Borel-Weil theorem*, Ann. of Math. (2) **74** (1961), 329–387, DOI 10.2307/1970237. MR142696

[KK] Bertram Kostant and Shrawan Kumar, *The nil Hecke ring and cohomology of G/P for a Kac-Moody group G*, Adv. in Math. **62** (1986), no. 3, 187–237, DOI 10.1016/0001-8708(86)90101-5. MR866159

[Kr] V. Kreiman, *Schubert classes in the equivariant K-theory and equivariant cohomology of the Grassmannian*, arXiv:math.AG/0512204, 2006.

[Ku] Shrawan Kumar, *Kac-Moody groups, their flag varieties and representation theory*, Progress in Mathematics, vol. 204, Birkhäuser Boston, Inc., Boston, MA, 2002, DOI 10.1007/978-1-4612-0105-2. MR1923198

[LRS] V. Lakshmibai, K. N. Raghavan, and P. Sankaran, *Equivariant Giambelli and determinantal restriction formulas for the Grassmannian*, Pure Appl. Math. Q. **2** (2006), no. 3, Special Issue: In honor of Robert D. MacPherson., 699–717, DOI 10.4310/PAMQ.2006.v2.n3.a5. MR2252114

[LLM] L. Lapointe, A. Lascoux, and J. Morse, *Determinantal expressions for Macdonald polynomials*, Internat. Math. Res. Notices **18** (1998), 957–978, DOI 10.1155/S1073792898000579. MR1653783

[L] Alain Lascoux, *Puissances extérieures, déterminants et cycles de Schubert* (French), Bull. Soc. Math. France **102** (1974), 161–179. MR364274

[LS] Alain Lascoux and Marcel-Paul Schützenberger, *Polynômes de Schubert* (French, with English summary), C. R. Acad. Sci. Paris Sér. I Math. **294** (1982), no. 13, 447–450. MR660739

[LR] Changzheng Li and Vijay Ravikumar, *Equivariant Pieri rules for isotropic Grassmannians*, Math. Ann. **365** (2016), no. 1-2, 881–909, DOI 10.1007/s00208-015-1266-0. MR3498930

[M] I. MacDonald, *Notes on Schubert polynomials*, LACIM, Univ. du Québec à Montréal, Montréal, 1991.

[Ma] Augustin-Liviu Mare, *Polynomial representatives of Schubert classes in $QH^*(G/B)$*, Math. Res. Lett. **9** (2002), no. 5-6, 757–769, DOI 10.4310/MRL.2002.v9.n6.a5. MR1906076

[MS] Dinesh P. Mehta and Sartaj Sahni (eds.), *Handbook of data structures and applications*, Chapman & Hall/CRC Computer and Information Science Series, Chapman & Hall/CRC, Boca Raton, FL, 2005. MR2122148

[MNS] Leonardo C. Mihalcea, Hiroshi Naruse, and Changjian Su, *Left Demazure-Lusztig operators on equivariant (quantum) cohomology and K-theory*, Int. Math. Res. Not. IMRN **16** (2022), 12096–12147, DOI 10.1093/imrn/rnab049. MR4465997

[P] D. Peterson, Lecture notes on Schubert calculus, MIT, Spring 1997.

[Pr] Piotr Pragacz, *Multiplying Schubert classes*, Topics in cohomological studies of algebraic varieties, Trends Math., Birkhäuser, Basel, 2005, pp. 163–174, DOI 10.1007/3-7643-7342-3_6. MR2143076

[PR] Piotr Pragacz and Jan Ratajski, *A Pieri-type formula for even orthogonal Grassmannians*, Fund. Math. **178** (2003), no. 1, 49–96, DOI 10.4064/fm178-1-2. MR1992238

[RYY] Colleen Robichaux, Harshit Yadav, and Alexander Yong, *The A·B·C·Ds of Schubert calculus*, Sém. Lothar. Combin. **85** ([2020), Art. B85a, 12. MR4226798

[Sp] T. A. Springer, *Linear algebraic groups*, Progress in Mathematics, vol. 9, Birkhäuser, Boston, MA, 1981. MR632835

[S] John R. Stembridge, *Tight quotients and double quotients in the Bruhat order*, Electron. J. Combin. **11** (2004/06), no. 2, Research Paper 14, 41, DOI 10.37236/1871. MR2120109

[Tam] Harry Tamvakis, *Double eta polynomials and equivariant Giambelli formulas*, J. Lond. Math. Soc. (2) **94** (2016), no. 1, 209–229, DOI 10.1112/jlms/jdw032. MR3532170

[TW] Harry Tamvakis and Elizabeth Wilson, *Double theta polynomials and equivariant Giambelli formulas*, Math. Proc. Cambridge Philos. Soc. **160** (2016), no. 2, 353–377, DOI 10.1017/S0305004115000754. MR3458957

[T] Julianna S. Tymoczko, *Permutation representations on Schubert varieties*, Amer. J. Math. **130** (2008), no. 5, 1171–1194, DOI 10.1353/ajm.0.0018. MR2450205

[T2] Julianna S. Tymoczko, *Permutation actions on equivariant cohomology of flag varieties*, Toric topology, Contemp. Math., vol. 460, Amer. Math. Soc., Providence, RI, 2008, pp. 365–384, DOI 10.1090/conm/460/09030. MR2428368

[T3] Julianna S. Tymoczko, *An introduction to equivariant cohomology and homology, following Goresky, Kottwitz, and MacPherson*, Snowbird lectures in algebraic geometry,

Contemp. Math., vol. 388, Amer. Math. Soc., Providence, RI, 2005, pp. 169–188, DOI 10.1090/conm/388/07264. MR2182897

Department of Mathematical Sciences, Smith College, Northampton Massachusetts 01063

Email address: jtymoczko@smith.edu

Contemporary Mathematics
Volume **837**, 2026
https://doi.org/10.1090/conm/837/16790

Exotic Spaltenstein varieties

Daniele Rosso and Neil Saunders

ABSTRACT. We define a new family of algebraic varieties, called exotic Spaltenstein varieties. These generalise the notion of Spaltenstein varieties (which are the partial flag analogues to classical Springer fibres) to the case of exotic Springer fibres. We show that, for self-adjoint nilpotent endomorphisms of order two, the top-dimensional irreducible components are in bijection with semi-standard Young bitableaux, via constructing an explicit map. Moreover, we are able to give a combinatorial formula for this top dimension. We conjecture that this description of the irreducible components holds for nilpotent endomorphisms of arbitrary order. Finally, we mention some connections to the Robinson-Schensted-Knuth correspondence.

1. Introduction

1.1. Classical Springer fibres and Spaltenstein varieties. Spaltenstein varieties are generalisations of the famous Springer fibres and can be defined for any algebraic group. Let G be a complex reductive algebraic group and let x be any element in the nilpotent cone $\mathcal{N}$ which is a singular subvariety of the Lie algebra $\mathrm{Lie}(G)$. Let B be a Borel subgroup of G and let $\mathfrak{b} = \mathfrak{h} \oplus \mathfrak{n}$ be the Lie algebra of B, where $\mathfrak{h}$ is the Cartan subalgebra and $\mathfrak{n} = [\mathfrak{b}, \mathfrak{b}]$. There is a resolution of singularities of $\mathcal{N}$, called the Springer resolution, and over each point $x \in \mathcal{N}$ its Springer fibre can be defined as $\mathcal{F}_x := \{gB \in G/B \mid gxg^{-1} \in \mathfrak{n}\}$, which are subvarieties of the full flag variety $\mathcal{B} := G/B$.

In Type A, for $G = \mathrm{GL}_n$, using the realisation of $\mathcal{B}$ as the variety of *complete flags* of vector spaces

$$0 = F_0 \subset F_1 \subset F_2 \subset \cdots \subset F_n = \mathbb{C}^n$$

where $\dim F_i = i$, we may identify the Springer fibre $\mathcal{F}_x$ as the subvariety of $\mathcal{B}$ stabilised by x, that is we have $x(F_i) \subset F_{i-1}$.

The geometry of these varieties is of great interest and plays a role in modern representation theory. For example, Springer [**20**], showed that the top degree cohomology of $\mathcal{F}_x$ has an action of the symmetric group of degree n, which makes it isomorphic to the Specht module $S^{\lambda^{tr}}$, where λ is the Jordan type of the nilpotent element x and λ^{tr} is the transpose partition. Moreoever, De Concini and Procesi [**4**]

2020 *Mathematics Subject Classification.* Primary 14M15, 14L35, 05E10, 05E14.

The first author was supported in this research by a Summer Faculty Fellowship and a Grant-in-Aid of research from Indiana University Northwest. The second author was supported by the London Mathematical Society's Scheme 4 *Research in Pairs* grant, ref. No. 41939.

showed that the total cohomology ring $H^*(\mathcal{F}_x, \mathbb{C})$ is isomorphic to a quotient of the polynomial ring $\mathbb{C}[x_1, \dots, x_n]$ by a certain ideal of elementary symmetric functions, for which there is an action of the symmetric group coming from an equivariant surjection from the cohomology of the flag variety.

Spaltenstein varieties arise from replacing the Borel subgroup by a parabolic subgroup P, and choosing a suitable Levi decomposition of Lie(P). Thus in type A, for a fixed nilpotent element x and parabolic P, the Spaltenstein variety corresponding to the triple $(G = \mathrm{GL}_n, P, x)$ is defined to be:

$$X_{P,x} = \{0 = F_0 \subset F_{\alpha_1} \subset F_{\alpha_2} \subset \cdots \subset F_{\alpha_d} = \mathbb{C}^n \mid \dim(F_{\alpha_i}/F_{\alpha_{i-1}}) = \alpha_i \text{ and } x(F_{\alpha_i}) \subset F_{\alpha_{i-1}}\}$$

and where $(\alpha_1, \dots, \alpha_d)$ is a composition of n (that is, a sequence of non-negative integers whose sum is n, see Definition 2.14).

In type A, Spaltenstein [**22**] proved that the Springer fibres and Spaltenstein varieties are pure dimensional - that is, all of the irreducible components have the same dimension. Moreover, he showed that the irreducible components are in bijection with semi-standard Young tableaux. Generalising De Concini and Procesi's results, Brundan and Ostrik [**2**] give a presentation for the cohomology of the Spaltenstein varieties, which can be viewed as an analogue of the Springer fibres sitting inside the partial flag variety.

Outside of type A, little is known about the pure dimensionality of the $X_{P,x}$ - Spaltenstein gave an example of an $X_{P,x}$ which is not pure dimensional for an $x \in \mathfrak{so}_8$. However, more recently in [**12**, Theorem A], Li proved that for a classical group G and fixed parabolic subgroup P and nilpotent element x, the Spaltenstein varieties $X_{P,x}$ are pure dimensional if the Jordan type of x is an even or odd partition - that is, of the form $(1^{m_1}3^{m_3}5^{m_5}\dots)$ or $(2^{m_2}4^{m_4}6^{m_6}\dots)$. By showing that the $X_{P,x}$ are Lagrangian in a certain partial resolution of a Slodowy slice, he is able to deduce further [**12**, Theorem C] that:

$$\dim(X_{P,x}) = \tfrac{1}{2}(\dim T^*(G/P) - \dim \mathcal{O}_x)$$

where $T^*(G/P)$ is the cotangent bundle of G/P and $\mathcal{O}_x$ is the nilpotent orbit of x.

1.2. Kato's exotic nilpotent cone and exotic Springer fibres. We continue with the notation as above. The combinatorics of the Springer correspondence in Type A are particularly nice in the sense that the orbits of GL_n on the nilpotent cone are in bijection with partitions of n, which parametrise the irreducible representations of its Weyl group (the symmetric group). This allowed Spaltenstein to show that there is a surjective map from $\mathcal{F}_x$ to std(λ) (standard Young tableaux of shape λ), which implies that the irreducible components of $\mathcal{F}_x$ are in bijection with std(λ).

For an algebraic group outside of type A, the action on the nilpotent cone of its Lie algebra has disconnected stabilisers, which means that the Springer correspondence needs the extra data of certain local systems on the nilpotent orbits, and therefore the combinatorics becomes more complicated when trying to match with the representations of its Weyl group.

However, Kato's exotic nilpotent cone $\mathfrak{N}$ (see Definition 2.6 and [**10**]) is a variation on the Type C setting which retains some of the combinatorial features seen in type A. In particular, the orbits of the action by the symplectic group Sp_{2n} on $\mathfrak{N}$ have connected stabilisers and are in bijection with the set $\mathcal{Q}_n$ of bipartitions of n, which parametrises the irreducible representations of the Weyl group of Type C (signed permutations). This enabled the authors in [**15**] to define an explicit map from the exotic Springer fibres (defined from a resolution of singularities of $\mathfrak{N}$) to standard Young bitableaux (SYB), which induced a bijection between the irreducible components of the exotic Springer fibres and the set of SYB ([**15**, Theorem 2.12]). One consequence of this explicit map, was to recover Kato's original result that showed that the exotic Springer fibres were of pure dimension. In addition to Kato's results, the authors were able to further describe the geometric structure of these fibres in low dimension, as either being projective spaces, or projectivised line bundles over projective spaces, and also combinatorially describe when and how the closures of these irreducible components intersected ([**15**, Section 8]).

Using a similar construction to Steinberg in [**19**], where he demonstrated that the relative position of two Springer fibres sitting in the flag variety was determined by the classical Robinson-Schensted correspondence, the authors further exploited the map between the exotic Springer fibres and SYB to define an *exotic Robinson-Schensted correspondence* that is a recursive bijection between pairs of SYB of the same shape and elements of the Weyl group of type C ([**16**, Section 3]).

In this paper, we define *exotic Spaltenstein varieties*, which are the partial flags/parabolic subgroups version of the exotic Spring fibres of Kato, and we study their geometry. These varieties are not pure dimensional in general (see Example 2.21) but in analogy to the results of [**22**] and [**15**] we conjecture (see Conjecture 2.19) that the top dimensional irreducible components are in bijection with semi-standard Young bitableaux, and we give a combinatorial formula for this dimension.

The main result of the paper is that we are able to prove that this is true in the cases where the nilpotent endomorphism x is such that $x^2 = 0$, see Theorem 2.23. We also provide some remarks about how a proof might proceed in general (see Section 6).

Our bijection between top-dimensional components of exotic Spaltenstein varieties and semi-standard Young bitableaux would give rise to an *exotic Robinson-Schensted-Knuth correspondence*, which should be a generalisation of our exotic Robinson-Schensted algorithm from [**16**], in a way that is analogous to the Type A case studied in [**18**] (see Section 7).

Finally, we briefly make some remarks about how these results fit into the wider context of studying varieties related to (the irreducible components of) Springer fibres corresponding to nilpotent endomorphisms of a given order, or corresponding to certain tableaux (or Young diagrams). It turns out that there are many results related to Springer fibres corresponding to nilpotent elements of degree 2, and so our work can be seen to be extending this area of research. While determining certain topological properties such as smoothness of irreducible components of Springer fibres and how the closures of these components intersect remain extremely difficult questions for these varieties, much work has been carried out for nilpotent endomorphisms corresponding to two-column tableaux, i.e. nilpotent endomorphisms of degree 2. In [**6**], Fresse considers irreducible components corresponding to a nilpotent endomorphism u of degree 2 that are fixed by the action of a standard torus relative to a Jordan basis of u. He is able to use that result to give necessary and sufficient conditions for the singularity of the components in that Springer fibre. He then extends and solves the question of describing components that are fixed by a torus to the case of nilpotent elements corresponding to hook, two-row and two-column tableaux in [**8**]. Moreover, with Melnikov in [**7**], they are able to give a criterion for when a component is singular in terms its Poincaré polynomial. Perrin and Smirnov in [**17**] show that irreducible components of Springer fibres of types A and D for nilpotent elements of degree 2 are normal varieties, and very recently, Manour in [**13**] gives a combinatorial formula counting the number of two-column tableaux corresponding to smooth components of a Springer fibre associated to a degree two nilpotent element.

For nilpotent elements of degree greater than 2, there are several papers that consider the number of Jordan blocks it has, or rows that the corresponding tableaux possesses. Im, Lai and Wilbert in [**9**] gave a description for the irreducible components of two-row Springer fibres in all classical types via cup and cap diagrams. In an analogous setting to the two-row Springer fibres for the exotic case, Wilbert and the second author in [**21**] studied exotic Springer fibres corresponding to 1-row bipartitions and gave a cup and cap description for their irreducible components. Finally, results for irreducible components and smoothness properties thereof, of so-called two-column Δ-Springer fibres have also recently been obtained in work of Connor, Griffin and Purohit in [**5**] and by Lacabanne, Vaz and Wilbert in [**11**] for two-row Δ-Springer varieties. Thus our work in defining and studying exotic Spaltenstein varieties fits well in the context of the existing literature and research in this area of Springer theory.

2. Background and notation

2.1. Symplectic spaces and isotropic Grassmannians. We let $\mathbb{N} = \{0, 1, 2, \ldots,\}$ be the set of nonnegative integers. All vector spaces will be assumed to be over the field $\mathbb{C}$ of complex numbers. Throughout the paper, we will have $n \in \mathbb{N}$ and V will be a $2n$-dimensional symplectic vector space, i.e. it is equipped with a nondegenerate, skew-symmetric bilinear form $\langle\ ,\ \rangle : V \times V \to \mathbb{C}$. If $W \subset V$, we denote by $W^\perp$ its perpendicular space with respect to the form $\langle\ ,\ \rangle$.

We define the symplectic group as the group of invertible linear transformations of V preserving the form

$$\mathrm{Sp}_{2n} = \mathrm{Sp}(V) := \{g \in \mathrm{GL}(V) \mid \langle gv, gw\rangle = \langle v, w\rangle,\ \forall v, w \in V\}.$$

Definition 2.1. For $0 \leq k \leq n$, we define the *isotropic Grassmannian variety* of all k-dimensional isotropic subspaces of V by

$$\mathrm{Gr}^\perp_{k,2n} = \mathrm{Gr}^\perp_k(V) = \{F \subset V \mid \dim(F) = k,\ F \subset F^\perp\}.$$

Remark 2.2. Notice that Sp_{2n} acts transitively on $\mathrm{Gr}^\perp_{k,2n}$, and it follows from [**3**, Prop 4.4] that $\mathrm{Gr}^\perp_{k,2n}$ is an irreducible projective variety of dimension $k(2n - k) - \frac{k(k-1)}{2}$.

If we restrict the bilinear form to a subspace $W \subset V$, in general it might be degenerate. So we will also need to consider degenerate skew-symmetric forms.

Definition 2.3. Let W be a vector space of dimension d, with a skew-symmetric bilinear form $\langle\ ,\ \rangle$, possibly degenerate. Consider the radical $\mathrm{rad}\langle\ ,\ \rangle = \{w \in W \mid \langle w, u\rangle = 0,\ \forall u \in W\}$. Then $W/\mathrm{rad}\langle\ ,\ \rangle$ is a symplectic space with the induced form on the quotient hence even dimensional. We say that $2r = \dim(W/\mathrm{rad}\langle\ ,\ \rangle)$ is the *rank* of the bilinear form, clearly $0 \leq r \leq \frac{d}{2}$.

In this case, for $0 \leq k \leq d-r$, we can define the (degenerate) isotropic Grassmannian

$$\mathrm{Gr}_{k,d}^{\perp r} = \mathrm{Gr}_k^{\perp}(W) = \{F \subset W \mid \dim(F) = k,\ F \subset F^{\perp}\}.$$

Note that for $d = 2n$ and $r = n$, we recover the previous Definition 2.1.

LEMMA 2.4. *The degenerate isotropic Grassmannian* $\mathrm{Gr}_{k,d}^{\perp r}$, *for all* $0 \leq r \leq \frac{d}{2}$ *and* $0 \leq k \leq d-r$ *is an irreducible projective variety of dimension* $k(d-k) - \frac{k(k-1)}{2} + \frac{(k-r)(k-r-1)}{2}$, *with the last term being zero when* $k \leq r$.

PROOF. We use the same idea as [**14**, §4.8]. Let $\{e_1, \ldots, e_d\}$ be a basis of W with the property that $e_1, \ldots, e_{d-2r}$ is a basis of $\mathrm{rad}\langle\ ,\ \rangle$, and such that $\langle e_i, e_j\rangle = \delta_{i+j,2d-2r+1}$ for all $1 \leq i \leq d-r$. We embed W in a symplectic vector space $\tilde{W}$, of dimension $2d-2r$, with basis $\{e_1, \ldots, e_{2d-2r}\}$ and non degenerate skew-symmetric bilinear form $\langle\ ,\ \rangle_\sim$ defined by $\langle e_i, e_j\rangle_\sim = \delta_{i+j,2d-2r+1}$ for all $1 \leq i \leq d-r$. In this way, the degenerate bilinear form on W becomes the restriction of the symplectic form $\langle\ ,\ \rangle_\sim$. Then the degenerate isotropic Grassmannian $\mathrm{Gr}_k^{\perp}(W)$ can be identified with a subvariety of the isotropic Grassmannian in $\tilde{W}$, more precisely

$$\mathrm{Gr}_k^{\perp}(W) = \{F \in \mathrm{Gr}_k^{\perp}(\tilde{W}) \mid F \subset W\}. \tag{2.1}$$

As in [**14**, §4.8], this is a Schubert variety in $\mathrm{Gr}_k^{\perp}(\tilde{W})$, i.e. the closure of a Schubert cell. We can distinguish two cases. When $k \leq r$, then (2.1) is the Schubert variety $X_{i_1,\ldots,i_k}$ for the index set $i_1 = d-k+1, \ldots, i_k = d$. When $r < k \leq d-r$, it's the Schubert variety for the index set $i_1 = d-r-k+1, \ldots, i_{k-r} = d-2r$, $i_{k-r+1} = d-r+1, \ldots, i_k = d$. In both cases, since the Schubert cell is isomorphic to an affine space, it is irreducible and so it its closure.

To conclude the proof we just need to compute the dimensions of the Schubert cells in the two cases. To do this, we will instead use the formulas from the proof of [**3**, Prop 4.4] (it would also be possible to use [**3**, Prop 4.4] directly by first translating the index set into a partition). Each index i_j gives a dimension of $i_j - j - \#\{\ell < j \mid i_\ell + i_j > 2d-2r+1\}$, hence

$$\begin{aligned}\dim X_{i_1,\ldots,i_k} &= \sum_{j=1}^{k}(i_j - j - \#\{\ell < j \mid i_\ell + i_j > 2d-2r+1\}) \\ &= \left(\sum_{j=1}^{k} i_j\right) - \frac{k(k+1)}{2} \\ &\quad - \#\{(\ell, j) \mid 1 \leq \ell < j \leq k,\ i_\ell + i_j > 2d-2r+1\}.\end{aligned}$$

When $k \leq r$, then $i_j = d-k+j$, so $\#\{(\ell, j) \mid 1 \leq \ell < j \leq k,\ i_\ell + i_j > 2d-2r+1\} = \binom{k}{2} = \frac{k(k-1)}{2}$, and we get indeed that

$$\dim X_{i_1,\ldots,i_k} = k(d-k) - \frac{k(k-1)}{2}.$$

When $k > r$, then $i_j = d-r-k+j$ for $1 \leq j \leq k-r$, and $i_j = d-k+j$ for $k-r+1 \leq j \leq k$, so $\#\{(\ell, j) \mid 1 \leq \ell < j \leq k,\ i_\ell + i_j > 2d-2r+1\} = \binom{r}{2} = \frac{r(r-1)}{2}$ and with a short computation we get indeed that

$$\dim X_{i_1,\ldots,i_k} = k(d-k) - \frac{k(k-1)}{2} + \frac{(k-r)(k-r-1)}{2}.$$

□

REMARK 2.5. Notice that if $k = r + 1$, the term $\frac{(k-r)(k-r-1)}{2}$ is again equal to zero, so in the formula of Lemma 2.4, the third term can be ignored whenever $k \leq r + 1$.

2.2. Exotic nilpotent cone. Let V be a $2n$-dimensional symplectic space V, with symplectic group $\mathrm{Sp}_{2n} = \mathrm{Sp}(V)$, we identify its Lie algebra as follows

$$\mathfrak{sp}_{2n} := \mathrm{Lie}(\mathrm{Sp}_{2n}) = \{x \in \mathrm{End}(V) \mid \langle xv, w\rangle + \langle v, xw\rangle = 0, \ \forall v, w \in V\}.$$

The adjoint action of Sp_{2n} on $\mathfrak{sp}_{2n}$ can be interpreted as the restriction of the Sp_{2n}-action on $\mathfrak{gl}_{2n} = \mathrm{End}(V)$ given by conjugation. This action gives a direct sum decomposition of Sp_{2n}-modules $\mathfrak{gl}_{2n} = \mathfrak{sp}_{2n} \oplus \mathcal{S}$, where $\mathcal{S}$ can also be described directly as

$$\mathcal{S} = \{x \in \mathrm{End}(V) \mid \langle xv, w\rangle = \langle v, xw\rangle, \ \forall v, w \in V\}.$$

Finally we let $\mathcal{N} := \{x \in \mathrm{End}(V) \mid x^k = 0, \text{ for some } k\}$ be the *nilpotent cone* of the Lie algebra of $\mathrm{GL}_{2n} = \mathrm{GL}(V)$.

DEFINITION 2.6. The *exotic nilpotent cone* is the (singular) variety $\mathfrak{N} := V \times (\mathcal{S} \cap \mathcal{N})$.

When it is not clear from the context, we might specify the underlying vector space and write $\mathcal{S}(V)$, $\mathcal{N}(V)$ or $\mathfrak{N}(V)$.

If $k \in \mathbb{N}$, a *composition* of k is a finite sequence $\alpha = (\alpha_1, \alpha_2, \ldots, \alpha_m)$ such that for all $1 \leq i \leq m$, α_i is a positive integer, and $\sum_{i=1}^{m} \alpha_i = k$. We denote this by $|\alpha| = k$ or $\alpha \vDash k$. If $\alpha = (\alpha_1, \alpha_2, \ldots, \alpha_m)$, then $\ell(\alpha) := m$, the *length* of the composition is the number of its parts.

A *partition* of k is a composition of k where the integers are weakly decreasing, that is $\lambda = (\lambda_1, \ldots, \lambda_m)$ is such that $\lambda_1 \geq \ldots \geq \lambda_m > 0$ and $\sum_{i=1}^{k} \lambda_i = k$. We denote it by $\lambda \vdash k$ or $|\lambda| = k$. On occasion it will be convenient for us to consider a partition to have some (or even infinitely many) parts of size zero appended to its end, this does not change the length of the partition which is the number of nonzero parts.

DEFINITION 2.7. If W is a vector space and $x \in \mathrm{End}(W)$ is a nilpotent transformation, its Jordan canonical form gives a partition of $\dim(W)$ by considering the sizes of the Jordan blocks (in weakly decreasing order). We denote this partition by $\mathrm{J}(x)$ and we call it the *Jordan type* of x.

If we have two partitions μ, ν (of potentially different integers), we define two new partitions $\mu+\nu := (\mu_1+\nu_1, \mu_2+\nu_2, \ldots)$, and $\mu \cup \nu$ which is the unique partition obtained by reordering the sequence $(\mu_1, \ldots, \mu_{\ell(\mu)}, \nu_1, \ldots, \nu_{\ell(\nu)})$ to make it weakly decreasing. A *bipartition* of n is a pair (μ, ν) of partitions such that $|\mu| + |\nu| = n$. We denote the set of all bipartitions of n by $\mathcal{Q}_n$.

The set $\mathcal{Q}_n$ is important for us because of the following result.

THEOREM 2.8 ([**1**, Thm 6.1]). *The orbits of the symplectic group* Sp_{2n} *on the exotic nilpotent cone* $\mathfrak{N}$ *are in bijection with* $\mathcal{Q}_n$.

More precisely, following Section 2 and Section 6 of [**1**] we can say that, given a bipartition $(\mu, \nu) \in \mathcal{Q}_n$, the corresponding orbit $\mathbb{O}_{(\mu,\nu)}$ contains the point (v, x) if and only if there is a 'normal' basis of V given by

$$\{v_{ij}, v_{ij}^* \mid 1 \leq i \leq \ell(\mu+\nu), 1 \leq j \leq \mu_i + \nu_i\},$$

with $\langle v_{ij}, v^*_{i'j'}\rangle = \delta_{i,i'}\delta_{j,j'}$, $v = \sum_{i=1}^{\ell(\mu)} v_{i,\mu_i}$ and such that the action of x on this basis is as follows:

$$xv_{ij} = \begin{cases} v_{i,j-1} & \text{if} \quad j \geq 2 \\ 0 & \text{if} \quad j = 1 \end{cases} \qquad xv^*_{ij} = \begin{cases} v^*_{i,j+1} & \text{if} \quad j \leq \mu_i + \nu_i - 1 \\ 0 & \text{if} \quad j = \mu_i + \nu_i \end{cases},$$

in particular the Jordan type of x is $(\mu + \nu) \cup (\mu + \nu)$.

DEFINITION 2.9. Similarly to Definition 2.7, if $(v, x) \in \mathbb{O}_{(\mu,\nu)}$, as defined above, we say that the bipartition (μ, ν) is the *exotic Jordan type* of (v, x) and we denote it by $\mathrm{eJ}(v, x) = (\mu, \nu)$.

Associated to a partition $\lambda = (\lambda_i)_i$ there is a Young diagram consisting of λ_i boxes on row i. We say that λ is the *shape* of the diagram. In the same way, a bipartition gives a pair of diagrams. If a Young diagram is filled with positive integers, we call it a Young tableau, and similarly, if we fill a pair of diagrams with positive integers we call it a bitableau.

EXAMPLE 2.10. The bipartition $(\mu, \nu) = ((3, 1), (2, 2, 1))$ corresponds to the pair of Young diagrams 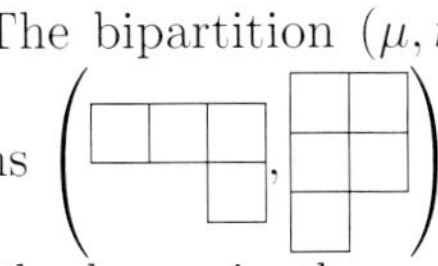, notice that we follow the usual convention of left-justifying the boxes in the second diagram, however we instead right-justify the boxes in the first diagram (equivalenty, we apply a vertical reflexion to a usual diagram). The reason for this comes from the look of normal bases, as in Example 2.11. An example of a bitableau in that shape would be 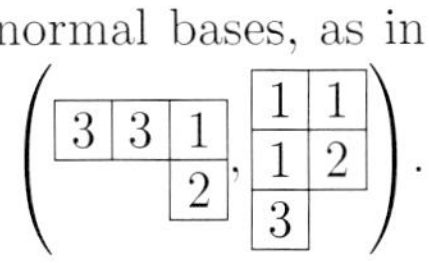

.

EXAMPLE 2.11. Consider the bipartition $(\mu, \nu) = ((3, 1), (2, 2, 1))$, and a point $(v, x) \in \mathbb{O}_{(\mu,\nu)}$. Then we can represent the 'normal' basis as

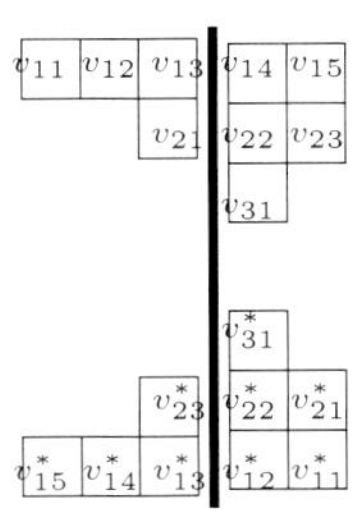

with the action of x given by moving one block left (and zero if there is nothing further left), and $v = v_{13} + v_{21}$ is given by the sum of the boxes just left of the dividing wall on the upper half of the diagram. Notice that the wall divides the two diagrams corresponding to the partitions that form the bipartition (the one on the left of the wall is facing backwards) and that the bipartition is repeated twice since $\mathrm{J}(x) = (\mu + \nu) \cup (\mu + \nu)$.

2.3. Exotic Spaltenstein varieties. We are now going to define the main object of study of this article and state our main conjecture about the irreducible components of exotic Spaltenstein varieties, which we will be able to prove in the case where $x^2 = 0$.

Let $\alpha = (\alpha_1, \alpha_2, \ldots, \alpha_m) \vDash n$ be a composition, we define a composition of $2n$ with $2m$ parts, $\hat{\alpha} = (\alpha_1, \alpha_2, \ldots, \alpha_m, \alpha_m, \alpha_{m-1}, \ldots, \alpha_1) \vDash 2n$. We also define,

$\check{\alpha}_i = \sum_{j=1}^{i} \hat{\alpha}_j$ for $i = 0, \dots, 2m$. We denote by $\mathcal{F}^\alpha(V)$ the variety of *partial symplectic flags* of type α in V, that is each $F_\bullet \in \mathcal{F}^\alpha(V)$ is a sequence of subspaces

$$F_\bullet = (0 = F_0 \subseteq F_{\check{\alpha}_1} \subseteq \cdots \subseteq F_{\check{\alpha}_m} \subseteq \cdots \subseteq F_{\check{\alpha}_{2m-1}} \subseteq F_{\check{\alpha}_{2m}} = V)$$

such that $\dim(F_{\check{\alpha}_i}/F_{\check{\alpha}_{i-1}}) = \hat{\alpha}_i$ and $F_{\check{\alpha}_i}^\perp = F_{\check{\alpha}_{2m-i}}$ for all $1 \leq i \leq 2m$.

Notice that $F_{\check{\alpha}_m} = F_n = F_n^\perp$ is a maximal isotropic subspace of V.

DEFINITION 2.12. Given $(v, x) \in \mathfrak{N}$, and $\alpha = (\alpha_1, \dots, \alpha_m) \vDash n$, we define the *exotic Spaltenstein variety*

$$\mathcal{C}^\alpha_{(v,x)} := \{F_\bullet = (0 \subseteq F_{\check{\alpha}_1} \subseteq \cdots \subseteq F_{\check{\alpha}_{2m}} = V) \in \mathcal{F}^\alpha(V) \mid v \in F_n,\ x(F_{\check{\alpha}_i}) \subseteq F_{\check{\alpha}_{i-1}}\}.$$

REMARK 2.13. Notice that for certain choices of v, x and α the variety could be empty. It is clear that if $\mathrm{eJ}(v, x) = \mathrm{eJ}(v', x')$, then there is an isomorphism of varieties $\mathcal{C}^\alpha_{(v,x)} \cong \mathcal{C}^\alpha_{(v',x')}$ given by the Sp_{2n}-action. When $\alpha = 1^n$, then the partial flags are actually complete flags and this variety is an exotic Springer fibre, as defined by Kato in [**10**] and studied by the authors in [**15, 16**].

Let $(v, x) \in \mathfrak{N}$, if $F_\bullet \in \mathcal{C}^\alpha_{(v,x)}$, for all $i = 0, \dots, m$, since $F_{\check{\alpha}_i}$ and $F_{\check{\alpha}_i}^\perp = F_{\check{\alpha}_{2m-i}}$ are both invariant under x, we can consider the restriction of x to $F_{\check{\alpha}_i}^\perp/F_{\check{\alpha}_i}$, which is a symplectic vector space of dimension $2(n - \check{\alpha}_i)$ (since $F_{\check{\alpha}_i}$ is isotropic, the bilinear form descends to the quotient and is nondegenerate because we restrict to $F_{\check{\alpha}_i}^\perp$). It can be easily verified that $x|_{F_{\check{\alpha}_i}^\perp/F_{\check{\alpha}_i}} \in \mathcal{N}(F_{\check{\alpha}_i}^\perp/F_{\check{\alpha}_i}) \cap \mathcal{S}(F_{\check{\alpha}_i}^\perp/F_{\check{\alpha}_i})$.

DEFINITION 2.14. Let $\alpha = (\alpha_1, \alpha_2, \dots, \alpha_m) \vDash n$ be a composition. Let $(v, x) \in \mathfrak{N}$, $\mathrm{eJ}(v, x) = (\mu, \nu)$. Define the following maps,

$$\Phi^i : \mathcal{C}^\alpha_{(v,x)} \to \mathcal{Q}_{n-\check{\alpha}_i} \qquad F_\bullet \mapsto \mathrm{eJ}\left(v + F_{\check{\alpha}_i}, x|_{F_{\check{\alpha}_i}^\perp/F_{\check{\alpha}_i}}\right) \qquad \forall\, 0 \leq i \leq m;$$

$$\Phi : \mathcal{C}^\alpha_{(v,x)} \to \prod_{i=m}^{0} \mathcal{Q}_{n-\check{\alpha}_i}$$

$$F_\bullet \mapsto \left((\emptyset, \emptyset) = \Phi^m(F_\bullet), \Phi^{m-1}(F_\bullet), \dots, \Phi^1(F_\bullet), \Phi^0(F_\bullet) = (\mu, \nu)\right).$$

This map assigns to each partial flag in the exotic Spaltenstein variety a sequence of bipartitions, such that the total size increases by α_i boxes at each step. Notice that in general there is no guarantee that the shape of $\Phi^i(F_\bullet)$ will contain the shape of $\Phi^{i+1}(F_\bullet)$. If the sequence of bipartitions is nested at each step, we can identify the sequence with a bitableau of shape (μ, ν), where at each step $i = m-1, m-2, \dots, 0$ we fill the boxes that have been added with the number $m - i$.

EXAMPLE 2.15.

(1) Let $n = 2$, $(\mu, \nu) = (1^2, \emptyset) = \left(\begin{array}{|c|}\hline \\ \hline \\ \hline \end{array}, \emptyset\right)$, and $(v, x) \in \mathbb{O}_{(\mu,\nu)}$, then we have a basis $\{v_{11}, v_{21}, v_{21}^*, v_{11}^*\}$ for V, with $v = v_{11} + v_{21}$. If $\alpha = (2)$, and $F_\bullet = (0 \subseteq F_2 \subseteq V)$ with $F_2 = \mathbb{C}v_{11} + \mathbb{C}v_{21}$ (or any subspace of V containing v), then $F_\bullet \in \mathcal{C}^\alpha_{(v,x)}$, and since $F_2^\perp/F_2 = 0$ we have $\Phi(F_\bullet) = \left((\emptyset, \emptyset), \left(\begin{array}{|c|}\hline \\ \hline \\ \hline \end{array}, \emptyset\right)\right)$. This sequence is nested and corresponds to the bitableau $\left(\begin{array}{|c|}\hline 1 \\ \hline 1 \\ \hline \end{array}, \emptyset\right)$.

(2) With (μ, ν) as above, if $\alpha = (1,1)$ and $F_\bullet = (0 \subseteq F_1 \subseteq F_2 \subseteq F_1^\perp \subseteq V)$ with $F_1 = \mathbb{C}v$, $F_2 = \mathbb{C}v_{11} + \mathbb{C}v_{21}$, then since $v \in F_1$ we have that $v + F_1$ is zero in the quotient $F_1^\perp/F_1$, which is a 2 dimensional space, while $F_2^\perp/F_2 = 0$, so $\Phi(F_\bullet) = \left((\emptyset,\emptyset),(\emptyset,\square),\left(\begin{smallmatrix}\square\\\square\end{smallmatrix},\emptyset\right)\right)$. This sequence of bipartitions is not nested, hence it does not correspond to a bitableau.

(3) Now let $(\mu,\nu) = (\emptyset, 2) = (\emptyset, \square\square)$, and $(v,x) \in \mathbb{O}_{(\mu,\nu)}$, then we have a basis $\{v_{11}, v_{12}, v_{12}^*, v_{11}^*\}$ for V, and $v = 0$. If $\alpha = (2)$, and $F_\bullet = (0 \subseteq F_2 \subseteq V)$ with $F_2 = \mathbb{C}v_{11} + \mathbb{C}v_{12}^*$, then $F_\bullet \in \mathcal{C}^\alpha_{(v,x)}$, and since $F_2^\perp/F_2 = 0$ we have $\Phi(F_\bullet) = ((\emptyset,\emptyset),(\emptyset,\square\square))$, which is a nested sequence corresponding to the bitableau $(\emptyset, \boxed{1}\boxed{1})$.

Definition 2.16. Let $\alpha = (\alpha_1, \dots, \alpha_m) \vDash n$, we say that a bitableau of shape $(\mu, \nu) \in \mathcal{Q}_n$ is *semistandard* with content α if for all $i = 1, \dots, m$ it contains α_i boxes with the number i, such that the numbers are strictly increasing along rows (right to left in the first tableau, left to right in the second tableau) and weakly increasing down columns. Equivalently, if we consider the bitableau as a nested sequence of bipartitions, it is semistandard if at each step we add a vertical strip to each of the two Young diagrams (no two boxes added are in the same row of the same diagram).

We denote by $\mathcal{T}^\alpha_{\mu,\nu}$ be the set of all semistandard bitableaux of shape (μ,ν) and content α.

Remark 2.17. Notice that our convention is the transpose of the usual convention for semistandard tableaux (usually they are taken to be increasing strictly down columns and weakly along rows) but it matches the convention used in [**18**]. In Example 2.15, the bitableau in case (1) is semistandard, while the one in case (3) is not.

Let $\lambda = (\lambda_i)_i \vdash n$ be a partition, we define $N(\lambda) = \sum_i (i-1)\lambda_i$. If $(\mu,\nu) \in \mathcal{Q}_n$ is a bipartition and $\alpha = (\alpha_1, \dots, \alpha_m) \vDash n$ is a composition, we define

$$d^\alpha_{(\mu,\nu)} := 2N(\mu + \nu) + |\nu| - \frac{1}{2}\sum_{i=1}^{m}(\alpha_i^2 - \alpha_i). \tag{2.2}$$

Remark 2.18. Recall that $N(\lambda)$ is the dimension of the Springer fibre over a nilpotent element of Jordan type λ, while when $\alpha = 1^n$, we have $d^{(1^n)}_{\mu,\nu} = 2N(\mu + \nu) + |\nu|$ is the dimension of the exotic Springer fibre over a pair $(v,x) \in \mathcal{N}$ with $\mathrm{eJ}(v,x) = (\mu,\nu)$.

Conjecture 2.19. *Let $T \in \mathcal{T}^\alpha_{\mu,\nu}$ be a semistandard bitableau of shape (μ,ν) and content α, and let $(v,x) \in \mathbb{O}_{(\mu,\nu)}$. Then $\overline{\Phi^{-1}(T)}$ is an irreducible component of $\mathcal{C}^\alpha_{(v,x)}$ of dimension $d^\alpha_{\mu,\nu}$ and all irreducible components of dimension $d^\alpha_{\mu,\nu}$ are of this form. All other irreducible components of $\mathcal{C}^\alpha_{(v,x)}$ have dimension strictly lower than $d^\alpha_{\mu,\nu}$.*

Remark 2.20. In the case of complete flags, when $\alpha = 1^n$, the variety is an exotic Springer fibre and the result is true (see [**15**]). In that case all irreducible components have the same dimension, but in general $\mathcal{C}^\alpha_{(v,x)}$ is not pure dimensional, as the following example shows.

EXAMPLE 2.21. Let $n = 3$, $(\mu, \nu) = (\emptyset, 21)$, $\alpha = (1, 2)$, then $v = 0$ and we have $F_\bullet \in \mathcal{C}^\alpha_{(v,x)}$ if $F_\bullet = (0 \subset F_1 \subset F_3 = F_3^\perp \subset F_1^\perp \subset V)$ with $F_1 \in \ker(x)$ and $x(F_3) \subset F_1$. The subvariety of $\mathcal{C}^\alpha_{(v,x)}$ where $F_1 \subset \operatorname{Im}(x)$ is an irreducible component of dimension $d^\alpha_{(\mu,\nu)} = 2(1) + 3 - \frac{1}{2}(1^2 - 1) - \frac{1}{2}(2^2 - 2) = 2 + 3 - 0 - 1 = 4$, in fact it equals $\Phi^{-1}\left(\emptyset, \begin{array}{|c|c|}\hline 1 & 2 \\ \hline 1 \\ \cline{1-1}\end{array}\right)$. The subvariety of $\mathcal{C}^\alpha_{(v,x)}$ where $F_1 \not\subset \operatorname{Im}(x)$ however is an irreducible locally closed subvariety, whose closure is an irreducible component of dimension 3, for these flags the corresponding sequence of bipartitions is still nested,

$$\Phi(F_\bullet) = \left((\emptyset, \emptyset), \left(\emptyset, \begin{array}{|c|c|}\hline & \\ \hline\end{array}\right), \left(\emptyset, \begin{array}{|c|c|}\hline & \\ \hline \\ \cline{1-1}\end{array}\right)\right) = \left(\emptyset, \begin{array}{|c|c|}\hline 1 & 1 \\ \hline 2 \\ \cline{1-1}\end{array}\right)$$

but the resulting bitableau is not semistandard.

2.4. Strategy for the proof of Conjecture 2.19. We induct on m, the number of parts of the composition $\alpha = (\alpha_1, \alpha_2, \ldots, \alpha_m) \vDash n$. For $m = 0$, then $n = 0$ as well and the statement is trivial (all the varieties involved are one single point), so we can assume that $m \geq 1$.

Let $(\mu, \nu) \in \mathcal{Q}_n$, $\alpha = (\alpha_1, \alpha_2, \ldots, \alpha_m) \vDash n$, $(v, x) \in \mathbb{O}_{(\mu,\nu)}$, and let $B = ((\mu^{[m]}, \nu^{[m]}), \ldots, (\mu^{[0]}, \nu^{[0]}))$ be a sequence of bipartitions such that $(\mu^{[i]}, \nu^{[i]}) \in \mathcal{Q}_{n-\check{\alpha}_i}$, for $i = m, \ldots, 0$ and $(\mu^{[0]}, \nu^{[0]}) = (\mu, \nu)$. Then if $\emptyset \neq \Phi^{-1}(B) \subseteq \mathcal{C}^\alpha_{v,x}$ we have a map

$$\begin{aligned} p : \Phi^{-1}(B) &\longrightarrow \operatorname{Gr}^\perp_{\alpha_1}(\ker(x) \cap (\mathbb{C}[x]v)^\perp); \\ F_\bullet &\mapsto F_{\alpha_1}. \end{aligned}$$

Define $(\mu', \nu') = (\mu^{[1]}, \nu^{[1]}) = \Phi^1(F_\bullet) = \operatorname{eJ}\left(v + F_{\alpha_1}, x|_{F^\perp_{\alpha_1}/F_{\alpha_1}}\right)$, $\alpha' = (\alpha_2, \ldots, \alpha_m)$, and $B' = ((\mu^{[m]}, \nu^{[m]}), \ldots, (\mu^{[1]}, \nu^{[1]}))$. We let

$$\mathcal{B}^{((\mu,\nu),\alpha)}_{((\mu',\nu'),\alpha')} := \operatorname{im} p = \{F \in \operatorname{Gr}^\perp_{\alpha_1}(\ker(x) \cap (\mathbb{C}[x]v)^\perp) \mid \operatorname{eJ}\left(v + F, x|_{F^\perp/F}\right) = (\mu', \nu')\}.$$

For any $F \in \mathcal{B}^{((\mu,\nu),\alpha)}_{((\mu',\nu'),\alpha')}$, we then have that

$$\begin{aligned} p^{-1}(F) &= \{F'_\bullet \in \Phi^{-1}(B) \mid F'_{\alpha_1} = F\} \\ &\cong \left\{\overline{F}_\bullet \in \mathcal{C}^{\alpha'}_{(v+F, x|_{F^\perp/F})} \subseteq \mathcal{F}^{\alpha'}(F^\perp/F) \mid (\Phi^{m-1}(\overline{F}_\bullet), \ldots, \Phi^0(\overline{F}_\bullet)) = B'\right\} \end{aligned} \tag{2.3}$$

where the isomorphism is given by

$$\begin{gathered} (F'_0, F'_1, \ldots, F'_{2m-1}, F'_{2m}) \mapsto (\overline{F}_0, \overline{F}_1, \ldots, \overline{F}_{2m-3}, \overline{F}_{2m-2}); \\ \overline{F}_i = F'_{i+1}/F \quad \forall i = 0, \ldots, 2m-2. \end{gathered}$$

In conclusion, the map p is a fibre bundle, and from (2.3), it follows that the fibres are $p^{-1}(F) \cong \Phi^{-1}(B')$, which by inductive hypothesis is an irreducible variety of dimension lesser or equal to $d^{\alpha'}_{\mu',\nu'}$, with equality if and only if the sequence B' is actually a semistandard tableau. Thus Conjecture 2.19 would be a consequence of the following result.

CONJECTURE 2.22. *For all $(\mu, \nu) \in \mathcal{Q}_n$ and for all $(\mu', \nu') \in \mathcal{Q}_{n-\alpha_1}$ the variety $\mathcal{B}^{((\mu,\nu),\alpha)}_{((\mu',\nu'),\alpha')}$ is irreducible, and we have*

$$\dim \mathcal{B}^{((\mu,\nu),\alpha)}_{((\mu',\nu'),\alpha')} \leq d^\alpha_{(\mu,\nu)} - d^{\alpha'}_{(\mu',\nu')},$$

with equality if and only if (μ', ν') is obtained from (μ, ν) by removing α_1 boxes in a vertical strip in each diagram (no two of the boxes are in the same row of the same diagram).

In Section 5, we will prove that Conjecture 2.22 holds when $x^2 = 0$, hence our main result for this paper is the following.

THEOREM 2.23. *Let $\alpha \vDash n$, $(\mu, \nu) \in \mathcal{Q}_n$, and $(v, x) \in \mathbb{O}_{(\mu,\nu)}$, with $x^2 = 0$. For all $T \in \mathcal{T}^{\alpha}_{\mu,\nu}$, $\overline{\Phi^{-1}(T)}$ is an irreducible component of $\mathcal{C}^{\alpha}_{(v,x)}$ of dimension $d^{\alpha}_{\mu,\nu}$ and all irreducible components of dimension $d^{\alpha}_{\mu,\nu}$ are of this form. All other irreducible components of $\mathcal{C}^{\alpha}_{(v,x)}$ have dimension strictly lower than $d^{\alpha}_{\mu,\nu}$.*

3. Computing Jordan types

It will be important for us to compute the Jordan type of a linear transformation when restricting to (or quotienting by) invariant subspaces. Recall that if V is any vector space (not necessarily symplectic) and $x \in \operatorname{End}(V)$ is a nilpotent linear transformation, then $\mathrm{J}(x) = \lambda$, where for all $j \geq 1$, $\sum_{k=1}^{j} \lambda^t_j = \dim(\ker(x^j))$. Here λ^t denotes the transposed partition, with $\lambda^t_j = \#\{k \mid \lambda_k \geq j\}$. Equivalently, we have that the number of boxes in the j-th column of the Young diagram of $\mathrm{J}(x)$ is $\lambda^t_j = \dim(\ker(x^j)) - \dim(\ker(x^{j-1}))$ for all $j \geq 1$.

LEMMA 3.1. *Let $0 \subseteq U \subseteq W \subseteq V$ be subspaces, and let $x \in \operatorname{End}(V)$ be a nilpotent linear transformation with $x(U) \subseteq U$, $x(W) \subseteq W$. Then we can consider the restriction to the subquotient, and $\mathrm{J}(x|_{W/U}) = \mu$ where*

$$\mu^t_i = \dim(x^i(W) \cap U) + \dim(\ker(x^i) \cap W) - \dim(x^{i-1}(W) \cap U) - \dim(\ker(x^{i-1}) \cap W)$$

PROOF. We know that $\mu^t_i = \dim\left(\ker(x|_{W/U})^i\right) - \dim\left(\ker(x|_{W/U})^{i-1}\right)$. Since $x^i(U) \subseteq U$, we have that $(x^i)^{-1}(U) \supseteq U$. It follows that

$$\ker(x|_{W/U})^i \simeq \left((x^i)^{-1}(U) \cap W\right)/U$$

so

$$\dim(\ker(x|_{W/U})^i) = \dim\left((x^i)^{-1}(U) \cap W\right) - \dim(U).$$

But, considering the map $x^i : (x^i)^{-1}(U) \cap W \to U$ we get an isomorphism

$$\left((x^i)^{-1}(U) \cap W\right)/(\ker(x^i) \cap W) \simeq x^i(W) \cap U$$

hence

$$\dim\left((x^i)^{-1}(U) \cap W\right) = \dim(x^i(W) \cap U) + \dim(\ker(x^i) \cap W).$$

In conclusion, we have

$$\begin{aligned}
\mu^t_i &= \dim\left(\ker(x|_{W/U})^i\right) - \dim\left(\ker(x|_{W/U})^{i-1}\right) \\
&= \left(\dim\left((x^i)^{-1}(U) \cap W\right) - \dim(U)\right) - \left(\dim\left((x^{i-1})^{-1}(U) \cap W\right) - \dim(U)\right) \\
&= \dim\left((x^i)^{-1}(U) \cap W\right) - \dim\left((x^{i-1})^{-1}(U) \cap W\right) \\
&= \dim(x^i(W) \cap U) + \dim(\ker(x^i) \cap W) - \dim(x^{i-1}(W) \cap U) \\
&\quad - \dim(\ker(x^{i-1}) \cap W).
\end{aligned}$$

$\square$

LEMMA 3.2. *Let x be a nilpotent transformation on a vector space V, and let $F \subseteq \ker(x)$. Define*

$$a_j = \dim(F \cap \mathrm{Im}(x^j)) \quad j \geq 0. \tag{3.1}$$

Then $\mathrm{J}(x|_{V/F})$ is obtained from $\mathrm{J}(x)$ by removing $(a_{j-1} - a_j)$ boxes at the bottom of column j for all $j \geq 1$. (Notice that these boxes form a vertical strip i.e. no two of them are in the same row)

PROOF. This follows directly from applying Lemma 3.1 to $U = F$, $W = V$ and comparing the result to $\dim(\ker(x^j)) - \dim(\ker(x^{j-1}))$. □

LEMMA 3.3. *Let x be a nilpotent transformation on a vector space V, and let $F \supseteq \mathrm{Im}(x)$. Define*

$$b_j = \dim\left(F + \ker(x^j)\right), \quad j \geq 0. \tag{3.2}$$

Then $\mathrm{J}(x|_F)$ is obtained from $\mathrm{J}(x)$ by removing $(b_j - b_{j-1})$ boxes at the bottom of column j for all $j \geq 1$. (Notice that these boxes form a vertical strip i.e. no two of them are in the same row)

PROOF. As in the previous proof, this follows directly from applying Lemma 3.1 to $U = 0$, $W = F$ and using that $\dim(F + \ker(x^j)) = \dim(F) + \dim(\ker(x^j)) - \dim(F \cap \ker(x^j))$. □

LEMMA 3.4. *Let V be a $2n$-dimensional symplectic space, and $x \in \mathrm{End}(V)$ such that $x \in \mathcal{S}$. Then $\mathrm{Im}(x^j) = \ker(x^j)^\perp$ for all $j \geq 0$.*

PROOF. Let $w \in \mathrm{Im}(x^j)$, so $w = x^j u$ for some $u \in V$. If $z \in \ker(x^j)$, then $\langle w, z\rangle = \langle x^j u, z\rangle = \langle u, x^j z\rangle = 0$ so $\mathrm{Im}(x^j) \subseteq \ker(x^j)^\perp$ and equality follows from the fact that they have the same dimension. □

LEMMA 3.5. *Let V be a $2n$-dimensional symplectic space, and $x \in \mathcal{S} \cap \mathcal{N}$. Suppose $F \subseteq \ker(x)$, which is equivalent to $F^\perp \supseteq \mathrm{Im}(x)$, then*

$$\mathrm{J}(x|_{V/F}) = \mathrm{J}(x|_{F^\perp}).$$

PROOF. Let $(a_j)_j$ be defined as in (3.1) for F, and $(b_j)_j$ be defined as in (3.2) for the space $F^\perp$. Then $(F \cap \mathrm{im}(x^j))^\perp = F^\perp + \ker(x^j)$ hence we have

$$\begin{aligned}
a_{j-1} - a_j &= \dim(F \cap \mathrm{im}(x^{j-1})) - \dim(F \cap \mathrm{im}(x^j)) \\
&= (2n - \dim(F^\perp + \ker(x^{j-1}))) - (2n - \dim(F^\perp + \ker(x^j))) \\
&= b_j - b_{j-1}
\end{aligned}$$

and the result then follows from Lemma 3.2 and Lemma 3.3. □

If $0 \subseteq F \subseteq F^\perp \subseteq V$ is an isotropic subspace of V, and $(v, x) \in \mathfrak{N}$, in order to compute $\mathrm{eJ}\left(v + F, x|_{F^\perp/F}\right)$ we will need to use the following result due to Travkin [**23**, Theorem 1 and Corollary 1] and Achar-Henderson [**1**, Theorem 6.1].

THEOREM 3.6. *If $(v, x) \in \mathfrak{N}(W)$, then $\mathrm{eJ}(v, x) = (\mu, \nu)$ if and only if*

$$\mathrm{J}\left(x|_W\right) = (\mu + \nu) \cup (\mu + \nu) = (\mu_1 + \nu_1, \mu_1 + \nu_1, \mu_2 + \nu_2, \mu_2 + \nu_2, \dots)$$

and

$$\mathrm{J}\left(x|_{W/\mathbb{C}[x]v}\right) = (\mu_1 + \nu_1, \mu_2 + \nu_1, \mu_2 + \nu_2, \mu_3 + \nu_2, \dots).$$

We want to apply this theorem to $W = F^{\perp}/F$, so we will need the next two lemmas, that tell us how to find $\mathrm{J}(x|_W) = \mathrm{J}\left(x|_{F^{\perp}/F}\right)$ and $\mathrm{J}\left(x|_{W/\mathbb{C}[x]v}\right) = \mathrm{J}\left(x|_{F^{\perp}/(F+\mathbb{C}[x]v)}\right)$.

LEMMA 3.7. *Let V be a symplectic vector space, $x \in \mathcal{S} \cap \mathcal{N}$, and $0 \subseteq F \subseteq F^{\perp} \subseteq V$ with $x(F) = 0$ and $x(F^{\perp}) \subseteq F^{\perp}$. Let*

$$a'_j = \dim(F \cap x^j(F^{\perp})), \quad j \geq 0$$

then $\mathrm{J}(x|_{F^{\perp}/F})$ is obtained from $\mathrm{J}(x|_{V/F}) = \mathrm{J}(x|_{F^{\perp}})$ by removing $(a'_{j-1} - a'_j)$ boxes from the bottom of column j for all $j \geq 1$.

PROOF. We can either apply Lemma 3.2 to the quotient $F^{\perp} \twoheadrightarrow F^{\perp}/F$ and look at the dimensions

$$\dim(F \cap \mathrm{im}(x|_{F^{\perp}})^j) = \dim(F \cap x^j(F^{\perp}))$$

to get the result, or use Lemma 3.1 (with $U = F$, $W = F^{\perp}$). □

LEMMA 3.8. *Let V be a symplectic vector space, $(v, x) \in \mathfrak{N}$, and $0 \subseteq F \subseteq F^{\perp} \subseteq V$ with $x(F) = 0$, $x(F^{\perp}) \subseteq F^{\perp}$, $v \in F^{\perp}$. Let*

$$b'_j = \dim((F + \mathbb{C}[x]v) \cap x^j(F^{\perp})), \quad j \geq 0$$

then $\mathrm{J}(x|_{F^{\perp}/F+\mathbb{C}[x]v})$ is obtained from $\mathrm{J}(x|_{V/F}) = \mathrm{J}(x|_{F^{\perp}})$ by removing $(b'_{j-1} - b'_j)$ boxes from the bottom of column j for all $j \geq 1$.

PROOF. This follows by applying Lemma 3.1 to $U = \mathbb{F} + \mathbb{C}[x]v \subseteq F^{\perp} = W$. □

4. Computing $\mathrm{J}\left(x|_{F^{\perp}/F}\right)$, $\mathrm{J}\left(x|_{F^{\perp}/F+\mathbb{C}[x]v}\right)$ and $\mathrm{eJ}\left(v + F, x|_{F^{\perp}/F}\right)$

For this section, we fix $x \in \mathfrak{N} \cap \mathcal{S}$ and we apply the results of Section 3. In particular we will look at the special case where $x^2 = 0$. We start with some general lemmas that we will need later.

DEFINITION 4.1. For each $j \geq 0$, we define a map $\mathrm{Im}(x^j) \times \mathrm{Im}(x^j) \to \mathbb{C}$ denoted by $\langle\langle\ ,\ \rangle\rangle_j$, as follows. If $u, w \in \mathrm{Im}(x^j)$, let $z \in V$ such that $u = x^j z$, then

$$\langle\langle u, w\rangle\rangle_j := \langle z, w\rangle.$$

LEMMA 4.2. *For each $j \geq 0$, the pairing $\langle\langle\ ,\ \rangle\rangle_j$ is well defined, and it is a non-degenerate skew-symmetric bilinear form on $\mathrm{Im}(x^j)$ (which is an even dimensional space).*

PROOF. To see that this pairing is well defined, let $u, w \in \mathrm{Im}(x^j)$ and suppose $z_1, z_2 \in V$ are such that $u = x^j z_1 = x^j z_2$. We also let $y \in V$ be such that $x^j y = w$. Then

$$\langle z_1, w\rangle = \langle z_1, x^j y\rangle = \langle x^j z_1, y\rangle = \langle x^j z_2, y\rangle = \langle z_2, x^j y\rangle = \langle z_2, w\rangle.$$

The fact that $\langle\langle\ ,\ \rangle\rangle_j$ is a skew-symmetric bilinear follows directly from the definition because if $u, w, z, y \in V$, with $x^j z = u$ and $x^j y = w$, then

$$\langle\langle u, w\rangle\rangle_j = \langle z, w\rangle = -\langle w, z\rangle = -\langle x^j y, z\rangle = -\langle y, x^j z\rangle = -\langle y, u\rangle = -\langle\langle w, u\rangle\rangle_j;$$

also, for $\alpha, \beta \in \mathbb{C}$ we have $x^j(\alpha z) = \alpha u$ and $x^j(\beta y) = \beta w$, so

$$\langle\langle \alpha u, \beta w\rangle\rangle_j = \langle \alpha z, \beta w\rangle = \alpha\beta\langle z, w\rangle = \alpha\beta\langle\langle u, w\rangle\rangle_j.$$

Finally, suppose that there is $u \in \mathrm{Im}(x^j)$ such that $\langle\langle u, w\rangle\rangle_j = 0$ for all $w \in \mathrm{Im}(x^j)$, and let $z \in V$ with $x^j z = u$. Then $0 = \langle\langle u, w\rangle\rangle_j = \langle z, w\rangle$ for all $w \in \mathrm{Im}(x^j)$ so

$z \in \left(\mathrm{Im}(x^j)\right)^{\perp} = \ker(x^j)$ (by Lemma 3.4), hence $u = x^j z = 0$. This proves that the form is indeed nondegenerate. □

In particular, $\langle\langle\ ,\ \rangle\rangle_0 = \langle\ ,\ \rangle$, and when $j = 1$ we might omit the index and just denote $\langle\langle\ ,\ \rangle\rangle_1 = \langle\langle\ ,\ \rangle\rangle$. We denote the perpendicular with respect to the form $\langle\langle\ ,\ \rangle\rangle_j$ by $\perp\!\!\!\perp_j$. So, if $W \subset \mathrm{Im}(x^j)$, then

$$W^{\perp\!\!\!\perp_j} = \{u \in \mathrm{Im}(x^j) \mid \langle\langle u, w\rangle\rangle_j = 0, \text{ for all } w \in W\}.$$

In particular $\perp\!\!\!\perp_0 = \perp$ is the perpendicular with respect to $\langle\ ,\ \rangle$, and when $j = 1$ we write $\perp\!\!\!\perp = \perp\!\!\!\perp_1$.

LEMMA 4.3. *Let $F \subset V$ be a subspace, then for all $j \geq 0$, $x^j(F^{\perp}) = (F \cap \mathrm{Im}(x^j))^{\perp\!\!\!\perp_j}$.*

PROOF. Let $u \in x^j(F^{\perp})$, then there is $z \in F^{\perp}$ such that $x^j z = u$, hence for all $w \in F \cap \mathrm{Im}(x^j)$ we have $\langle\langle u, w\rangle\rangle_j = \langle z, w\rangle = 0$, so $u \in (F \cap \mathrm{Im}(x^j))^{\perp\!\!\!\perp_j}$. Viceversa, if $u \in (F \cap \mathrm{Im}(x^j))^{\perp\!\!\!\perp_j}$, then $u \in \mathrm{Im}(x^j)$, so there is $z \in V$ with $u = x^j z$, and we have that for all $w \in F \cap \mathrm{Im}(x^j)$, $0 = \langle\langle u, w\rangle\rangle_j = \langle z, w\rangle$, so $z \in (F \cap \mathrm{Im}(x^j))^{\perp} = F^{\perp} + \mathrm{Im}(x^j)^{\perp} = F^{\perp} + \ker(x^j)$. This shows that $(F \cap \mathrm{Im}(x^j))^{\perp\!\!\!\perp_j} \subseteq x^j(F^{\perp} + \ker(x^j)) = x^j(F^{\perp})$ which concludes the proof. □

The next two lemmas will specifically be relevant in the case where $x^2 = 0$.

LEMMA 4.4. *If $v \in \mathrm{Im}(x)$, $v \notin F$, then $F \cap x(F^{\perp}) = (F + \mathbb{C}v) \cap x(F^{\perp})$ if and only if*

$$F \cap x(F^{\perp}) \not\subset (\mathbb{C}v)^{\perp\!\!\!\perp}.$$

PROOF. Since $F \cap x(F^{\perp}) \subseteq (F + \mathbb{C}v) \cap x(F^{\perp})$, we have equality exactly when the two spaces have the same dimension. Notice that $F \cap x(F^{\perp}) = F \cap \mathrm{Im}(x) \cap x(F^{\perp})$ and, since $v \in \mathrm{Im}(x)$,

$$(F + \mathbb{C}v) \cap x(F^{\perp}) = (F + \mathbb{C}v) \cap \mathrm{Im}(x) \cap x(F^{\perp}) = ((F \cap \mathrm{Im}(x)) + \mathbb{C}v) \cap x(F^{\perp}).$$

Hence $F \cap x(F^\perp) = (F + \mathbb{C}v) \cap x(F^\perp)$ if and only if

$$
\begin{aligned}
&\dim\left(F \cap x(F^\perp)\right) = \dim\left((F + \mathbb{C}v) \cap x(F^\perp)\right)\\
\iff &\dim\left(F \cap \mathrm{Im}(x) \cap x(F^\perp)\right) = \dim\left(((F \cap \mathrm{Im}(x)) + \mathbb{C}v) \cap x(F^\perp)\right)\\
\iff &\dim\left(\left(F \cap \mathrm{Im}(x) \cap x(F^\perp)\right)^{\perp\!\!\perp}\right) = \dim\left(\left(((F \cap \mathrm{Im}(x)) + \mathbb{C}v) \cap x(F^\perp)\right)^{\perp\!\!\perp}\right)\\
\iff &\dim\left((F \cap \mathrm{Im}(x))^{\perp\!\!\perp} + \left(x(F^\perp)\right)^{\perp\!\!\perp}\right)\\
&= \dim\left(((F \cap \mathrm{Im}(x)) + \mathbb{C}v))^{\perp\!\!\perp} + \left(x(F^\perp)\right)^{\perp\!\!\perp}\right)\\
\iff &\dim\left(x(F^\perp) + (F \cap \mathrm{Im}(x))\right)\\
&= \dim\left(((F \cap \mathrm{Im}(x))^{\perp\!\!\perp} \cap (\mathbb{C}v)^{\perp\!\!\perp}\right) + (F \cap \mathrm{Im}(x)))\\
\iff &\dim\left(x(F^\perp)\right) + \dim\left(F \cap \mathrm{Im}(x)\right) - \dim\left(F \cap \mathrm{Im}(x) \cap x(F^\perp)\right)\\
&= \dim\left(\left(x(F^\perp) \cap (\mathbb{C}v)^{\perp\!\!\perp}\right) + (F \cap \mathrm{Im}(x))\right)\\
\iff &\dim\left(x(F^\perp)\right) + \dim\left(F \cap \mathrm{Im}(x)\right) - \dim\left(F \cap x(F^\perp)\right) = \dim\left(x(F^\perp) \cap (\mathbb{C}v)^{\perp\!\!\perp}\right)\\
&+ \dim\left(F \cap \mathrm{Im}(x)\right) - \dim\left(x(F^\perp) \cap (\mathbb{C}v)^{\perp\!\!\perp} \cap F \cap \mathrm{Im}(x)\right)\\
\iff &\dim\left(x(F^\perp)\right) + \dim\left(F \cap \mathrm{Im}(x)\right) - \dim\left(F \cap x(F^\perp)\right) = \dim\left(x(F^\perp) \cap (\mathbb{C}v)^{\perp\!\!\perp}\right)\\
&+ \dim\left(F \cap \mathrm{Im}(x)\right) - \dim\left(x(F^\perp) \cap (\mathbb{C}v)^{\perp\!\!\perp} \cap F \cap \mathrm{Im}(x)\right)
\end{aligned}
$$

(4.1)

$$
\begin{aligned}
\iff &\dim\left(x(F^\perp)\right) - \dim\left(x(F^\perp) \cap (\mathbb{C}v)^{\perp\!\!\perp}\right)\\
&= \dim\left(F \cap x(F^\perp)\right) - \dim\left(F \cap x(F^\perp) \cap (\mathbb{C}v)^{\perp\!\!\perp}\right)
\end{aligned}
$$

Now, since $(\mathbb{C}v)^{\perp\!\!\perp}$ is a codimension 1 subspace of $\mathrm{Im}(x)$ we have that each of the two sides of the inequality are either equal to 0 or 1. Since $v \notin F$, hence $v \notin F \cap \mathrm{Im}(x)$, we have that

$$
\begin{aligned}
(F \cap \mathrm{Im}(x)) + \mathbb{C}v &\neq F \cap \mathrm{Im}(x)\\
((F \cap \mathrm{Im}(x)) + \mathbb{C}v)^{\perp\!\!\perp} &\neq (F \cap \mathrm{Im}(x))^{\perp\!\!\perp}\\
((F \cap \mathrm{Im}(x))^{\perp\!\!\perp} \cap (\mathbb{C}v)^{\perp\!\!\perp} &\neq (F \cap \mathrm{Im}(x))^{\perp\!\!\perp}\\
x(F^\perp) \cap (\mathbb{C}v)^{\perp\!\!\perp} &\neq x(F^\perp),
\end{aligned}
$$

hence the LHS of (4.1) is 1. Therefore, the RHS equals the LHS if and only if

$$F \cap x(F^\perp) \neq F \cap x(F^\perp) \cap (\mathbb{C}v)^{\perp\!\!\perp}$$

which is equivalent to $F \cap x(F^\perp) \not\subset (\mathbb{C}v)^{\perp\!\!\perp}$. □

LEMMA 4.5. *If $v \notin \mathrm{Im}(x)$, $v \notin F$, then $F \cap x(F^\perp) = (F + \mathbb{C}v) \cap x(F^\perp)$ if and only if*

$$(v + F) \cap x(F^\perp) = \emptyset.$$

PROOF. Notice that $v \notin \mathrm{Im}(x)$, so in particular $v \notin x(F^\perp)$. Since additionally $v \notin F$, we have that $(F + \mathbb{C}v) \cap x(F^\perp) = F \cap x(F^\perp)$ if and only if for all $u \in F$, $u + v \notin x(F^\perp)$. □

4.1. Computing $\mathrm{J}\left(x|_{F_k^\perp/F_k}\right)$. We consider now any isotropic subspace $F_k \in \mathrm{Gr}_k^\perp(\ker(x)\cap(\mathbb{C}[x]v)^\perp)$, using the subscript to emphasize the dimension $k=\dim(F_k)$. We assume that $x^2=0$ and we separate two cases.

4.1.1. $x=0$. In this case, $\mathrm{J}(x)=(1^{2n})$, and $\mathrm{Im}(x^j)=0$ when $j>0$, hence, using Lemma 3.2 we have $a_0=k$, $a_j=0$ for $j>0$ and so $\mathrm{J}(x|_{V/F_k})=\mathrm{J}(x|_{F_k^\perp})=(1^{2n-k})$. Further, using Lemma 3.7 we have $a_0'=k$, $a_j'=0$ for $j>0$, and so $\mathrm{J}\left(x|_{F_k^\perp/F_k}\right)=(1^{2n-2k})$.

4.1.2. $x\neq 0$, $x^2=0$. In this case, since $x\in\mathcal{S}$, $\mathrm{J}(x)=\lambda\cup\lambda$, with $\lambda=(2^{n_2}1^{n_1})$, $2n_2+n_1=n$, we then have $\mathrm{J}(x)=(2^{2n_2}1^{2n_1})$. Let $k_2=\dim(F_k\cap\mathrm{Im}(x))$, $k_1=\dim\left(F_k/(F_k\cap\mathrm{Im}(x))\right)$, so $k=k_1+k_2$. Using Lemma 3.2 we have $a_0=k=k_1+k_2$, $a_1=k_2$, $a_j=0$ for $j>1$. We remove $a_0-a_1=k_1$ boxes from the first column and $a_1-a_2=k_2$ boxes from the second column, which gives $\mathrm{J}(x|_{V/F_k})=\mathrm{J}(x|_{F_k^\perp})=(2^{2n_2-k_2}1^{2n_1-k_1+k_2})$.

Then, to use Lemma 3.7 we need to know $\dim(F_k\cap x(F_k^\perp))$ and, by Lemma 4.3, $F_k\cap x(F_k^\perp)$ is the radical of the bilinear form $\langle\langle\ ,\ \rangle\rangle$ restricted to $F_k\cap\mathrm{Im}(x)$. Therefore, $F_k\cap\mathrm{Im}(x)/F_k\cap x(F_k^\perp)$ is a symplectic space with form $\langle\langle\ ,\ \rangle\rangle$, hence even dimensional, say $\dim\left(F_k\cap\mathrm{Im}(x)/F_k\cap x(F_k^\perp)\right)=2h$. We then have from Lemma 3.7 that $a_0'=k=k_1+k_2$, $a_1'=k_2-2h$, $a_j'=0$ for $j>1$, so we need to remove k_1+2h boxes from the first column and k_2-2h boxes from the second column which results in $\mathrm{J}\left(x|_{F_k^\perp/F_k}\right)=(2^{2n_2-2k_2+2h}1^{2n_1-2k_1+2k_2-4h})$.

4.2. Computing $\mathrm{J}\left(x|_{F_k^\perp/F_k+\mathbb{C}[x]v}\right)$. We consider F_k as above, computing the Jordan type will then give us, using Theorem 3.6, $\mathrm{eJ}\left(v+F_k,x|_{F_k^\perp/F_k}\right)$. There are six cases to consider.

4.2.1. $x=0$, $v=0$. In this case, $\mu=\emptyset$, $\nu=\lambda=1^n$, so we have $\mathrm{J}\left(x|_{F_k^\perp/F_k+\mathbb{C}[x]v}\right)=\mathrm{J}\left(x|_{F_k^\perp/F_k}\right)=(1^{2n-2k})$ and hence $\mathrm{eJ}\left(v+F_k,x|_{F_k^\perp/F_k}\right)=(\emptyset,1^{n-k})$ by Theorem 3.6.

4.2.2. $x=0$, $v\neq 0$. In this case, $\mu=\lambda=1^n$, $\nu=\emptyset$.

(a) If $v\in F_k$, then $b_0'=k$, $b_j'=0$ for $j>0$, so, by Lemma 3.8, $\mathrm{J}\left(x|_{F_k^\perp/F_k+\mathbb{C}[x]v}\right)=\mathrm{J}\left(x|_{F_k^\perp/F_k}\right)=(1^{2n-2k})$, and $\mathrm{eJ}\left(v+F_k,x|_{F_k^\perp/F_k}\right)=(\emptyset,1^{n-k})$.

(b) If $v\notin F_k$, then $b_0'=k+1$, $b_j'=0$, so by Lemma 3.8, $\mathrm{J}\left(x|_{F_k^\perp/F_k+\mathbb{C}[x]v}\right)=(1^{2n-2k-1})$, and $\mathrm{eJ}\left(v+F_k,x|_{F_k^\perp/F_k}\right)=(1^{n-k},\emptyset)$.

4.2.3. $x\neq 0$, $x^2=0$, $v=0$. In this case, $\mu=\emptyset$, $\nu=\lambda=2^{n_2}1^{n_1}$, so we have $\mathrm{J}\left(x|_{F_k^\perp/F_k+\mathbb{C}[x]v}\right)=\mathrm{J}\left(x|_{F_k^\perp/F_k}\right)=(2^{2n_2-2k_2+2h}1^{2n_1-2k_1+2k_2-4h})$ and hence $\mathrm{eJ}\left(v+F_k,x|_{F_k^\perp/F_k}\right)=(\emptyset,2^{n_2-k_2+h}1^{n_1-k_1+k_2-2h})$ by Theorem 3.6.

4.2.4. $x\neq 0$, $x^2=0$, $0\neq v\in\mathrm{Im}(x)$. In this case, $\mu=1^{n_2}$, $\nu=1^{n_2+n_1}$, with $n_2>0$.

(a) If $v\in F_k$, then $b_j'=a_j'$, for all $j\geq 0$, so, by Lemma 3.8,
$$\mathrm{J}\left(x|_{F_k^\perp/F_k+\mathbb{C}[x]v}\right)=\mathrm{J}\left(x|_{F_k^\perp/F_k}\right)=(2^{2n_2-2k_2+2h}1^{2n_1-2k_1+2k_2-4h}),$$
and

$\mathrm{eJ}\left(v + F_k, x|_{F_k^\perp/F_k}\right) = (\emptyset, 2^{n_2-k_2+h}1^{n_1-k_1+k_2-2h})$.

(b) If $v \notin F_k$, and $(F_k + \mathbb{C}v) \cap x(F_k^\perp) \supsetneq F_k \cap x(F_k^\perp)$ (by Lemma 4.4 this means that $F_k \cap x(F_k^\perp) \subset (\mathbb{C}v)^{\perp\perp}$), then we have $b_0' = k+1 = a_0'+1$, and $b_1' = k_2 - 2h+1 = a_1'+1$. Hence $\mathrm{J}(x|_{F_k^\perp/F_k+\mathbb{C}[x]v}) = (2^{2n_2-2k_2+2h-1}1^{2n_1-2k_1+2k_2-4h+1})$ and we now demonstrate how to obtain that

$$\mathrm{eJ}\left(v + F_k, x|_{F_k^\perp/F_k}\right) = (1^{n_2-k_2+h}, 1^{n_1+n_2-k_1-h}),$$

by using Theorem 3.6.

We want to find the exotic Jordan type $\mathrm{eJ}\left(v + F, x|_{F^\perp/F}\right) = (\mu', \nu')$. Since $\mathrm{J}\left(x|_{F_k^\perp/F_k}\right) = (2^{2n_2-2k_2+2h}1^{2n_1-2k_1+2k_2-4h}) = (\mu' + \nu') \cup (\mu' + \nu')$, we have $\mu'_{n_1+n_2-k_1-h+1} + \nu'_{n_1+n_2-k_1-h+1} = 0$, hence $\mu'_{n_1+n_2-k_1-h+1} = 0$. Then, from $\mathrm{J}(x|_{F_k^\perp/F_k+\mathbb{C}[x]v}) = (2^{2n_2-2k_2+2h-1}1^{2n_1-2k_1+2k_2-4h+1}) = (\mu_1' + \nu_1', \mu_2' + \nu_1', \ldots)$ we get that $\mu'_{n_1+n_2-k_1-h+1} + \nu'_{n_1+n_2-k_1-h} = 1$, hence $\nu'_{n_1+n_2-k_1-h} = 1$. Inductively we have $\mu_j' = 0$ for all $j > n_2 - k_2 + h$, and $\nu_j' = 1$ for $n_2 - k_2 + h \le j \le n_1 + n_2 - k_1 - h + 1$. Then, since $\mu'_{n_2-k_2+h} + \nu'_{n_2-k_2+h} = 2$, we get $\mu'_{n_2-k_2+h} = 1$ and then inductively $\mu_j' = \nu_j' = 1$ for all $1 \le j \le n_2 - k_2 + h$.

(c) If $v \notin F_k$, and $(F_k + \mathbb{C}v) \cap x(F_k^\perp) = F_k \cap x(F_k^\perp)$ (by Lemma 4.4 this means that $F_k \cap x(F_k^\perp) \not\subset (\mathbb{C}v)^{\perp\perp}$), then we have $b_0' = k + 1 = a_0' + 1$, and $b_1' = k_2 - 2h = a_1'$. Hence $\mathrm{J}(x|_{F_k^\perp/F_k+\mathbb{C}[x]v}) = (2^{2n_2-2k_2+2h}1^{2n_1-2k_1+2k_2-4h-1})$ and with Theorem 3.6 and a similar inductive argument as above, which we will omit, we get $\mathrm{eJ}\left(v + F, x|_{F^\perp/F}\right) = (1^{n_1+n_2-k_1-h}, 1^{n_2-k_2+h})$.

4.2.5. $x \neq 0$, $x^2 = 0$, $v \in \ker(x) \setminus \mathrm{Im}(x)$. In this case, $\mu = 1^{n_2+n_1}$, $\nu = 1^{n_2}$, with $n_1, n_2 > 0$. The various possibilities for $\mathrm{eJ}\left(v + F, x|_{F^\perp/F}\right)$ in this case are the same as in Section 4.2.4 and depend in the exact same way on whether $v \in F_k$ and whether $(F_k + \mathbb{C}v) \cap x(F_k^\perp) = F_k \cap x(F_k^\perp)$ (with the difference that we will use Lemma 4.5 instead of Lemma 4.4 to verify the condition).

4.2.6. $x \neq 0$, $x^2 = 0$, $v \notin \ker(x)$. In this case, $\mu = 2^{n_2}1^{n_1}$, $\nu = \emptyset$, with $n_2 > 0$. Since $F_k \subset \ker(x)$, we always have that $v \notin F_k$ and since $v \in F_n \subset F_k^\perp$, $xv \in x(F_k^\perp)$. Also, since $v = v_{1,2} + \cdots + v_{n_2,2} + v_{n_2+1,1} \cdots + v_{n_1+n_2,1}$, and $xv = v_{1,1} + \cdots + v_{n_2,1}$, then

$$\dim(F_k + \mathbb{C}[x]v) = \dim(F_k + \mathbb{C}v + \mathbb{C}xv) = \begin{cases} k+1 & \text{if } xv \in F_k \\ k+2 & \text{if } xv \notin F_k \end{cases}.$$

(a) If $xv \in F_k$, then $\dim((F_k + \mathbb{C}xv + \mathbb{C}v) \cap x(F_k^\perp)) = \dim((F_k + \mathbb{C}v) \cap x(F_k^\perp)) = \dim(F_k \cap x(F_k^\perp)) = k_2 - 2h$ because for any $c \neq 0$, $(cv + \ker(x)) \cap \mathrm{Im}(x) = \emptyset$. It follows that in this case $b_0' = k + 1$, $b_1' = k_2 - 2h$, so, like in 4.2.4(c), $\mathrm{J}(x|_{F_k^\perp/F_k+\mathbb{C}[x]v}) = (2^{2n_2-2k_2+2h}1^{2n_1-2k_1+2k_2-4h-1})$ and $\mathrm{eJ}\left(v + F, x|_{F_k^\perp/F_k}\right) = (1^{n_1+n_2-k_1-h}, 1^{n_2-k_2+h})$.

(b) If $xv \notin F_k$, then by the modular property of lattices of vector spaces we have $(F_k + \mathbb{C}xv + \mathbb{C}v) \cap x(F_k^\perp) = ((F_k + \mathbb{C}v) \cap x(F_k^\perp)) + \mathbb{C}xv$, hence

$$\begin{aligned}\dim(F_k + \mathbb{C}[x]v) &= \dim((F_k + \mathbb{C}xv + \mathbb{C}v) \cap x(F_k^\perp)) \\ &= \dim(((F_k + \mathbb{C}v) \cap x(F_k^\perp)) + \mathbb{C}xv) \\ &= \dim(F_k \cap x(F_k^\perp) + \mathbb{C}xv) \\ &= k_2 - 2h + 1.\end{aligned}$$

It follows that in this case $b_0' = k+2$, $b_1' = k_2 - 2h + 1$, so,

$$\mathrm{J}(x|_{F_k^\perp/F_k+\mathbb{C}[x]v}) = (2^{2n_2-2k_2+2h-1}1^{2n_1-2k_1+2k_2-4h})$$

and by Theorem 3.6 and a similar inductive argument as above, we get $\mathrm{eJ}\left(v + F_k, x|_{F_k^\perp/F_k}\right) = (2^{n_2-k_2+h}1^{n_1-k_1+k_2-2h}, \emptyset)$.

5. Proof of Conjecture 2.22 when $x^2 = 0$

For notational simplicity, we let $\alpha_1 = k$, so to prove the conjecture we want to describe the variety

$$\mathcal{B}^{((\mu,\nu),\alpha)}_{((\mu',\nu'),\alpha')} = \{F_k \in \mathrm{Gr}_k^\perp\left(\ker(x) \cap (\mathbb{C}[x]v)^\perp\right) \mid \mathrm{eJ}\left(v + F_k, x|_{F_k^\perp/F_k}\right) = (\mu',\nu')\}$$

for all the possible cases of (μ,ν) and (μ',ν') that we found in Section 4. More precisely, we want to show its irreducibility and compute its dimension, and to do that we will need some more fibre bundles.

5.1. More fibre bundles. Given any $F_k \in \mathcal{B}^{((\mu,\nu),\alpha)}_{((\mu',\nu'),\alpha')}$, we have a filtration $F_k \cap x(F_k^\perp) \subseteq F_k \cap \mathrm{Im}(x) \subseteq F_k$.

Definition 5.1. For $\ell \geq 0$, we define two maps

$$\pi_1^\ell : \mathrm{Gr}_\ell^\perp(\ker(x) \cap (\mathbb{C}[x]v)^\perp) \to \coprod_{j\geq 0} \mathrm{Gr}_j(\ker(x) \cap (\mathbb{C}[x]v)^\perp \cap \mathrm{Im}(x))$$

$$\pi_1^\ell(F) = F \cap \mathrm{Im}(x);$$

$$\pi_2^\ell : \mathrm{Gr}_\ell(\ker(x) \cap (\mathbb{C}[x]v)^\perp \cap \mathrm{Im}(x)) \to \coprod_{j\geq 0} \mathrm{Gr}_j^{\perp\!\!\perp}(\ker(x) \cap (\mathbb{C}[x]v)^\perp \cap \mathrm{Im}(x))$$

$$\pi_2^\ell(U) = U \cap U^{\perp\!\!\perp}.$$

Remark 5.2. If $F_k \in \mathcal{B}^{((\mu,\nu),\alpha)}_{((\mu',\nu'),\alpha')}$, with $\dim(F_k \cap \mathrm{Im}(x)) = k_2$ as in Section 4.1.2, then by Lemma 4.3 we have

$$\pi_2^{k_2}(\pi_1^k(F_k)) = (F_k \cap \mathrm{Im}(x)) \cap (F_k \cap \mathrm{Im}(x))^{\perp\!\!\perp} = F_k \cap x(F_k^\perp).$$

If we have a subspace $U \subset \ker(x) \cap (\mathbb{C}[x]v)^\perp \cap \mathrm{Im}(x)$, $\dim(U) = u$, and $F \cap \mathrm{Im}(x) = U$, we can consider the following subspaces in the quotient space $\ker(x) \cap (\mathbb{C}[x]v)^\perp/U$: $\overline{F} = F/U$, and $\overline{\mathrm{Im}(x)} = ((\ker(x) \cap (\mathbb{C}[x]v)^\perp \cap \mathrm{Im}(x))/U$. We then have

$$\begin{aligned}(5.1)\qquad (\pi_1^\ell)^{-1}(U) &= \{F \in \mathrm{Gr}_\ell^\perp(\ker(x) \cap (\mathbb{C}[x]v)^\perp) \mid F \cap \mathrm{Im}(x) = U\} \\ &\cong \{\overline{F} \in \mathrm{Gr}_{\ell-u}^\perp((\ker(x) \cap (\mathbb{C}[x]v)^\perp)/U) \mid \overline{F} \cap \overline{\mathrm{Im}(x)} = 0\}\end{aligned}$$

which is empty when $\ell - u + \dim((\ker(x) \cap (\mathbb{C}[x]v)^{\perp} \cap \mathrm{Im}(x)) - u > \dim(\ker(x) \cap (\mathbb{C}[x]v)^{\perp}) - u$ and otherwise is an open dense subvariety of the isotropic Grassmannian $\mathrm{Gr}^{\perp}_{\ell-u}((\ker(x) \cap (\mathbb{C}[x]v)^{\perp})/U)$. This shows that the map π_1^{ℓ} is a fibre bundle over any given $\mathrm{Gr}_j(\ker(x) \cap (\mathbb{C}[x]v)^{\perp} \cap \mathrm{Im}(x))$.

Similarly, if we have an isotropic subspace $S \subset S^{\perp\!\!\perp}$, with respect to $\langle\langle\ ,\ \rangle\rangle$ restricted to $\ker(x) \cap (\mathbb{C}[x]v)^{\perp} \cap \mathrm{Im}(x)$, with $\dim(S) = s$, and if $U \subseteq \ker(x) \cap (\mathbb{C}[x]v)^{\perp} \cap \mathrm{Im}(x)$, with $U \cap U^{\perp\!\!\perp} = S$, then $U^{\perp\!\!\perp} + U = S^{\perp\!\!\perp}$. We have then

$$
\begin{aligned}
(\pi_2^{\ell})^{-1}(S) &= \{U \in \mathrm{Gr}_{\ell}(\ker(x) \cap (\mathbb{C}[x]v)^{\perp} \cap \mathrm{Im}(x)) \mid U \cap U^{\perp\!\!\perp} = S\} \\
&= \{U \in \mathrm{Gr}_{\ell}(\ker(x) \cap (\mathbb{C}[x]v)^{\perp} \cap \mathrm{Im}(x)) \mid S \subseteq U \subseteq S^{\perp\!\!\perp}\} \qquad (5.2)\\
&\cong \mathrm{Gr}_{\ell-s}(S^{\perp\!\!\perp}/S)
\end{aligned}
$$

This fibre is empty when $\ell - s > \dim((\ker(x) \cap (\mathbb{C}[x]v)^{\perp} \cap \mathrm{Im}(x))) - 2s$ and otherwise is isomorphic to a Grassmannian. This shows that the map π_2^{ℓ} is a fibre bundle over any given $\mathrm{Gr}_j^{\perp\!\!\perp}(\ker(x) \cap (\mathbb{C}[x]v)^{\perp} \cap \mathrm{Im}(x))$.

REMARK 5.3. In the computations of $d^{\alpha}_{(\mu,\nu)}$ it will be helpful to use the fact that $N(1^n) = n(n-1)/2$ and $N(2^{n_2}1^{n_1}) = (n_2+n_1)(n_2+n_1-1)/2 + n_2(n_2-1)/2$, also since $k = \alpha_1$, we have $\frac{1}{2}\sum_{i\geq 1}(\alpha_i^2 - \alpha_i) = \frac{1}{2}k(k-1) + \frac{1}{2}\sum_{i\geq 2}(\alpha_i^2 - \alpha_i)$.

We now examine all the different possible cases for the proof of Conjecture 2.22 when $x^2 = 0$.

5.2. $(\mu, \nu) = (\emptyset, 1^n)$**.** In this case we have $x = 0$ and $v = 0$, so $\ker(x) \cap (\mathbb{C}[x]v)^{\perp} = V$. By Section 4.2.1, no matter the choice of $F_k \in \mathrm{Gr}_k^{\perp}(V)$, we have $(\mu', \nu') = (\emptyset, 1^{n-k})$.

Therefore

$$\mathcal{B}^{((\emptyset,1^n),\alpha)}_{((\emptyset,1^{n-k}),\alpha')} = \mathrm{Gr}_k^{\perp}(\ker(x) \cap (\mathbb{C}[x]v)^{\perp}) = \mathrm{Gr}_k^{\perp}(V) \simeq \mathrm{Gr}^{\perp}_{k,2n}$$

is an irreducible variety of dimension $k(2n-k) - \frac{1}{2}k(k-1)$ by Remark 2.2. We also have

$$
\begin{aligned}
d^{\alpha}_{(\mu,\nu)} - d^{\alpha'}_{(\mu',\nu')} &= 2N(1^n) + |1^n| - \frac{1}{2}\sum_{i\geq 1}(\alpha_i^2 - \alpha_i) \\
&\quad - \left(2N(1^{n-k}) + |1^{n-k}| - \frac{1}{2}\sum_{i\geq 2}(\alpha_i^2 - \alpha_i)\right) \\
&= n(n-1) + n - (n-k)(n-k-1) - (n-k) - \frac{1}{2}k(k-1) \\
&= n^2 - (n-k)^2 - \frac{1}{2}k(k-1) \\
&= k(2n-k) - \frac{1}{2}k(k-1)
\end{aligned}
$$

and the shape $(\emptyset, 1^{n-k})$ is obtained from $(\emptyset, 1^n)$ by removing k-boxes that are all in different rows.

EXAMPLE 5.4. We show the two diagrams with $n = 5$ and $k = 3$, as well as the boxes removed marked with an x.

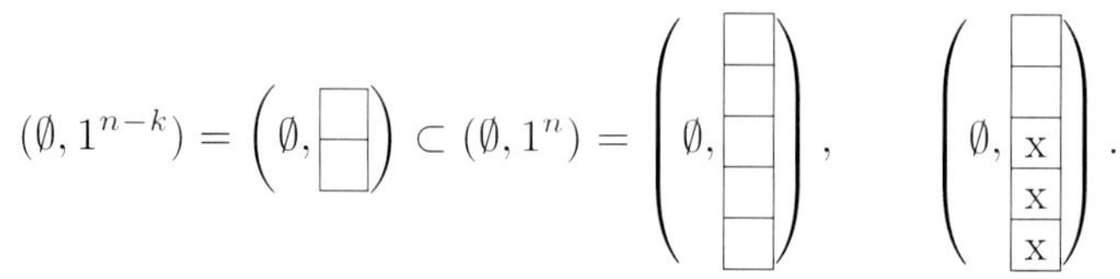

5.3. $(\mu, \nu) = (1^n, \emptyset)$. In this case we have $x = 0$ and $v \neq 0$, so $\ker(x) = V$ and $(\mathbb{C}[x]v)^\perp = (\mathbb{C}v)^\perp$ is a $2n-1$ dimensional subspace. By Section 4.2.2, we have $(\mu', \nu') = (\emptyset, 1^{n-k})$ when $v \in F_k$, and $(\mu', \nu') = (1^{n-k}, \emptyset)$ when $v \notin F_k$.

5.3.1. Suppose that $(\mu', \nu') = (\emptyset, 1^{n-k})$, so that $v \in F_k$ (which implies $k \geq 1$), then $F_k/\mathbb{C}v$ is a $k-1$-dimensional isotropic subspace of $\ker(x) \cap (\mathbb{C}[x]v)^\perp/(\mathbb{C}v) = (\mathbb{C}v)^\perp/(\mathbb{C}v)$, which the restriction of $\langle , \rangle$ makes into a symplectic space of dimension $2n-2$. Hence

$$\begin{aligned}
\mathcal{B}^{((1^n,\emptyset),\alpha)}_{((\emptyset,1^{n-k}),\alpha')} &= \{F_k \subset \ker(x) \cap (\mathbb{C}[x]v)^\perp \mid F_k \subset F_k^\perp,\ v \in F_k\} \\
&\simeq \{F_k/(\mathbb{C}v) \subset (\mathbb{C}v)^\perp/(\mathbb{C}v) \mid F_k/(\mathbb{C}v) \subset (F_k/(\mathbb{C}v))^\perp\} \\
&\simeq \mathrm{Gr}^\perp_{k-1}((\mathbb{C}v)^\perp/(\mathbb{C}v)) \simeq \mathrm{Gr}^\perp_{k-1,2n-2}
\end{aligned}$$

is an irreducible variety of dimension $(k-1)(2n-2-(k-1)) - \frac{1}{2}(k-1)(k-2) = (k-1)(2n-k-1) - \frac{1}{2}(k-1)(k-2)$ by Remark 2.2.

We then have

$$\begin{aligned}
&d^\alpha_{(1^n,\emptyset)} - d^{\alpha'}_{(\emptyset,1^{n-k})} - \dim \mathcal{B}^{((1^n,\emptyset),\alpha)}_{((\emptyset,1^{n-k}),\alpha')} \\
&= 2N(1^n) + 0 - \frac{1}{2}\sum_{i\geq 1}(\alpha_i^2 - \alpha_i) - \left(2N(1^{n-k}) + |1^{n-k}| - \frac{1}{2}\sum_{i\geq 2}(\alpha_i^2 - \alpha_i)\right) \\
&- \left((k-1)(2n-k-1) - \frac{1}{2}(k-1)(k-2)\right) \\
&= n(n-1) - (n-k)(n-k-1) - (n-k) - \frac{1}{2}k(k-1) - (k-1)(2n-k-1) \\
&\quad + \frac{1}{2}(k-1)(k-2) \\
&= n-k.
\end{aligned}$$

This shows that if $1 \leq k < n$, $\dim \mathcal{B}^{((1^n,\emptyset),\alpha)}_{((\emptyset,1^{n-k}),\alpha')} < d^\alpha_{(1^n,\emptyset)} - d^{\alpha'}_{(\emptyset,1^{n-k})}$ and in fact the shape $(\emptyset, 1^{n-k})$ is not contained in $(1^n, \emptyset)$. In the case when $k = n$ (which means that $m = 1$ and there is a single space in the flag) then

$$\dim \mathcal{B}^{((1^n,\emptyset),\alpha)}_{((\emptyset,1^{n-k}),\alpha')} = \dim \mathcal{B}^{((1^n,\emptyset),\alpha)}_{((\emptyset,\emptyset),\alpha')} = d^\alpha_{(1^n,\emptyset)} - d^{\alpha'}_{(\emptyset,\emptyset)}$$

and $(\emptyset, 1^{n-k}) = (\emptyset, \emptyset)$ is indeed obtained from $(1^n, \emptyset)$ by removing k boxes in different rows.

EXAMPLE 5.5. We show the two diagrams with $n = 5$ and $k = 3$, where the diagrams are not nested, and the case with $n = k = 5$, with the boxes removed marked with an x.

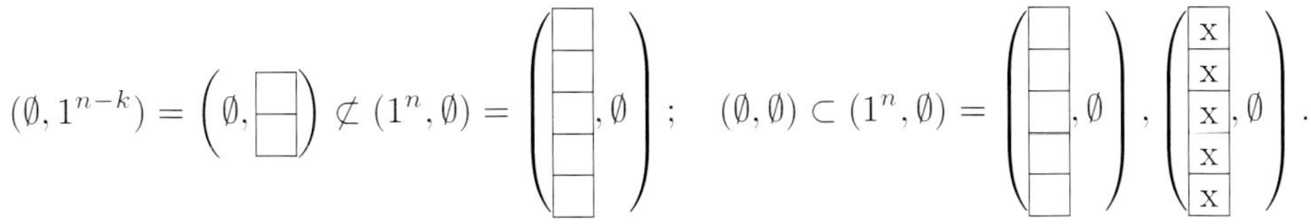

5.3.2. Suppose that $(\mu',\nu') = (1^{n-k},\emptyset)$, so that $v \notin F_k$, which implies $k < n$. Then F_k is a k-dimensional isotropic subspace of $\ker(x) \cap (\mathbb{C}[x]v)^{\perp} = (\mathbb{C}v)^{\perp}$ which is a $2n-1$-dimensional space with the restriction of the form $\langle,\rangle$ having rank $2n-2$, so $r = n-1$. We then have

$$\mathcal{B}^{((1^n,\emptyset),\alpha)}_{((1^{n-k},\emptyset),\alpha')} = \{F_k \subset (\mathbb{C}v)^{\perp} \mid F_k \subset F_k^{\perp},\ v \notin F_k\}$$

which is an open dense subvariety of

$$\mathrm{Gr}^{\perp}_k\left((\mathbb{C}v)^{\perp}\right) \simeq \mathrm{Gr}^{n-1}_{k,2n-1}$$

hence it is irreducible of dimension $k(2n-1-k) - \frac{1}{2}k(k-1)$ by Lemma 2.4 (there is no third term because $k \leq n-1$).

In this case, we have:

$$\begin{aligned} d^{\alpha}_{(1^n,\emptyset)} - d^{\alpha'}_{(1^{n-k},\emptyset)} &= 2N(1^n) - \frac{1}{2}\sum_{i\geq 1}(\alpha_i^2 - \alpha_i) - (2N(1^{n-k}) - \frac{1}{2}\sum_{i\geq 2}(\alpha_i^2 - \alpha_i)) \\ &= n(n-1) - \frac{1}{2}(k^2-k) - (n-k)(n-k-1) \\ &= k(2n-k-1) - \frac{1}{2}(k^2-k) \end{aligned}$$

which agrees with the dimension of the variety and the shape $(1^{n-k},\emptyset)$ is obtained from $(1^n,\emptyset)$ by removing k-boxes that are all in different rows.

EXAMPLE 5.6. We show the two diagrams with $n = 5$ and $k = 3$, as well as the boxes removed marked with an x.

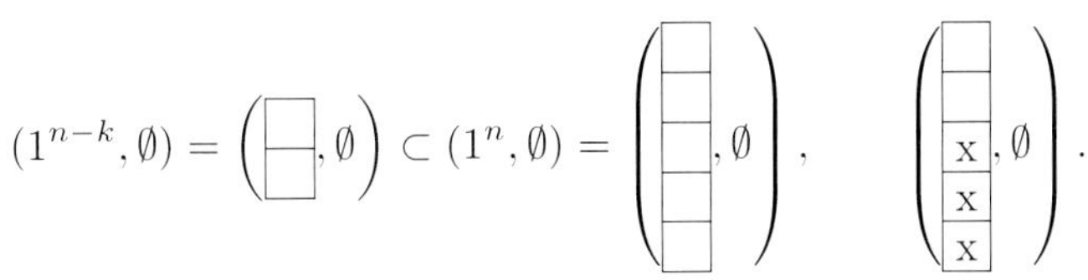

5.4. $\mu = \emptyset$, $\nu = 2^{n_2}1^{n_1}$. In this case we have $x \neq 0$, $x^2 = 0$ and $v = 0$, so $(\mathbb{C}[x]v)^{\perp} = V$ and $\ker(x)$ is a $2n_1 + 2n_2$-dimensional space where the restriction of $\langle,\rangle$ has rank $2n_1$ ($r = n_1$). We define k_1, k_2, and h as in Section 4.1.2, by Section 4.2.3 we have that $(\mu',\nu') = (\emptyset, 2^{n_2-k_2+h}1^{n_1-k_1+k_2-2h})$. To describe the variety $\mathcal{B}^{((\mu,\nu),\alpha)}_{((\mu',\nu'),\alpha')}$ we use the ideas from Section 5.1. We start by describing the variety

$$\mathcal{X}(k_2,h) := \{F_k \in \mathrm{Gr}^{\perp}_k(\ker(x)) \mid \dim(F_k \cap \mathrm{Im}(x)) = k_2, \dim(F_k \cap x(F_k^{\perp})) = k_2 - 2h\}$$

LEMMA 5.7. *For $0 \leq k \leq n_1 + 2n_2$, the variety $\mathcal{X}(k_2,h)$ is nonempty if and only if the following inequalities are satisfied:*

$$\max\{0, k-n_1\} \leq k_2 \leq \min\{k, 2n_2\}; \qquad \max\{0, k_2 - n_2\} \leq h \leq \frac{k_2}{2}. \tag{5.3}$$

PROOF. If the variety is nonempty, then for all $F_k \in \mathcal{X}(k_2,h)$ we have

$$0 \le \dim(F_k \cap \operatorname{Im}(x)) = k_2 \le \dim(\operatorname{Im}(x)) = 2n_2,$$

$$k_1 = \dim(F_k/F_k \cap \operatorname{Im}(x)) \ge 0, \text{ so } k_2 = k - k_1 \le k,$$

$$F_k/(F_k \cap \operatorname{Im}(x)) \in \operatorname{Gr}_{k_1}^{\perp n_1}(\ker(x)/F_k \cap \operatorname{Im}(x)), \text{ so } k_1 \le n_1, \text{ and } k_2 = k - k_1 \ge k - n_1,$$

$$0 \le \dim(F_k \cap \operatorname{Im}(x)/F_k \cap x(F_k^{\perp})) = 2h \le \dim(F_k \cap \operatorname{Im}(x)) = k_2,$$

Since $F_k \cap \operatorname{Im}(x) \subset (F_k \cap x(F_k^{\perp}))^{\perp\!\!\perp}$, we have $k_2 \le 2n_2 - (k_2 - 2h)$, which implies

$$0 \le n_2 - k_2 + h, \text{ so } h \ge k_2 - n_2.$$

Viceversa, if all the inequalities are satisfied, we can define $F_k = \operatorname{span}\{u_1, \dots, u_k\}$. Here we choose $u_1, \dots, u_{k_2-2h}$ to be linearly independent vectors in $\operatorname{Im}(x)$ such that $\langle\langle u_i, u_j\rangle\rangle = 0$ for all $1 \le i, j, \le k_2 - 2h$, then we choose $u_{k_2-2h}, \dots, u_{k_2}$ to be linearly independent vectors in $(\operatorname{span}\{u_1, \dots, u_{k_2-2h}\}^{\perp\!\!\perp}) \setminus \operatorname{span}\{u_1, \dots, u_{k_2-2h}\}$, and finally we choose $u_{k_2+1}, \dots, u_k$ to be linearly independent vectors that are not in $\operatorname{Im}(x)$, such that $\langle u_i, u_j\rangle = 0$ for all $k_2 + 1 \le i, j, \le k$. Then $F_k \in \mathcal{X}(k_2, h) \ne \emptyset$. □

Observe that

$$\begin{aligned}
\mathcal{X}(k_2,h) &:= \{F_k \in \operatorname{Gr}_k^{\perp}(\ker(x)) \mid \dim(F_k \cap \operatorname{Im}(x)) = k_2, \\
&\qquad \dim(F_k \cap x(F_k^{\perp})) = k_2 - 2h\} \\
&= (\pi_1^k)^{-1}(\pi_2^{k_2})^{-1}\left(\operatorname{Gr}_{k_2-2h}^{\perp\!\!\perp}(\ker(x) \cap (\mathbb{C}[x]v)^{\perp} \cap \operatorname{Im}(x))\right) \\
&= (\pi_1^k)^{-1}(\pi_2^{k_2})^{-1}\left(\operatorname{Gr}_{k_2-2h}^{\perp\!\!\perp}(\operatorname{Im}(x))\right).
\end{aligned}$$

The base $\operatorname{Gr}_{k_2-2h}^{\perp\!\!\perp}(\operatorname{Im}(x))$ is an isotropic Grassmannian, which (since $\dim(\operatorname{Im}(x)) = 2n_2$), is irreducible of dimension

$$(k_2 - 2h)(2n_2 - k_2 + 2h) - \tfrac{1}{2}(k_2 - 2h)(k_2 - 2h - 1).$$

Over every point $S \in \operatorname{Gr}_{k_2-2h}^{\perp\!\!\perp}(\operatorname{Im}(x))$, we have $(\pi_2^{k_2})^{-1}(S) \simeq \operatorname{Gr}_{2h}(S^{\perp\!\!\perp}/S)$ by (5.2), so each fibre is irreducible of dimension

$$2h(2n_2 - 2k_2 + 2h)$$

hence $(\pi_2^{k_2})^{-1}\left(\operatorname{Gr}_{k_2-2h}^{\perp\!\!\perp}(\operatorname{Im}(x))\right)$ is irreducible, being a fibre bundle with irreducible fibres over an irreducible base.

Again, for each $U \in (\pi_2^{k_2})^{-1}\left(\operatorname{Gr}_{k_2-2h}^{\perp\!\!\perp}(\operatorname{Im}(x))\right)$, we consider $(\pi_1^k)^{-1}(U)$ which by (5.1) is isomorphic to an open dense subvariety of $\operatorname{Gr}_{k_1}^{\perp}(\ker(x)/U)$. Notice that $\ker(x)/U$ is a vector space of dimension $2n_1 + 2n_2 - k_2$ and the restriction of $\langle , \rangle$ has rank $2n_1$, hence each fibre $(\pi_1^k)^{-1}(U)$ is an irreducible subvariety of dimension

$$\dim \operatorname{Gr}_{k_1, 2n_1+2n_2-k_2}^{\perp n_1} = k_1(2n_2 + 2n_1 - k_2 - k_1) - \tfrac{1}{2}k_1(k_1 - 1).$$

In conclusion $\mathcal{X}(k_2, h)$ is an irreducible variety of dimension

$$\begin{aligned}
&(k_2 - 2h)(2n_2 - k_2 + 2h) - \tfrac{1}{2}(k_2 - 2h)(k_2 - 2h - 1) + 2h(2n_2 - 2k_2 + 2h) \\
&+ k_1(2n_2 + 2n_1 - k_2 - k_1) - \tfrac{1}{2}k_1(k_1 - 1).
\end{aligned} \tag{5.4}$$

For $(\mu, \nu) = (\emptyset, 2^{n_2}1^{n_1})$, $(\mu', \nu') = (\emptyset, 2^{n_2-k_2+h}1^{n_1-k_1+k_2-2h})$ we compute the difference :

$$
\begin{aligned}
(5.5)\quad & d^{\alpha}_{(\mu,\nu)} - d^{\alpha'}_{(\mu',\nu')} - \dim \mathcal{X}(k_2, h) \\
&= 2N(2^{n_2}1^{n_1}) + 2n_2 + n_1 - \frac{1}{2}\sum_{i\geq 1}(\alpha_i^2 - \alpha_i) - 2N(2^{n_2-k_2+h}1^{n_1-k_1+k_2-2h}) \\
&\quad - (2(n_2 - k_2 + h) + n_1 - k_1 + k_2 - 2h) + \frac{1}{2}\sum_{i\geq 2}(\alpha_i^2 - \alpha_i) - (5.4) \\
&= (n_1 + n_2)(n_1 + n_2 - 1) + n_2(n_2 - 1) + 2n_2 + n_1 - \tfrac{1}{2}k(k-1) \\
&\quad - (n_1 + n_2 - k_1 - h)(n_1 + n_2 - k_1 - h - 1) \\
&\quad - (n_2 - k_2 + h)(n_2 - k_2 + h - 1) \\
&\quad - (2(n_2 - k_2 + h) + n_1 - k_1 + k_2 - 2h) - (5.4) \\
&= h(2n_1 - 2k_1 + 1).
\end{aligned}
$$

Since $h \geq 0$ and $n_1 \geq k_1$, this difference is always ≥ 0 and is equal to zero if and only if $h = 0$.

Notice that for fixed $0 \leq n_1' \leq n_1 + n_2$, $0 \leq n_2' \leq n_2$ such that $n_1' + 2n_2' + k = n_1 + 2n_2$, we have

$$
\begin{aligned}
\mathcal{B}^{((\emptyset,2^{n_2}1^{n_1}),\alpha)}_{((\emptyset,2^{n_2'}1^{n_1'}),\alpha')} &= \coprod_{\substack{\max\{0,k-n_1\}\leq k_2\leq \min\{k,2n_2\} \\ \max\{0,k_2-n_2\}\leq h\leq \frac{k_2}{2} \\ k_2-h=n_2-n_2'}} \mathcal{X}(k_2, h) \\
&= \coprod_{\substack{\max\{0,n_2-n_1'-n_2'\}\leq h \\ h\leq \min\{k-(n_2-n_2'),n_2-n_2'\}}} \mathcal{X}(h + n_2 - n_2', h).
\end{aligned}
$$

Where the inequalities follow from (5.3), together with the fact that $n_2' = n_2 - k_2 + h$ and $n_1' = n_1 - k_1 + k_2 - 2h$ by Section 4.2.3.

Lemma 5.8. *Suppose that $\mathcal{X}(k_2, h) \neq \emptyset \neq \mathcal{X}(k_2+1, h+1)$, then $\mathcal{X}(k_2+1, h+1) \subset \overline{\mathcal{X}(k_2, h)}$.*

Proof. Let $F_k \in \mathcal{X}(k_2 + 1, h + 1) \neq \emptyset$, such that $F_k = \mathrm{span}\{u_1, \ldots, u_k\}$, with $F_k \cap x(F_k^{\perp}) = \mathrm{span}\{u_1, \ldots, u_{k_2-2h-1}\}$, $F_k \cap \mathrm{Im}(x) = \mathrm{span}\{u_1, \ldots, u_{k_2+1}\}$. Since $\mathcal{X}(k_2, h) \neq \emptyset$, we have $k - n_1 \leq k_2$, so $k - (k_2 + 1) < n_1$, it follows that there exists a vector $w \notin \mathrm{Im}(x)$, $w \in F_k^{\perp} \notin F_k$. Also, $\mathcal{X}(k_2, h) \neq \emptyset$ implies that $h \geq 0$, hence $2h + 2 > 0$ and $u_{k_2+1} \in F_k \cap \mathrm{Im}(x) \setminus F_k \cap x(F_k^{\perp})$. For all $\alpha \in \mathbb{C}$, define a new space F_k^{α} to be spanned by the same vectors as F_k except that u_{k_2+1} is replaced by $u_{k_2+1} + \alpha w$. Then for all $\alpha \neq 0$ we have $F_k^{\alpha} \in \mathcal{X}(k_2, h)$, which shows that $F_k = F_k^0 \in \overline{\mathcal{X}(k_2, h)}$. □

By (5.5), since $h(2n_1 - 2k_1 + 1) = h(2n_1 + 2n_2 - 2n_2' - 2k + 2h + 1)$, and by Lemma 5.8, we have that for $h' > h$, $\mathcal{X}(h' + n_2 - n_2', h')$ is contained in the closure of $\mathcal{X}(h + n_2 - n_2', h)$ as a positive codimension subvariety. Therefore $\mathcal{B}^{((\mu,\nu),\alpha)}_{((\mu',\nu'),\alpha')} = \overline{\mathcal{X}(h + n_2 - n_2', h)}$, for the minimal possible value of h, i.e. $h = \max\{0, n_2 - n_1' - n_2'\}$. This proves that $\mathcal{B}^{((\mu,\nu),\alpha)}_{((\mu',\nu'),\alpha')}$ is an irreducible variety of dimension $\dim \mathcal{X}(h + n_2 - n_2', h)$ with $h = \max\{0, n_2 - n_1' - n_2'\}$.

Now, the minimal h is $h = 0$ if and only if $n_2 \leq n_1' + n_2'$ if and only if the shape $(\emptyset, 2^{n_2'}1^{n_1'})$ is obtained from $(\emptyset, 2^{n_2}1^{n_1})$ by removing boxes in a vertical strip. In conclusion, $\dim \mathcal{B}^{((\mu,\nu),\alpha)}_{((\mu',\nu'),\alpha')} = d^{\alpha}_{(\mu,\nu)} - d^{\alpha'}_{(\mu',\nu')}$ exactly when the shape is obtained by removing boxes from different rows and is strictly lower otherwise.

EXAMPLE 5.9. Here $n_1 = 2$, $n_2 = 3$, $k = 4$. In the first example $n_1' + n_2' = 0 + 2 < 3 = n_2$ so some boxes are removed from the same row, while in the second example $n_1' + n_2' = 2 + 1 = 3 \geq n_2$ so the boxes are removed from different rows.

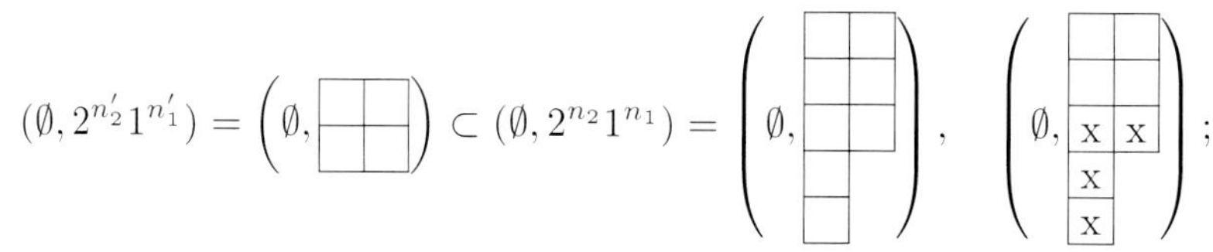

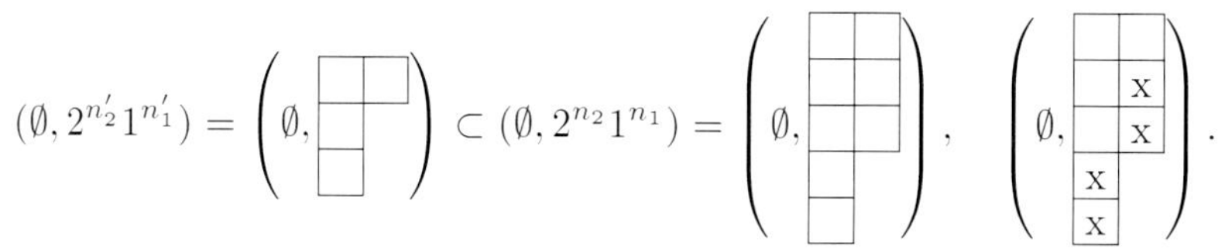

5.5. $\mu = 1^{n_2}$**,** $\nu = 1^{n_1+n_2}$**.** In this case we have $x \neq 0$, $x^2 = 0$ and $0 \neq v \in \mathrm{Im}(x)$, so $\ker(x) \subset (\mathbb{C}[x]v)^{\perp}$ and $\ker(x) \cap (\mathbb{C}[x]v)^{\perp} = \ker(x)$ is a $2n_1 + 2n_2$-dimensional space where the restriction of $\langle , \rangle$ has rank $2n_1$ ($r = n_1$). We define k_1, k_2 and h as in Section 4.1.2. According to Section 4.2.4, there are three possibilities for (μ', ν'), which we now examine separately.

5.5.1. Suppose that $(\mu', \nu') = (\emptyset, 2^{n_2'}1^{n_1'}) = (\emptyset, 2^{n_2-k_2+h}1^{n_1-k_1+k_2-2h})$. Then by Section 4.2.4, we have $v \in F_k \cap \mathrm{Im}(x)$, which implies that $k_2 \geq 1$. Notice that there are two possibilities, either $v \in F_k \cap x(F_k^{\perp})$, or $v \in (F_k \cap \mathrm{Im}(x)) \setminus (F_k \cap x(F_k^{\perp}))$. We define the following varieties

$$\mathcal{X}^1(k_2, h) = \{F_k \in \mathrm{Gr}_k^{\perp}(\ker(x)) \mid \dim(F_k \cap \mathrm{Im}(x)) = k_2,$$
$$\dim(F_k \cap x(F_k^{\perp})) = k_2 - 2h, v \in F_k \cap x(F_k^{\perp})\};$$

$$\mathcal{Y}(k_2, h) = \{F_k \in \mathrm{Gr}_k^{\perp}(\ker(x)) \mid \dim(F_k \cap \mathrm{Im}(x)) = k_2,$$
$$\dim(F_k \cap x(F_k^{\perp})) = k_2 - 2h, v \in (F_k \cap \mathrm{Im}(x)) \setminus (F_k \cap x(F_k^{\perp}))\}.$$

If $v \in F_k \cap x(F_k^{\perp})$, then $k_2 - 2h \geq 1$, and $(F_k \cap x(F_k^{\perp}))/\mathbb{C}v \in \mathrm{Gr}^{\perp\!\!\perp}_{k_2-2h-1}(\mathbb{C}(v)^{\perp\!\!\perp}/\mathbb{C}v)$. It follows that we have a description as an iterated vector bundle

$$\mathcal{X}^1(k_2, h) = (\pi_1^k)^{-1}(\pi_2^{k_2})^{-1}\left(\mathrm{Gr}^{\perp\!\!\perp}_{k_2-2h-1}(\mathbb{C}(v)^{\perp\!\!\perp}/\mathbb{C}v)\right).$$

The base is an isotropic Grassmannian, hence an irreducible variety of dimension

$$(k_2 - 2h - 1)(2n_2 - k_2 + 2h - 1) - \tfrac{1}{2}(k_2 - 2h - 1)(k_2 - 2h - 2). \tag{5.6}$$

Over every point $S \in \mathrm{Gr}^{\perp\!\!\perp}_{k_2-2h-1}(\mathbb{C}(v)^{\perp\!\!\perp}/\mathbb{C}v)$, we have $(\pi_2^{k_2})^{-1}(S) \simeq \mathrm{Gr}_{2h}(S^{\perp\!\!\perp}/S)$ by (5.2), so each fibre is irreducible of dimension

$$2h(2n_2 - 2k_2 + 2h).$$

Finally over a point $U \in (\pi_2^{k_2})^{-1}\left(\mathrm{Gr}^{\perp\!\!\perp}_{k_2-2h-1}(\mathbb{C}(v)^{\perp\!\!\perp}/\mathbb{C}v)\right)$, we have that $(\pi_1^k)^{-1}(U)$ is isomorphic to an open subvariety of $\mathrm{Gr}^{\perp}_{k_1}(\ker(x)/U)$. Exactly as in the previous section, then, each fibre $(\pi_1^k)^{-1}(U)$ is an irreducible subvariety of dimension

$$\dim \mathrm{Gr}^{\perp n_1}_{k_1,2n_1+2n_2-k_2} = k_1(2n_2+2n_1-k_2-k_1) - \tfrac{1}{2}k_1(k_1-1).$$

In conclusion $\mathcal{X}^1(k_2,h)$ is an irreducible variety of dimension

$$(5.7)\quad (k_2-2h-1)(2n_2-k_2+2h-1) - \tfrac{1}{2}(k_2-2h-1)(k_2-2h-2) \\ + 2h(2n_2-2k_2+2h) + k_1(2n_2+2n_1-k_2-k_1) - \tfrac{1}{2}k_1(k_1-1).$$

For $(\mu,\nu) = (1^{n_2}, 1^{n_1+n_2})$, $(\mu',\nu') = (\emptyset, 2^{n_2-k_2+h}1^{n_1-k_1+k_2-2h})$ we compute the difference:

$$\begin{aligned}
(5.8)\quad & d^{\alpha}_{(\mu,\nu)} - d^{\alpha'}_{(\mu',\nu')} - \dim \mathcal{X}^1(k_2,h) \\
&= 2N(2^{n_2}1^{n_1}) + n_2 + n_1 - \frac{1}{2}\sum_{i\geq 1}(\alpha_i^2-\alpha_i) - 2N(2^{n_2-k_2+h}1^{n_1-k_1+k_2-2h}) \\
&\quad - (2(n_2-k_2+h)+n_1-k_1+k_2-2h) + \frac{1}{2}\sum_{i\geq 2}(\alpha_i^2-\alpha_i) - (5.7) \\
&= (n_1+n_2)(n_1+n_2-1) + n_2(n_2-1) + n_2 + n_1 - \tfrac{1}{2}k(k-1) \\
&\quad - (n_1+n_2-k_1-h)(n_1+n_2-k_1-h-1) \\
&\quad - (n_2-k_2+h)(n_2-k_2+h-1) \\
&\quad - (2(n_2-k_2+h)+n_1-k_1+k_2-2h) - (5.7) \\
&= 2h(n_1-k_1+1) + n_2 - k_2 + h.
\end{aligned}$$

Since $h \geq 0$, $n_1 \geq k_1$, and $n_2-k_2+h \geq 0$, this difference is always ≥ 0 and is equal to zero if and only if $h=0$ and $n_2-k_2+h=0$ (which implies that $n_2=k_2$).

Now we want to describe $\mathcal{Y}(k_2,h)$, notice that if $v \in (F_k \cap \mathrm{Im}(x)) \setminus (F_k \cap x(F_k^{\perp}))$, that means $h \geq 1$. Since $F_k \cap x(F_k^{\perp}) = (F_k \cap \mathrm{Im}(x))^{\perp\!\!\perp}$, if $v \in (F_k \cap \mathrm{Im}(x)) \setminus (F_k \cap x(F_k^{\perp}))$, then $F_k \cap x(F_k^{\perp}) \subseteq (\mathbb{C}v)^{\perp\!\!\perp}$, with $\mathbb{C}v \not\subset F_k \cap x(F_k^{\perp})$. Also the bilinear form $\langle\langle\ ,\ \rangle\rangle$ restricted to the $(2n_2-1)$-dimensional space $(\mathbb{C}v)^{\perp\!\!\perp} \subset \mathrm{Im}(x)$ has rank $2n_2-2$ ($r=n_2-1$). It follows that $F_k \cap x(F_k^{\perp}) \in Z := \{S \in \mathrm{Gr}^{\perp\!\!\perp}_{k_2-2h}((\mathbb{C}v)^{\perp\!\!\perp}) \mid S \cap \mathbb{C}v = 0\}$ which is an open dense subvariety of $\mathrm{Gr}^{\perp\!\!\perp}_{k_2-2h}((\mathbb{C}v)^{\perp\!\!\perp})$, hence irreducible of dimension

$$\dim \mathrm{Gr}^{\perp n_2-1}_{k_2-2h,2n_2-1} = (k_2-2h)(2n_2-1-k_2+2h) - \tfrac{1}{2}(k_2-2h)(k_2-2h-1).$$

(There is no third term because $k-r = k_2-2h-(n_2-1) = -(n_2-k_2+h)-h+1 \leq 0$). It follows that

$$\mathcal{Y}(k_2,h) = \{F_k \in (\pi_1^k)^{-1}(\pi_2^{k_2})^{-1}(Z) \mid v \in \pi_1^k(F_k)\}.$$

Over every point $S \in Z$, we consider only the points $U \in (\pi_2^{k_2})^{-1}(S)$ such that $v \in U$, i.e. we are choosing $U/(S+\mathbb{C}v) \in \mathrm{Gr}_{2h-1}(S^{\perp\!\!\perp}/(S+\mathbb{C}v))$, so each fibre is irreducible of dimension

$$\begin{aligned}
\dim \mathrm{Gr}_{2h-1,2n_2-2(k_2-2h)-1} &= (2h-1)(2n_2-2(k_2-2h)-1-(2h-1)) \\
&= (2h-1)(2n_2-2k_2+2h).
\end{aligned}$$

Finally, over a point $U \in (\pi_2^{k_2})^{-1}(Z)$, (with $v \in U$) we have that $(\pi_1^k)^{-1}(U)$ is isomorphic to an open dense subvariety of $\mathrm{Gr}^{\perp}_{k_1}(\ker(x)/U)$ so, as in previous sections, it is irreducible of dimension

$$\dim \mathrm{Gr}^{\perp n_1}_{k_1,2n_1+2n_2-k_2} = k_1(2n_2+2n_1-k_2-k_1) - \tfrac{1}{2}k_1(k_1-1).$$

In conclusion $\mathcal{Y}(k_2, h)$ is an irreducible variety of dimension

$$(5.9)\quad (k_2 - 2h)(2n_2 - 1 - k_2 + 2h) - \tfrac{1}{2}(k_2 - 2h)(k_2 - 2h - 1) + (2h - 1)(2n_2 - 2k_2 + 2h) + k_1(2n_2 + 2n_1 - k_2 - k_1) - \tfrac{1}{2}k_1(k_1 - 1).$$

For $(\mu, \nu) = (1^{n_2}, 1^{n_1+n_2})$, $(\mu', \nu') = (\emptyset, 2^{n_2-k_2+h}1^{n_1-k_1+k_2-2h})$ we compute the difference:

$$\begin{aligned}(5.10)\quad & d^{\alpha}_{(\mu,\nu)} - d^{\alpha'}_{(\mu',\nu')} - \dim \mathcal{Y}(k_2, h)\\
&= 2N(2^{n_2}1^{n_1}) + n_2 + n_1 - \frac{1}{2}\sum_{i\geq 1}(\alpha_i^2 - \alpha_i) - 2N(2^{n_2-k_2+h}1^{n_1-k_1+k_2-2h})\\
&\quad - (2(n_2 - k_2 + h) + n_1 - k_1 + k_2 - 2h) + \frac{1}{2}\sum_{i\geq 2}(\alpha_i^2 - \alpha_i) - (5.9)\\
&= (n_1 + n_2)(n_1 + n_2 - 1) + n_2(n_2 - 1) + n_2 + n_1 - \tfrac{1}{2}k(k-1)\\
&\quad - (n_1 + n_2 - k_1 - h)(n_1 + n_2 - k_1 - h - 1)\\
&\quad - (n_2 - k_2 + h)(n_2 - k_2 + h - 1)\\
&\quad - (2(n_2 - k_2 + h) + n_1 - k_1 + k_2 - 2h) - (5.9)\\
&= 2h(n_1 - k_1) + n_2 - k_2 + h.\end{aligned}$$

Since $h \geq 1$, $n_1 \geq k_1$, and $n_2 - k_2 + h \geq 0$, this difference is always ≥ 0 and is equal to zero if and only if $n_1 = k_1$ and $n_2 - k_2 + h = 0$. For fixed $0 \leq n_1' \leq n_1 + n_2$, $0 \leq n_2' \leq n_2$ such that $n_1' + 2n_2' + k = n_1 + 2n_2$, we have

$$\begin{aligned}\mathcal{B}^{((1^{n_2},1^{n_1+n_2}),\alpha)}_{((\emptyset,2^{n_2'}1^{n_1'}),\alpha')} &= \left(\coprod_{\substack{\max\{1,k-n_1\}\leq k_2\leq \min\{k,2n_2\}\\ \max\{0,k_2-n_2\}\leq h<\frac{k_2}{2}\\ k_2-h=n_2-n_2'}} \mathcal{X}^1(k_2, h)\right)\\
&\quad \coprod \left(\coprod_{\substack{\max\{1,k-n_1\}\leq k_2\leq \min\{k,2n_2\}\\ \max\{1,k_2-n_2\}\leq h\leq\frac{k_2}{2}\\ k_2-h=n_2-n_2'}} \mathcal{Y}(k_2, h)\right)\\
&= \left(\coprod_{\substack{\max\{0,n_2-n_1'-n_2'\}\leq h\\ 1-n_2+n_2'\leq h\\ h\leq \min\{k-(n_2-n_2'),n_2-n_2'-1\}}} \mathcal{X}^1(h + n_2 - n_2', h)\right)\\
&\quad \coprod \left(\coprod_{\substack{\max\{1,n_2-n_1'-n_2'\}\leq h\\ h\leq \min\{k-(n_2-n_2'),n_2-n_2'\}}} \mathcal{Y}(h + n_2 - n_2', h)\right);\end{aligned}$$

where the inequalities follow from the same reasoning as Lemma 5.7, plus the observations that $k_2 > 0$, in $\mathcal{X}^1(k_2, h)$ we have $k_2 - 2h > 0$, and in $\mathcal{Y}(k_2, h)$ we have $h > 0$.

LEMMA 5.10. *If $\mathcal{X}^1(k_2, h) \neq \emptyset \neq \mathcal{X}^1(k_2+1, h+1)$, then $\mathcal{X}^1(k_2+1, h+1) \subset \overline{\mathcal{X}^1(k_2, h)}$; if $\mathcal{Y}(k_2, h) \neq \emptyset \neq \mathcal{Y}(k_2+1, h+1)$, then $\mathcal{Y}(k_2+1, h+1) \subset \overline{\mathcal{Y}(k_2, h)}$; if $\mathcal{X}^1(k_2, h) \neq \emptyset \neq \mathcal{Y}(k_2+1, h+1)$, then $\mathcal{Y}(k_2+1, h+1) \subset \overline{\mathcal{X}^1(k_2, h)}$; if $\mathcal{X}^1(k_2, h) \neq \emptyset \neq \mathcal{Y}(k_2, h)$, then $\mathcal{X}^1(k_2, h) \subset \overline{\mathcal{Y}(k_2, h)}$.*

PROOF. This is entirely analogous to the proof of Lemma 5.8, by choosing for each case an appropriate basis and a vector to be deformed, details are omitted. □

Lemma 5.10 then implies, like it happened in Section 5.4, that there is one irreducible piece (either $\mathcal{X}^1(k_2, h)$ or $\mathcal{Y}(k_2, h)$) of maximal dimension that contains all the other ones in its closure as positive codimension subvarieties, hence $\mathcal{B}^{((1^{n_2}, 1^{n_1+n_2}), \alpha)}_{((\emptyset, 2^{n_2'}1^{n_1'}), \alpha')}$ is irreducible. Notice that by (5.8) and (5.10) we have that if $n_2' = n_2 - k_2 + h > 0$, then even the piece of maximal dimension will be of dimension strictly less than $d^{\alpha}_{(\mu,\nu)} - d^{\alpha'}_{(\mu',\nu')}$, and the shapes are not nested $(\emptyset, 2^{n_2'}1^{n_1'}) \not\subset (1^{n_2}, 1^{n_1+n_2})$. When $n_2' = 0$, we have that the piece of maximal dimension is $\mathcal{X}^1(n_2, 0)$ if and only if $h = 0$ is the minimal h, if and only if $n_1' \geq n_2$, and otherwise it is $\mathcal{Y}(2n_2 - n_1', n_2 - n_1')$ if and only if $n_2 > n_1'$. If $n_1' \geq n_2$, then $\dim \mathcal{B}^{((1^{n_2}, 1^{n_1+n_2}), \alpha)}_{((\emptyset, 1^{n_1'}), \alpha')} = \dim \mathcal{X}(n_2, 0) = d^{\alpha}_{(\mu,\nu)} - d^{\alpha'}_{(\mu',\nu')}$. Otherwise, with $n_2' = 0$ and $n_2 > n_1'$ we have that the minimal h, is $h = n_2 - n_1'$, which gives

$$\begin{aligned}
h &= n_2 - n_1' \\
h &= n_2 - (n_1 - k_1) - (k_2 - 2h) \\
h &= n_2 - n_1 + k_1 - k_2 + 2h \\
0 &= n_2 - k_2 + h - n_1 + k_1 \\
0 &= -n_1 + k_1 \\
n_1 &= k_1,
\end{aligned}$$

so that indeed $\dim \mathcal{B}^{((1^{n_2}, 1^{n_1+n_2}), \alpha)}_{((\emptyset, 1^{n_1'}), \alpha')} = \dim \mathcal{Y}(2n_2 - n_1', n_2 - n_1') = d^{\alpha}_{(\mu,\nu)} - d^{\alpha'}_{(\mu',\nu')}$.

EXAMPLE 5.11. Here $n_1 = 2$, $n_2 = 3$, $k = 4$. In the first example $n_1' = 2$, $n_2' = 1 > 0$, so the diagrams are not nested, in the second example $n_1' = 4$, $n_2' = 0$ so the diagrams are nested and the boxes are removed from different rows of the two tableaux.

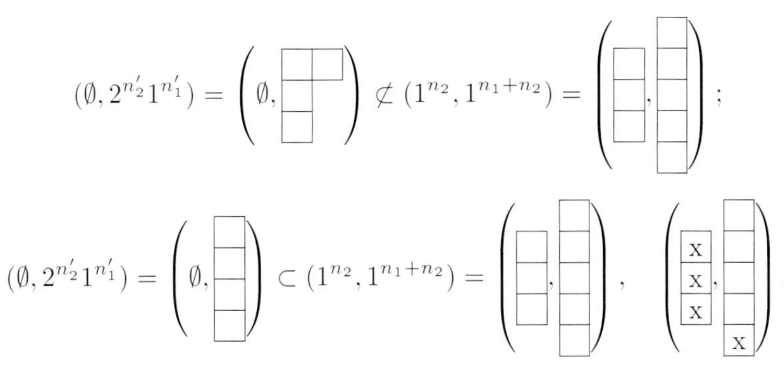

5.5.2. Suppose that $(\mu', \nu') = (1^{n_2'}, 1^{n_2'+n_1'}) = (1^{n_2-k_2+h}, 1^{n_1+n_2-k_1-h})$, with $n_2' = n_2 - k_2 + h > 0$. Then by Section 4.2.4, we have $v \notin F_k$, and $(F_k + \mathbb{C}v) \cap x(F_k^\perp) \supsetneq F_k \cap x(F_k^\perp)$, or equivalently that $F_k \cap x(F_k^\perp) \subset (\mathbb{C}v)^{\perp\!\!\perp}$, by Lemma 4.4. In this case we have that $k_2 < 2n_2$. We define the variety

$$\mathcal{X}^2(k_2, h) = \{F_k \in \mathrm{Gr}_k^\perp(\ker(x)) \mid \dim(F_k \cap \mathrm{Im}(x)) = k_2, \\ \dim(F_k \cap x(F_k^\perp)) = k_2 - 2h,\ v \notin F_k,\ F_k \cap x(F_k^\perp) \subset (\mathbb{C}v)^{\perp\!\!\perp}\}.$$

Notice that $(\mathbb{C}v)^{\perp\!\!\perp}$ is a $2n_2 - 1$-dimensional space, and the restriction of $\langle\langle, \rangle\rangle$ to it has rank $2n_2 - 2$ ($r = n_2 - 1$). It follows that $F_k \cap x(F_k^\perp) \in Z := \{S \in \mathrm{Gr}_{k_2-2h}^{\perp\!\!\perp}((\mathbb{C}v)^{\perp\!\!\perp}) \mid S \cap \mathbb{C}v = 0\}$, which is nonempty because $n_2 - k_2 + h > 0$ implies that $k_2 - 2h < 2n_2 - 1$, so it is an open dense subvariety of $\mathrm{Gr}_{k_2-2h}^{\perp\!\!\perp}((\mathbb{C}v)^{\perp\!\!\perp})$, hence irreducible of dimension

$$\dim \mathrm{Gr}_{k_2-2h, 2n_2-1}^{\perp n_2-1} = (k_2 - 2h)(2n_2 - 1 - k_2 + 2h) - \tfrac{1}{2}(k_2 - 2h)(k_2 - 2h - 1).$$

We then have

$$\mathcal{X}^2(k_2, h) = \{F_k \in (\pi_1^k)^{-1}(\pi_2^{k_2})^{-1}(Z) \mid v \notin \pi_1^k(F_k)\}$$

Over every point $S \in Z$, we consider only the points $U \in (\pi_2^{k_2})^{-1}(S)$ such that $v \notin U$, i.e. we are choosing $U/S \in \mathrm{Gr}_{2h}(S^{\perp\!\!\perp}/S)$, such that $U/S \cap \mathbb{C}v/S = 0$ (where $\mathbb{C}v/S$ is the image of $\mathbb{C}v$ under the projection onto $S^{\perp\!\!\perp}/S$) so each fibre, if nonempty, is an open subvariety of a Grassmannian, hence irreducible of dimension

$$\dim \mathrm{Gr}_{2h, 2n_2-2(k_2-2h)} = 2h(2n_2 - 2k_2 + 2h).$$

This fibre is nonempty if and only if it is possible to choose U/S that does not contain $\mathbb{C}v/S$, if and only if

$$\begin{aligned} 2h &< 2n_2 - 2(k_2 - 2h) \\ 2h &< 2n_2 - 2k_2 + 4h \\ 0 &< 2n_2 - 2k_2 + 2h \\ 0 &< n_2 - k_2 + h. \end{aligned}$$

Finally, over a point $U \in (\pi_2^{k_2})^{-1}(Z)$, (with $v \notin U$) we have that $(\pi_1^k)^{-1}(U)$ is isomorphic to an open subvariety of $\mathrm{Gr}_{k_1}^\perp(\ker(x)/U)$ so, as in previous sections, it is irreducible of dimension

$$\dim \mathrm{Gr}_{k_1, 2n_1+2n_2-k_2}^{\perp n_1} = k_1(2n_2 + 2n_1 - k_2 - k_1) - \tfrac{1}{2}k_1(k_1 - 1).$$

In conclusion $\mathcal{X}^2(k_2, h)$ is an irreducible variety of dimension

(5.11)
$$(k_2 - 2h)(2n_2 - 1 - k_2 + 2h) - \tfrac{1}{2}(k_2 - 2h)(k_2 - 2h - 1) + 2h(2n_2 - 2k_2 + 2h) \\ + k_1(2n_2 + 2n_1 - k_2 - k_1) - \tfrac{1}{2}k_1(k_1 - 1).$$

For $(\mu,\nu)=(1^{n_2},1^{n_1+n_2})$, $(\mu',\nu')=(1^{n_2-k_2+h},1^{n_1+n_2-k_1-h})$ we compute the difference :

$$\begin{aligned}
(5.12)\quad & d^{\alpha}_{(\mu,\nu)}-d^{\alpha'}_{(\mu',\nu')}-\dim\mathcal{X}^2(k_2,h)\\
&=2N(2^{n_2}1^{n_1})+n_2+n_1-\frac{1}{2}\sum_{i\geq 1}(\alpha_i^2-\alpha_i)-2N(2^{n_2-k_2+h}1^{n_1-k_1+k_2-2h})\\
&\quad-(n_1+n_2-k_1-h)+\frac{1}{2}\sum_{i\geq 2}(\alpha_i^2-\alpha_i)-(5.11)\\
&=(n_1+n_2)(n_1+n_2-1)+n_2(n_2-1)+n_2+n_1-\tfrac{1}{2}k(k-1)\\
&\quad-(n_1+n_2-k_1-h)(n_1+n_2-k_1-h-1)\\
&\quad-(n_2-k_2+h)(n_2-k_2+h-1)\\
&\quad-(n_1+n_2-k_1-h)-(5.11)\\
&=2h(n_1-k_1).
\end{aligned}$$

Since $h\geq 0$ and $n_1\geq k_1$, this difference is always greater or equal to zero and it equals zero if and only if $h=0$ or $n_1=k_1$.

For fixed $0\leq n_1'\leq n_1+n_2$, $0<n_2'\leq n_2$ such that $n_1'+2n_2'+k=n_1+2n_2$, we have

$$\begin{aligned}
\mathcal{B}^{((1^{n_2},1^{n_1+n_2}),\alpha)}_{((1^{n_2'},1^{n_1'+n_2'}),\alpha')} &= \coprod_{\substack{\max\{0,k-n_1\}\leq k_2\leq\min\{k,2n_2-1\}\\ \max\{0,k_2-n_2+1\}\leq h\leq\frac{k_2}{2}\\ k_2-h=n_2-n_2'}} \mathcal{X}^2(k_2,h)\\
&= \coprod_{\substack{\max\{0,n_2-n_1'-n_2'\}\leq h\\ h\leq\min\{k-(n_2-n_2'),n_2-n_2'\}}} \mathcal{X}^2(h+n_2-n_2',h).
\end{aligned}$$

Again, the inequalities follow from Lemma 5.7 and from the fact that $k_2<2n_2$ and $n_2-k_2+h>0$. As in previous sections, we have that if $\mathcal{X}^2(k_2,h)\neq\emptyset\neq\mathcal{X}^2(k_2+1,h+1)$, then $\mathcal{X}^2(k_2+1,h+1)\subset\overline{\mathcal{X}^2(k_2,h)}$, so there is one irreducible piece of maximal dimension that contains all the other ones in its closure as positive codimension subvarieties. By (5.12), the piece of maximal dimension is $\mathcal{X}^2(h+n_2-n_2',h)$ for minimal h, which equals zero if and only if $n_1'+n_2'\geq n_2$. Otherwise, if $n_1'+n_2'<n_2$, then the minimal h is $h=n_2-n_1'-n_2'$ which gives us

$$\begin{aligned}
h&=n_2-n_1'-n_2'\\
h&=n_2-(n_1-k_1)-(k_2-2h)-(n_2-k_2+h)\\
h&=n_2-n_1+k_1-k_2+2h-n_2+k_2-h\\
0&=-n_1+k_1\\
n_1&=k_1.
\end{aligned}$$

So in either case $h(n_1-k_1)=0$ when h is minimal and we have indeed that $\mathcal{B}^{((1^{n_2},1^{n_1+n_2}),\alpha)}_{((1^{n_2'},1^{n_1'+n_2'}),\alpha')}$ is an irreducible variety of dimension $d^{\alpha}_{(\mu,\nu)}-d^{\alpha'}_{(\mu',\nu')}$ and the diagram of $(1^{n_2'},1^{n_1'+n_2'})$ is always obtained by removing a vertical strip from $(1^{n_2},1^{n_1+n_2})$.

5.5.3. Suppose that $(\mu',\nu') = (1^{n_2'+n_1'}, 1^{n_2'}) = (1^{n_1+n_2-k_1-h}, 1^{n_2-k_2+h})$, with $n_1' = n_1 - k_1 + k_2 - 2h > 0$. Then by Section 4.2.4, we have $v \notin F_k$, and $(F_k + \mathbb{C}v) \cap x(F_k^\perp) = F_k \cap x(F_k^\perp)$, or equivalently that $F_k \cap x(F_k^\perp) \not\subset (\mathbb{C}v)^{\perp\!\!\perp}$, by Lemma 4.4. In this case we have that $k_2 < 2n_2$.

We define the variety

$$\mathcal{X}^3(k_2,h) = \{F_k \in \mathrm{Gr}_k^\perp(\ker(x)) \mid \dim(F_k \cap \mathrm{Im}(x)) = k_2, \\ \dim(F_k \cap x(F_k^\perp)) = k_2 - 2h,\ v \notin F_k,\ F_k \cap x(F_k^\perp) \not\subset (\mathbb{C}v)^{\perp\!\!\perp}\}.$$

Notice that $F_k \cap x(F_k^\perp) \in Z := \{S \in \mathrm{Gr}^{\perp\!\!\perp}_{k_2-2h}(\mathrm{Im}(x)) \mid S \not\subset (\mathbb{C}v)^{\perp\!\!\perp},\ S \cap \mathbb{C}v = 0\}$ which is impossible if $k_2 - 2h = 0$, and otherwise, when $k_2 - 2h > 0$, is an open dense subvariety of $\mathrm{Gr}^{\perp\!\!\perp}_{k_2-2h}(\mathrm{Im}(x))$, hence irreducible of dimension

$$(k_2-2h)(2n_2-k_2+2h) - \tfrac{1}{2}(k_2-2h)(k_2-2h-1).$$

We then have

$$\mathcal{X}^3(k_2,h) = \{F_k \in (\pi_1^k)^{-1}(\pi_2^{k_2})^{-1}(Z) \mid v \notin \pi_1^k(F_k)\} = (\pi_1^k)^{-1}(\pi_2^{k_2})^{-1}(Z)$$

because, if $S \in Z$, since $S \not\subset (\mathbb{C}v)^{\perp\!\!\perp}$, then $\mathbb{C}v \notin S^{\perp\!\!\perp}$, hence for all $U \in (\pi_2^{k_2})^{-1}(S) = \mathrm{Gr}_{2h}(S^{\perp\!\!\perp}/S)$ automatically $\mathbb{C}v \not\subset U$. So we get, for each $S \in Z$, that the fibre is irreducible and

$$\dim(\pi_2^{k_2})^{-1}(S) = \dim \mathrm{Gr}_{2h,2n_2-2(k_2-2h)} = 2h(2n_2-2k_2+2h).$$

Finally, over a point $U \in (\pi_2^{k_2})^{-1}(Z)$, we have that $(\pi_1^k)^{-1}(U)$ is isomorphic to an open dense subvariety of $\mathrm{Gr}^\perp_{k_1}(\ker(x)/U)$ so, as in previous sections, it is irreducible of dimension

$$\dim \mathrm{Gr}^{\perp n_1}_{k_1,2n_1+2n_2-k_2} = k_1(2n_2+2n_1-k_2-k_1) - \tfrac{1}{2}k_1(k_1-1).$$

In conclusion $\mathcal{X}^3(k_2,h)$ is an irreducible variety of dimension

$$(5.13)\quad (k_2-2h)(2n_2-k_2+2h) - \tfrac{1}{2}(k_2-2h)(k_2-2h-1) + 2h(2n_2-2k_2+2h) \\ + k_1(2n_2+2n_1-k_2-k_1) - \tfrac{1}{2}k_1(k_1-1).$$

For $(\mu,\nu) = (1^{n_2}, 1^{n_1+n_2})$, $(\mu',\nu') = (1^{n_1+n_2-k_1-h}, 1^{n_2-k_2+h})$ we compute the difference :

$$(5.14)\quad d^\alpha_{(\mu,\nu)} - d^{\alpha'}_{(\mu',\nu')} - \dim \mathcal{X}^3(k_2,h)$$
$$= 2N(2^{n_2}1^{n_1}) + n_2 + n_1 - \frac{1}{2}\sum_{i\geq 1}(\alpha_i^2 - \alpha_i) - 2N(2^{n_2-k_2+h}1^{n_1-k_1+k_2-2h})$$
$$-(n_2-k_2+h) + \frac{1}{2}\sum_{i\geq 2}(\alpha_i^2-\alpha_i) - (5.13)$$
$$= (n_1+n_2)(n_1+n_2-1) + n_2(n_2-1) + n_2 + n_1 - \tfrac{1}{2}k(k-1)$$
$$-(n_1+n_2-k_1-h)(n_1+n_2-k_1-h-1)$$
$$-(n_2-k_2+h)(n_2-k_2+h-1)$$
$$-(n_2-k_2+h) - (5.13)$$
$$(5.15)\quad = (2h+1)(n_1-k_1).$$

Since $h \geq 0$ and $n_1 \geq k_1$, this difference is always greater or equal to zero and it equals zero if and only if $n_1 = k_1$.

For fixed $0 < n_1' \leq n_1 + n_2$, $0 \leq n_2' \leq n_2$ such that $n_1' + 2n_2' + k = n_1 + 2n_2$, we have

$$\begin{aligned}
\mathcal{B}^{((1^{n_2},1^{n_1+n_2}),\alpha)}_{((1^{n_1'+n_2'},1^{n_2'}),\alpha')} &= \coprod_{\substack{\max\{0,k-n_1\}\leq k_2\leq\min\{k,2n_2-1\}\\ \max\{0,k_2-n_2\}\leq h<\frac{k_2}{2}\\ k_2-h=n_2-n_2'}} \mathcal{X}^3(k_2,h) \\
&= \coprod_{\substack{\max\{k-n_1,n_2-n_2'\}\leq k_2\\ k_2\leq\min\{k,2n_2-2n_2',2n_2-1\}}} \mathcal{X}^3(k_2,k_2-n_2+n_2').
\end{aligned}$$

As in previous sections, the inequalities follow from Lemma 5.7 plus the stricter inequalities that we found in this section, also if $\mathcal{X}^3(k_2,h) \neq \emptyset \neq \mathcal{X}^3(k_2+1,h+1)$, then $\mathcal{X}^3(k_2+1,h+1) \subset \overline{\mathcal{X}^3(k_2,h)}$. It follows that there is one irreducible piece of maximal dimension that contains all the other ones in its closure as positive codimension subvarieties and hence $\mathcal{B}^{((1^{n_2},1^{n_1+n_2}),\alpha)}_{((1^{n_1'+n_2'},1^{n_2'}),\alpha')}$ is an irreducible variety. By (5.14) we have that the piece of maximal dimension is $\mathcal{X}^3(k_2,k_2-n_2+n_2')$ for minimal k_2. Notice that the shape $(1^{n_1'+n_2'},1^{n_2'})$ is contained in the shape $(1^{n_2},1^{n_1+n_2})$ (and hence obtained by removing boxes appropriately) if and only if $n_2 \geq n_1' + n_2'$,

$$\begin{aligned}
n_2 &\geq n_1' + n_2' \\
n_2 - n_1' - n_2' &\geq 0 \\
2n_2 - n_2 + n_1 - n_1 - n_1' - 2n_2' + n_2' - k + k &\geq 0 \\
n - n_2 - n_1 - n + n_2' + k &\geq 0 \\
k - n_1 &\geq n_2 - n_2'
\end{aligned}$$

if and only if $\max\{k-n_1, n_2-n_2'\} = k - n_1$, if and only if the minimal k_2 is $k_2 = k - n_1$, if and only if

$$\begin{aligned}
k_2 &= k - n_1 \\
k_2 &= k_1 + k_2 - n_1 \\
0 &= k_1 - n_1 \\
n_1 &= k_1
\end{aligned}$$

and as a conclusion $\dim \mathcal{B}^{((1^{n_2},1^{n_1+n_2}),\alpha)}_{((1^{n_1'+n_2'},1^{n_2'}),\alpha')} = d^{\alpha}_{(\mu,\nu)} - d^{\alpha'}_{(\mu',\nu')}$ if and only if the shape is obtained by removing boxes as desired, and is strictly lower otherwise.

Example 5.12. Here $n_1 = 2$, $n_2 = 3$. In the first example $k = 2$, $n_1' + n_2' = 2 + 2 > 3 = n_2$, so the diagrams are not nested, in the second example $k = 5$, $n_1' + n_2' = 1 + 1 = 2 \leq n_2$, so the diagrams are nested and the boxes are removed from different rows of the two tableaux.

$(1^{n_1'+n_2'}, 1^{n_1'}) =$ 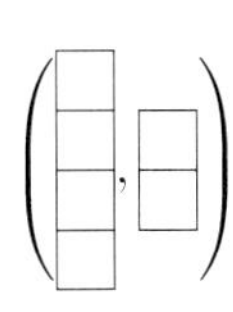 $\not\subset (1^{n_2}, 1^{n_1+n_2}) =$ 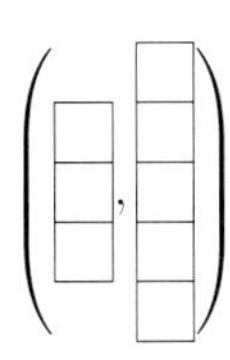 ;

$(1^{n_1'+n_2'}, 1^{n_1'}) =$ 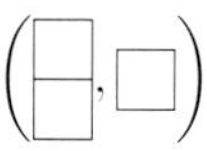 $\subset (1^{n_2}, 1^{n_1+n_2}) =$ 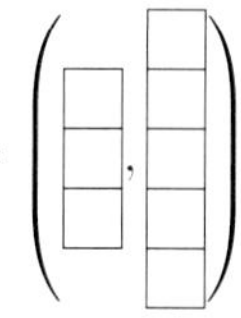 , 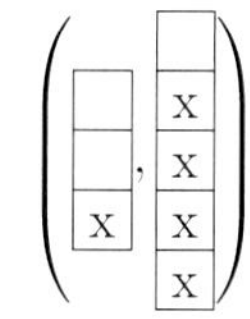

.

5.6. $\underline{\mu = 1^{n_1+n_2}, \nu = 1^{n_2}}$**.** Here $n_1, n_2 > 0$. In this case we have $x \neq 0$, $x^2 = 0$ and $v \in \ker(x) \setminus \mathrm{Im}(x)$, so $\ker(x) \cap (\mathbb{C}[x]v)^\perp = \ker(x) \cap (\mathbb{C}v)^\perp$ is a $2n_1 + 2n_2 - 1$-dimensional space where the restriction of $\langle , \rangle$ has rank $2n_1 - 2$ ($r = n_1 - 1$). We define k_1, k_2 and h as in Section 4.1.2. According to Section 4.2.5, there are three possibilities for (μ', ν'), which we now examine separately.

5.6.1. Suppose that $(\mu', \nu') = (\emptyset, 2^{n_2'} 1^{n_1'}) = (\emptyset, 2^{n_2 - k_2 + h} 1^{n_1 - k_1 + k_2 - 2h})$. Then by Section 4.2.4, we have $v \in F_k \setminus F_k \cap \mathrm{Im}(x)$, which implies that $k_1 \geq 1$. We define the following variety

$$\mathcal{X}^4(k_2, h) = \{F_k \in \mathrm{Gr}_k^\perp(\ker(x) \cap (\mathbb{C}[x]v)^\perp) \mid \dim(F_k \cap \mathrm{Im}(x)) = k_2,$$
$$\dim(F_k \cap x(F_k^\perp)) = k_2 - 2h, v \in F_k \setminus F_k \cap \mathrm{Im}(x)\}.$$

Then

$$\mathcal{X}^4(k_2, h) = \{F_k \in (\pi_1^k)^{-1}(\pi_2^{k_2})^{-1}\left(\mathrm{Gr}_{k_2-2h}^{\perp\!\!\perp}(\mathrm{Im}(x))\right) \mid v \in F_k\}.$$

The base of the iterated fibre bundle is the isotropic Grassmannian $\mathrm{Gr}_{k_2-2h}^{\perp\!\!\perp}(\mathrm{Im}(x))$ which is irreducible of dimension

$$(k_2 - 2h)(2n_2 - k_2 + 2h) - \tfrac{1}{2}(k_2 - 2h)(k_2 - 2h - 1).$$

Over each point $S \in \mathrm{Gr}_{k_2-2h}^{\perp\!\!\perp}(\mathrm{Im}(x))$, we have $(\pi_2^{k_2})^{-1}(S) \simeq \mathrm{Gr}_{2h}(S^{\perp\!\!\perp}/S)$, so each fibre is irreducible of dimension

$$2h(2n_2 - 2k_2 + 2h).$$

Finally, for each $U \in (\pi_2^{k_2})^{-1}\left(\mathrm{Gr}_{k_2-2h}^{\perp\!\!\perp}(\mathrm{Im}(x))\right)$, we only consider the points $F \in (\pi_1^k)^{-1}(U)$ such that $v \in F$, so we are choosing

$$F/(U + \mathbb{C}v) \in \mathrm{Gr}_{k_1-1}^\perp(\ker(x) \cap (\mathbb{C}[x]v)^\perp/(U + \mathbb{C}v)) \simeq \mathrm{Gr}_{k_1-1, 2n_1+2n_2-k_2-2}^{\perp, n_1-1}$$

which is irreducible of dimension $(k_1 - 1)(2n_2 + 2n_1 - k_2 - k_1 - 1) - \frac{1}{2}(k_1 - 1)(k_1 - 2)$.

In conclusion, $\mathcal{X}^4(k_2, h)$ is an irreducible variety of dimension

$$(5.16) \quad (k_2 - 2h)(2n_2 - k_2 + 2h) - \tfrac{1}{2}(k_2 - 2h)(k_2 - 2h - 1)$$
$$+ 2h(2n_2 - 2k_2 + 2h) + (k_1 - 1)(2n_2 + 2n_1 - k_2 - k_1 - 1) - \tfrac{1}{2}(k_1 - 1)(k_1 - 2).$$

For $(\mu,\nu) = (1^{n_1+n_2}, 1^{n_2})$, $(\mu',\nu') = (\emptyset, 2^{n_2-k_2+h}1^{n_1-k_1+k_2-2h})$ we compute the difference :

$$
\begin{aligned}
(5.17)\qquad & d^{\alpha}_{(\mu,\nu)} - d^{\alpha'}_{(\mu',\nu')} - \dim \mathcal{X}^4(k_2,h) \\
&= 2N(2^{n_2}1^{n_1}) + n_2 - \frac{1}{2}\sum_{i\geq 1}(\alpha_i^2-\alpha_i) - 2N(2^{n_2-k_2+h}1^{n_1-k_1+k_2-2h}) \\
&\quad - 2(n_2-k_2+h) - (n_1-k_1+k_2-2h) + \frac{1}{2}\sum_{i\geq 2}(\alpha_i^2-\alpha_i) - (5.16) \\
&= (n_1+n_2)(n_1+n_2-1) + n_2(n_2-1) + n_2 - \tfrac{1}{2}k(k-1) \\
&\quad - (n_1+n_2-k_1-h)(n_1+n_2-k_1-h-1) \\
&\quad - (n_2-k_2+h)(n_2-k_2+h-1) \\
&\quad - 2(n_2-k_2+h) - (n_1-k_1+k_2-2h) - (5.16) \\
&= (2h+1)(n_1-k_1) + (n_2-k_2+h).
\end{aligned}
$$

which is always greater or equal to zero and equals zero if and only if $n_1 = k_1$ and $n_2 - k_2 + h = 0$.

For fixed $0 \leq n_1' \leq n_1 + n_2$, $0 \leq n_2' \leq n_2$ such that $n_1' + 2n_2' + k = n_1 + 2n_2$, we have

$$
\begin{aligned}
\mathcal{B}^{((1^{n_1+n_2},1^{n_2}),\alpha)}_{((\emptyset,2^{n_2'}1^{n_1'}),\alpha')} &= \coprod_{\substack{\max\{0,k-n_1\}\leq k_2\leq \min\{k-1,2n_2\} \\ \max\{0,k_2-n_2\}\leq h\leq \frac{k_2}{2} \\ k_2-h=n_2-n_2'}} \mathcal{X}^4(k_2,h) \\
&= \coprod_{\substack{\max\{k-n_1,n_2-n_2'\}\leq k_2 \\ k_2\leq \min\{k-1,2n_2-2n_2'\}}} \mathcal{X}^4(k_2,k_2-n_2+n_2')
\end{aligned}
$$

As before, if $\mathcal{X}^4(k_2,h) \neq \emptyset \neq \mathcal{X}^4(k_2+1,h+1)$, then $\mathcal{X}^4(k_2+1,h+1) \subset \overline{\mathcal{X}^4(k_2,h)}$. It follows that there is one maximal dimensional piece that contains the others in its closure as positive codimension subvarieties, hence $\mathcal{B}^{((1^{n_1+n_2},1^{n_2}),\alpha)}_{((\emptyset,2^{n_2'}1^{n_1'}),\alpha')}$ is irreducible, and its dimension is the dimension of $\mathcal{X}^4(k_2,k_2-n_2+n_2')$ for minimal k_2. If $n_2' = n_2 - k_2 + h > 0$, then all pieces are strictly lower dimensional than $d^{\alpha}_{(\mu,\nu)} - d^{\alpha'}_{(\mu',\nu')}$ and the shape of (μ',ν') is not contained in the shape of (μ,ν). Otherwise, when $n_2' = 0$, then $n = n_1' + k$ so we have that the minimal k_2 is $k - n_1$ if and only if $k - n_1 \geq n_2$,

$$
\begin{aligned}
k - n_1 &\geq n_2 \\
k &\geq n_2 + n_1 \\
n - n_1' &\geq 2n_2 + n_1 - n_2 \\
n - n_1' &\geq n - n_2 \\
n_2 &\geq n_1'
\end{aligned}
$$

if and only if the shape (μ',ν') is contained in the shape of (μ,ν). Also notice that $k_2 = k - n_1$, is equivalent to $n_1 = k_1$ so that indeed when $n_1' \leq n_2$ we have $\dim \mathcal{B}^{((1^{n_1+n_2},1^{n_2}),\alpha)}_{((\emptyset,2^{n_2'}1^{n_1'}),\alpha')} = \dim \mathcal{X}^4(k-n_1,k-n_1-n_2) = d^{\alpha}_{(\mu,\nu)} - d^{\alpha'}_{(\mu',\nu')}$, and is strictly lower otherwise.

EXAMPLE 5.13. Here $n_1 = 2$, $n_2 = 3$. In the first example $k = 4$, $n_1' = 2$, $n_2' = 1 > 0$, in the second example $k = 4$ and, while $n_2' = 0$, $n_1' = 4 > 3 = n_2$, so in both cases the diagrams are not nested. In the third example $k = 6$, $n_2' = 0$ and $n_1' = 2 \leq 3 = n_2$, so the diagrams are nested and the boxes are removed from different rows of the two tableaux.

$(\emptyset, 2^{n_2'}1^{n_1'}) =$ 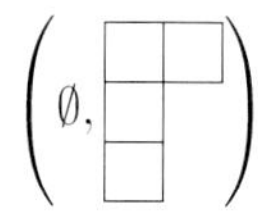 $\not\subset (1^{n_1+n_2}, 1^{n_2}) =$ 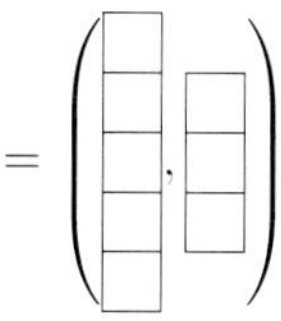 ;

$(\emptyset, 2^{n_2'}1^{n_1'}) =$ 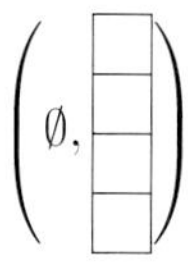 $\not\subset (1^{n_1+n_2}, 1^{n_2}) =$ 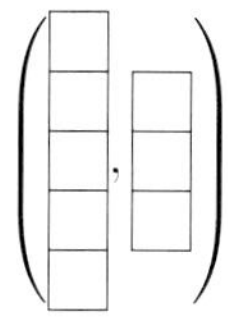 ;

$(\emptyset, 2^{n_2'}1^{n_1'}) =$ 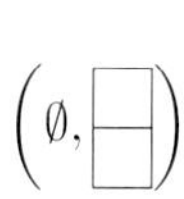 $\subset (1^{n_1+n_2}, 1^{n_2}) =$ 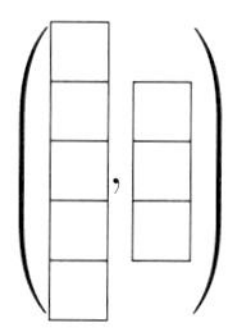 , 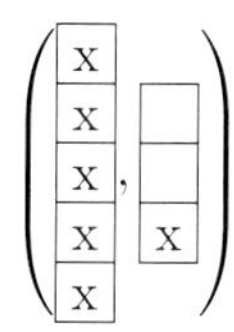 .

5.6.2. Suppose that $(\mu', \nu') = (1^{n_2'}, 1^{n_2'+n_1'}) = (1^{n_2-k_2+h}, 1^{n_1+n_2-k_1-h})$, with $n_2' = n_2 - k_2 + h > 0$. Then by Section 4.2.4, we have $v \notin F_k$, and $(F_k + \mathbb{C}v) \cap x(F_k^\perp) \supsetneq F_k \cap x(F_k^\perp)$, or equivalently that $(v + F_k) \cap x(F_k^\perp) \neq \emptyset$, by Lemma 4.5. In this case we have that $k_1 \geq 1$ (otherwise if $F_k \subset \mathrm{Im}(x)$, then $(v + F_k) \cap \mathrm{Im}(x) = \emptyset$). Notice that since $v \notin \mathrm{Im}(x)$, the condition that $v \notin F_k$, $(v + F_k) \cap x(F_k^\perp) \neq \emptyset$ is only possible if $x(F_k^\perp) \not\subset F_k$, equivalently,

$$\begin{aligned} x(F_k^\perp) \setminus F_k \neq \emptyset &\iff x(F_k^\perp) \setminus F_k \cap x(F_k^\perp) \neq \emptyset \\ &\iff \dim(x(F_k^\perp)) > \dim(F_k \cap x(F_k^\perp)). \end{aligned}$$

Since $x(F_k^\perp) = (F_k \cap \mathrm{Im}(x))^{\perp\!\!\perp}$ by Lemma 4.3, the dimension condition can be expressed as

$$2n_2 - k_2 > k_2 - 2h \iff 2n_2 - 2k_2 + 2h > 0 \iff n_2 - k_2 + h > 0$$

which has to be satisfied. We define the variety

$$\begin{aligned} \mathcal{X}^5(k_2, h) = \{F_k \in \mathrm{Gr}_k^\perp(\ker(x) \cap (\mathbb{C}[x]v)^\perp) \mid \dim(F_k \cap \mathrm{Im}(x)) = k_2, \\ \dim(F_k \cap x(F_k^\perp)) = k_2 - 2h,\ v \notin F_k,\ (v + F_k) \cap x(F_k^\perp) \neq \emptyset\} \end{aligned}$$

which we can write as

$$\begin{aligned} \mathcal{X}^5(k_2, h) = \{F_k \in (\pi_1^k)^{-1}(\pi_2^{k_2})^{-1}\left(\mathrm{Gr}_{k_2-2h}^{\perp\!\!\perp}(\mathrm{Im}(x))\right) \mid v \notin F_k, \\ (v + F_k) \cap (\pi_1^k(F_k))^{\perp\!\!\perp} \neq \emptyset\}. \end{aligned}$$

The base of the iterated fibre bundle is the isotropic Grassmannian $\mathrm{Gr}_{k_2-2h}^{\perp\!\!\perp}(\mathrm{Im}(x))$ which is irreducible of dimension

$$(k_2 - 2h)(2n_2 - k_2 + 2h) - \tfrac{1}{2}(k_2 - 2h)(k_2 - 2h - 1).$$

Over each point $S \in \mathrm{Gr}^{\perp\!\!\perp}_{k_2-2h}(\mathrm{Im}(x))$, we have $(\pi_2^{k_2})^{-1}(S) \simeq \mathrm{Gr}_{2h}(S^{\perp\!\!\perp}/S)$, so each fibre is irreducible of dimension

$$2h(2n_2 - 2k_2 + 2h).$$

Finally, for each $U \in (\pi_2^{k_2})^{-1}\left(\mathrm{Gr}^{\perp\!\!\perp}_{k_2-2h}(\mathrm{Im}(x))\right)$, we only consider the points $F \in (\pi_1^k)^{-1}(U)$ such that $v \notin F$, $(v+F) \cap (U)^{\perp\!\!\perp} \neq \emptyset$.

Suppose $L = (v+F) \cap (U)^{\perp\!\!\perp} \neq \emptyset$ and $w \in L$, then $w \in (U)^{\perp\!\!\perp} \setminus F \cap U^{\perp\!\!\perp} = (U)^{\perp\!\!\perp} \setminus U \cap U^{\perp\!\!\perp}$, viceversa given $w \in (U)^{\perp\!\!\perp} \setminus U \cap U^{\perp\!\!\perp}$, we have that $L \neq \emptyset$ if and only if $w = v + f$ with $f \in F$, if and only if $w - v \in F$. L (if nonempty) is an affine space of dimension equal to $\dim F \cap U^{\perp\!\!\perp} = \dim U \cap U^{\perp\!\!\perp}$ and all points in L are of the form $w+u$, with $w \in L$ and $u \in U \cap U^{\perp\!\!\perp}$. It follows that if $w, w' \in (U)^{\perp\!\!\perp} \setminus U \cap U^{\perp\!\!\perp}$, then $(U + \mathbb{C}(w-v))/U = (U + \mathbb{C}(w'-v))/U$ if and only if $w - w' \in U \cap U^{\perp\!\!\perp}$.

So, we have an isomorphism of varieties

$$\begin{aligned}\mathcal{D} := \{(w, D) \mid w \in ((U)^{\perp\!\!\perp} \setminus U \cap U^{\perp\!\!\perp})/(U \cap U^{\perp\!\!\perp}),\\ D \in \mathrm{Gr}^{\perp}_{k_1-1}(\ker(x) \cap (\mathbb{C}[x]v)^{\perp}/(U + \mathbb{C}(w-v)), v \notin D\}\end{aligned}$$

$$\simeq \{F \in (\pi_1^k)^{-1}(U) \mid v \notin F,\ (v+F) \cap (U)^{\perp\!\!\perp} \neq \emptyset\}$$

given by $(w, D) \mapsto \tilde{D}$, where $\tilde{D}$ is any pre-image of D under the projection $\ker(x) \cap (\mathbb{C}[x]v)^{\perp}/U \to \ker(x) \cap (\mathbb{C}[x]v)^{\perp}/(U + \mathbb{C}(w-v))$.

The variety $\mathcal{D}$ is by definition a fibre bundle with the map $(w, D) \mapsto w$, with base $((U)^{\perp\!\!\perp} \setminus U \cap U^{\perp\!\!\perp})/U \cap U^{\perp\!\!\perp} = ((U)^{\perp\!\!\perp} \setminus \pi_2^{k_2}(U))/\pi_2^{k_2}(U)$, which is an irreducible variety of dimension $\dim((U)^{\perp\!\!\perp}) - \dim(U \cap U^{\perp\!\!\perp}) = 2n_2 - k_2 - (k_2 - 2h) = 2n_2 - 2k_2 + 2h$ (isomorphic to a vector space minus the origin), and fibre isomorphic to an open dense subvariety (to guarantee that $v \notin D$) of

$$\mathrm{Gr}^{\perp}_{k_1-1}(\ker(x) \cap (\mathbb{C}[x]v)^{\perp}/(U + \mathbb{C}(w-v)) \simeq \mathrm{Gr}^{\perp, n_1-1}_{k_1-1, 2n_1+2n_2-k_2-2}$$

which is irreducible of dimension $(k_1-1)(2n_2+2n_1-k_2-k_1-1) - \frac{1}{2}(k_1-1)(k_1-2)$. In conclusion, $\mathcal{X}^5(k_2, h)$ is an irreducible variety of dimension

(5.18)
$$\begin{aligned}&(k_2 - 2h)(2n_2 - k_2 + 2h) - \tfrac{1}{2}(k_2 - 2h)(k_2 - 2h - 1) + 2h(2n_2 - 2k_2 + 2h)\\ &\quad + 2n_2 - 2k_2 + 2h + (k_1 - 1)(2n_2 + 2n_1 - k_2 - k_1 - 1) - \tfrac{1}{2}(k_1 - 1)(k_1 - 2).\end{aligned}$$

For $(\mu,\nu)=(1^{n_1+n_2},1^{n_2})$, $(\mu',\nu')=(1^{n_2-k_2+h},1^{n_1+n_2-k_1-h})$ we compute the difference:

$$\begin{aligned}(5.19)\qquad & d^{\alpha}_{(\mu,\nu)}-d^{\alpha'}_{(\mu',\nu')}-\dim\mathcal{X}^5(k_2,h)\\ &=2N(2^{n_2}1^{n_1})+n_2-\frac{1}{2}\sum_{i\geq 1}(\alpha_i^2-\alpha_i)-2N(2^{n_2-k_2+h}1^{n_1-k_1+k_2-2h})\\ &\quad-(n_1+n_2-k_1-h)+\frac{1}{2}\sum_{i\geq 2}(\alpha_i^2-\alpha_i)-(5.18)\\ &=(n_1+n_2)(n_1+n_2-1)+n_2(n_2-1)+n_2-\tfrac{1}{2}k(k-1)\\ &\quad-(n_1+n_2-k_1-h)(n_1+n_2-k_1-h-1)\\ &\quad-(n_2-k_2+h)(n_2-k_2+h-1)\\ &\quad-(n_1+n_2-k_1-h)-(5.18)\\ &=(2h+1)(n_1-k_1).\end{aligned}$$

which is always greater or equal to zero and equals zero if and only if $n_1=k_1$. For fixed $0\leq n_1'\leq n_1+n_2$, $0\leq n_2'\leq n_2$ such that $n_1'+2n_2'+k=n_1+2n_2$, we have

$$\begin{aligned}\mathcal{B}^{((1^{n_1+n_2},1^{n_2}),\alpha)}_{((1^{n_2'},1^{n_1'+n_2'}),\alpha')}&=\coprod_{\substack{\max\{0,k-n_1\}\leq k_2\leq\min\{k-1,2n_2\}\\ \max\{0,k_2-n_2\}\leq h\leq\frac{k_2}{2}\\ k_2-h=n_2-n_2'}}\mathcal{X}^5(k_2,h)\\ &=\coprod_{\substack{\max\{k-n_1,n_2-n_2'\}\leq k_2\\ k_2\leq\min\{k-1,2n_2-2n_2'\}}}\mathcal{X}^5(k_2,k_2-n_2+n_2')\end{aligned}$$

As before, if $\mathcal{X}^5(k_2,h)\neq\emptyset\neq\mathcal{X}^5(k_2+1,h+1)$, then $\mathcal{X}^5(k_2+1,h+1)\subset\overline{\mathcal{X}^5(k_2,h)}$. It follows that there is one irreducible piece of maximal dimension that contains all the other ones in its closure as positive codimension subvarieties, hence $\mathcal{B}^{((1^{n_1+n_2},1^{n_2}),\alpha)}_{((1^{n_2'},1^{n_1'+n_2'}),\alpha')}$ is an irreducible variety. By (5.19) we have that the piece of maximal dimension is $\mathcal{X}^5(k_2,k_2-n_2+n_2')$ for minimal k_2. Notice that the shape $(1^{n_2'},1^{n_1'+n_2'})$ is contained in the shape $(1^{n_1+n_2},1^{n_2})$ (and hence obtained by removing boxes appropriately) if and only if $n_2-n_1'-n_2'\geq 0$

$$\begin{aligned}n_2-n_1'-n_2'&\geq 0\\ 2n_2-n_2+n_1-n_1-n_1'-2n_2'+n_2'-k+k&\geq 0\\ n-n_2-n_1-n+n_2'+k&\geq 0\\ k-n_1&\geq n_2-n_2'\end{aligned}$$

if and only if $\max\{k-n_1,n_2-n_2'\}=k-n_1$, if and only if the minimum possible k_2 is $k_2=k-n_1$, if and only if

$$\begin{aligned}k_2&=k-n_1\\ k_2&=k_1+k_2-n_1\\ 0&=k_1-n_1\\ n_1&=k_1\end{aligned}$$

and as a conclusion $\dim \mathcal{B}^{((1^{n_2},1^{n_1+n_2}),\alpha)}_{((1^{n'_1+n'_2},1^{n'_2}),\alpha')} = d^{\alpha}_{(\mu,\nu)} - d^{\alpha'}_{(\mu',\nu')}$ if and only if the shape is obtained by removing boxes as desired and is strictly lower otherwise.

Example 5.14. Here $n_1 = 2$, $n_2 = 3$. In the first example $k = 2$, $n'_1 + n'_2 = 2 + 2 > 3 = n_2$, so the diagrams are not nested, in the second example $k = 5$, $n'_1 + n'_2 = 1 + 1 = 2 \leq n_2$, so the diagrams are nested and the boxes are removed from different rows of the two tableaux.

$(1^{n'_1}, 1^{n'_1+n'_2}) =$ 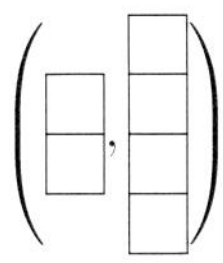 $\not\subset (1^{n_1+n_2}, 1^{n_2}) =$ 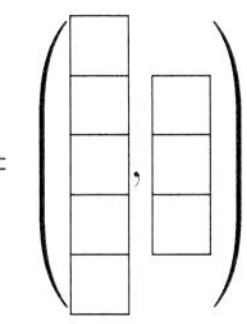 ;

$(1^{n'_1}, 1^{n'_1+n'_2}) =$ 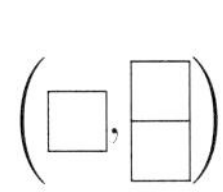 $\subset (1^{n_1+n_2}, 1^{n_2}) =$ 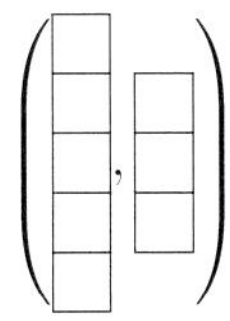 , 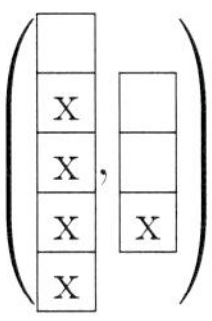

.

5.6.3. Suppose that $(\mu', \nu') = (1^{n'_2+n'_1}, 1^{n'_2}) = (1^{n_1+n_2-k_1-h}, 1^{n_2-k_2+h})$, with $n'_1 = n_1 - k_1 + k_2 - 2h > 0$. Then by Section 4.2.4, we have $v \notin F_k$, and $(F_k + \mathbb{C}v) \cap x(F_k^{\perp}) = F_k \cap x(F_k^{\perp})$, or equivalently that $(v + F_k) \cap x(F_k^{\perp}) = \emptyset$, by Lemma 4.5. We define the variety

$$\mathcal{X}^6(k_2, h) = \{F_k \in \mathrm{Gr}_k^{\perp}(\ker(x) \cap (\mathbb{C}[x]v)^{\perp}) \mid \dim(F_k \cap \mathrm{Im}(x)) = k_2, \\ \dim(F_k \cap x(F_k^{\perp})) = k_2 - 2h,\ (v + F_k) \cap x(F_k^{\perp}) = \emptyset\}$$

which we can write as

$$\mathcal{X}^6(k_2, h) = \{F_k \in (\pi_1^k)^{-1}(\pi_2^{k_2})^{-1}\left(\mathrm{Gr}^{\perp\!\!\perp}_{k_2-2h}(\mathrm{Im}(x))\right) \mid (v + F_k) \cap (\pi_1^k(F_k))^{\perp\!\!\perp} = \emptyset\}.$$

The base of the iterated fibre bundle is the isotropic Grassmannian $\mathrm{Gr}^{\perp\!\!\perp}_{k_2-2h}(\mathrm{Im}(x))$ which is irreducible of dimension

$$(k_2 - 2h)(2n_2 - k_2 + 2h) - \tfrac{1}{2}(k_2 - 2h)(k_2 - 2h - 1).$$

Over each point $S \in \mathrm{Gr}^{\perp\!\!\perp}_{k_2-2h}(\mathrm{Im}(x))$, we have $(\pi_2^{k_2})^{-1}(S) \simeq \mathrm{Gr}_{2h}(S^{\perp\!\!\perp}/S)$, so each fibre is irreducible of dimension

$$2h(2n_2 - 2k_2 + 2h).$$

Finally, for each $U \in (\pi_2^{k_2})^{-1}\left(\mathrm{Gr}^{\perp\!\!\perp}_{k_2-2h}(\mathrm{Im}(x))\right)$, we only consider the points $F \in (\pi_1^k)^{-1}(U)$ such that $(v + F) \cap (U)^{\perp\!\!\perp} = \emptyset$, which is an open dense subvariety of

$$\mathrm{Gr}_{k_1}^{\perp}(\ker(x) \cap (\mathbb{C}[x]v)^{\perp}/U) \simeq \mathrm{Gr}^{\perp, n_1-1}_{k_1, 2n_1+2n_2-1-k_2}$$

which is irreducible of dimension $k_1(2n_2 + 2n_1 - k_2 - k_1 - 1) - \frac{1}{2}k_1(k_1 - 1)$.

In conclusion, $\mathcal{X}^6(k_2, h)$ is an irreducible variety of dimension

$$(5.20)\quad (k_2 - 2h)(2n_2 - k_2 + 2h) - \tfrac{1}{2}(k_2 - 2h)(k_2 - 2h - 1) \\ + 2h(2n_2 - 2k_2 + 2h) + k_1(2n_2 + 2n_1 - k_2 - k_1 - 1) - \tfrac{1}{2}k_1(k_1 - 1).$$

For $(\mu,\nu)=(1^{n_1+n_2},1^{n_2})$, $(\mu',\nu')=(1^{n_1+n_2-k_1-h},1^{n_2-k_2+h})$ we compute the difference:

$$\begin{aligned}
(5.21)\qquad & d^{\alpha}_{(\mu,\nu)}-d^{\alpha'}_{(\mu',\nu')}-\dim\mathcal{X}^6(k_2,h)\\
&=2N(2^{n_2}1^{n_1})+n_2-\frac{1}{2}\sum_{i\geq 1}(\alpha_i^2-\alpha_i)-2N(2^{n_2-k_2+h}1^{n_1-k_1+k_2-2h})\\
&\quad-(n_2-k_2+h)+\frac{1}{2}\sum_{i\geq 2}(\alpha_i^2-\alpha_i)-(5.20)\\
&=(n_1+n_2)(n_1+n_2-1)+n_2(n_2-1)+n_2-\tfrac{1}{2}k(k-1)\\
&\quad-(n_1+n_2-k_1-h)(n_1+n_2-k_1-h-1)\\
&\quad-(n_2-k_2+h)(n_2-k_2+h-1)\\
&\quad-(n_2-k_2+h)-(5.20)\\
&=2h(n_1-k_1).
\end{aligned}$$

Since $h\geq 0$ and $n_1\geq k_1$, this difference is always greater or equal to zero and it equals zero if and only if $h=0$ or $n_1=k_1$.

For fixed $0<n_1'\leq n_1+n_2$, $0\leq n_2'\leq n_2$ such that $n_1'+2n_2'+k=n_1+2n_2$, we have

$$\begin{aligned}
\mathcal{B}^{((1^{n_1+n_2},1^{n_2}),\alpha)}_{((1^{n_1'+n_2'},1^{n_2'}),\alpha')} &= \coprod_{\substack{\max\{0,k-n_1\}\leq k_2\leq\min\{k,2n_2\}\\ \max\{0,k_2-n_2\}\leq h\leq\frac{k_2}{2}\\ k_2-h=n_2-n_2'}} \mathcal{X}^6(k_2,h)\\
&= \coprod_{\substack{\max\{0,n_2-n_1'-n_2'\}\leq h\\ h\leq\min\{k-(n_2-n_2'),n_2-n_2'\}}} \mathcal{X}^6(h+n_2-n_2',h).
\end{aligned}$$

As in previous cases, if $\mathcal{X}^6(k_2,h)\neq\emptyset\neq\mathcal{X}^6(k_2+1,h+1)$, then $\mathcal{X}^6(k_2+1,h+1)\subset\overline{\mathcal{X}^6(k_2,h)}$. It follows that there is one irreducible piece of maximal dimension that contains all the other ones in its closure as positive codimension subvarieties, hence $\mathcal{B}^{((1^{n_1+n_2},1^{n_2}),\alpha)}_{((1^{n_2'},1^{n_1'+n_2'}),\alpha')}$ is an irreducible variety. By (5.21), the piece of maximal dimension is $\mathcal{X}^6(h+n_2-n_2',h)$ for minimal h. This minimal h equals zero if and only if $n_1'+n_2'\geq n_2$. Otherwise, if $n_1'+n_2'<n_2$, then the minimal h is $h=n_2-n_1'-n_2'$ which gives us

$$\begin{aligned}
h&=n_2-n_1'-n_2'\\
h&=n_2-(n_1-k_1)-(k_2-2h)-(n_2-k_2+h)\\
h&=n_2-n_1+k_1-k_2+2h-n_2+k_2-h\\
0&=-n_1+k_1\\
n_1&=k_1.
\end{aligned}$$

So we always have that $\mathcal{B}^{((1^{n_1+n_2},1^{n_2}),\alpha)}_{((1^{n_1'+n_2'},1^{n_2'}),\alpha')}$ is of dimension $d^{\alpha}_{(\mu,\nu)}-d^{\alpha'}_{(\mu',\nu')}$.

5.7. $\underline{\mu=2^{n_2}1^{n_1},\nu=\emptyset}$**.** Here $n_2>0$. In this case we have $x\neq 0$, $x^2=0$ and $v\notin\ker(x)$, so $\ker(x)\cap(\mathbb{C}[x]v)^{\perp}=\ker(x)\cap(\mathbb{C}v)^{\perp}$ is a $2n_1+2n_2-1$-dimensional space where the restriction of $\langle,\rangle$ has rank $\max\{2n_1-2,0\}$ ($r=\max\{n_1-1,0\}$).

We define k_1, k_2 and h as in Section 4.1.2. According to Section 4.2.6, there are two possibilities for (μ', ν'), which we now examine separately.

5.7.1. Suppose that $(\mu', \nu') = (1^{n_2'+n_1'}, 1^{n_2'}) = (1^{n_1+n_2-k_1-h}, 1^{n_2-k_2+h})$. Then by Section 4.2.6, we have $0 \neq xv \in F_k$, and since we know that $xv \in x(F_k^\perp)$ we necessarily have that $xv \in F_k \cap x(F_k^\perp)$, hence we have $k_2 - 2h > 0$ and $n_1' = n_1 - k_1 + k_2 - 2h > 0$.

We define the variety

$$\begin{aligned}\mathcal{X}^7(k_2, h) = \{F_k \in \mathrm{Gr}_k^\perp(\ker(x) \cap (\mathbb{C}[x]v)^\perp) \mid \dim(F_k \cap \mathrm{Im}(x)) = k_2,\\ \dim(F_k \cap x(F_k^\perp)) = k_2 - 2h, xv \in F_k \cap x(F_k^\perp)\}\end{aligned}$$

If $xv \in F_k \cap x(F_k^\perp)$, then $(F_k \cap x(F_k^\perp))/\mathbb{C}xv \in \mathrm{Gr}_{k_2-2h-1}^{\perp\!\!\perp}(\mathbb{C}(xv)^{\perp\!\!\perp}/\mathbb{C}xv)$. It follows that we have a description as an iterated vector bundle

$$\mathcal{X}^7(k_2, h) = (\pi_1^k)^{-1}(\pi_2^{k_2})^{-1}\left(\mathrm{Gr}_{k_2-2h-1}^{\perp\!\!\perp}((\mathbb{C}xv)^{\perp\!\!\perp}/\mathbb{C}xv)\right).$$

The base is an isotropic Grassmannian, hence an irreducible variety of dimension

$$(k_2 - 2h - 1)(2n_2 - k_2 + 2h - 1) - \tfrac{1}{2}(k_2 - 2h - 1)(k_2 - 2h - 2). \tag{5.22}$$

Over every point $S \in \mathrm{Gr}_{k_2-2h}^{\perp\!\!\perp}((\mathbb{C}xv)^{\perp\!\!\perp}/\mathbb{C}xv)$, we have $(\pi_2^{k_2})^{-1}(S) \simeq \mathrm{Gr}_{2h}(S^{\perp\!\!\perp}/S)$, so each fibre is irreducible of dimension

$$2h(2n_2 - 2k_2 + 2h).$$

Finally over a point $U \in (\pi_2^{k_2})^{-1}\left(\mathrm{Gr}_{k_2-2h-1}^{\perp\!\!\perp}((\mathbb{C}xv)^{\perp\!\!\perp}/\mathbb{C}xv)\right)$, we have that, when $n_1 \geq 1$, then $(\pi_1^k)^{-1}(U)$ is isomorphic to an open dense subvariety of

$$\mathrm{Gr}_{k_1}^\perp(\ker(x) \cap (\mathbb{C}[x]v)^\perp/U) \simeq \mathrm{Gr}_{k_1, 2n_1+2n_2-1-k_2}^{\perp, n_1-1}$$

which is irreducible of dimension $k_1(2n_2 + 2n_1 - k_2 - k_1 - 1) - \frac{1}{2}k_1(k_1 - 1)$. Notice that if $n_1 = 0$, then necessarily $k_1 = 0$ and $(\pi_1^k)^{-1}(U)$ is a single point, so the formula for the dimension works in this case as well. In conclusion, $\mathcal{X}^7(k_2, h)$ is an irreducible variety of dimension

$$\begin{aligned}(k_2 - 2h - 1)(2n_2 - k_2 + 2h - 1) - \tfrac{1}{2}(k_2 - 2h - 1)(k_2 - 2h - 2)\\ + 2h(2n_2 - 2k_2 + 2h) + k_1(2n_2 + 2n_1 - k_2 - k_1 - 1) - \tfrac{1}{2}k_1(k_1 - 1).\end{aligned} \tag{5.23}$$

For $(\mu,\nu) = (2^{n_2}1^{n_1},\emptyset)$, $(\mu',\nu') = (1^{n_1+n_2-k_1-h}, 1^{n_2-k_2+h})$ we compute the difference:

$$\begin{aligned}
(5.24)\qquad & d^{\alpha}_{(\mu,\nu)} - d^{\alpha'}_{(\mu',\nu')} - \dim \mathcal{X}^7(k_2,h) \\
&= 2N(2^{n_2}1^{n_1}) - \frac{1}{2}\sum_{i\geq 1}(\alpha_i^2 - \alpha_i) - 2N(2^{n_2-k_2+h}1^{n_1-k_1+k_2-2h}) \\
&\quad - (n_2 - k_2 + h) + \frac{1}{2}\sum_{i\geq 2}(\alpha_i^2 - \alpha_i) - (5.23) \\
&= (n_1+n_2)(n_1+n_2-1) + n_2(n_2-1) - \tfrac{1}{2}k(k-1) \\
&\quad - (n_1+n_2-k_1-h)(n_1+n_2-k_1-h-1) \\
&\quad - (n_2-k_2+h)(n_2-k_2+h-1) \\
&\quad - (n_2-k_2+h) - (5.23) \\
(5.25)\qquad &= h(2n_1 - 2k_1 + 1) + n_2 - k_2 + h.
\end{aligned}$$

Since $h \geq 0$, $n_1 \geq k_1$, and $n_2 - k_2 + h \geq 0$ this difference is always greater or equal to zero and it equals zero if and only if $h = 0$ and $n_2 - k_2 + h = 0$, (which means $n_2 = k_2$).

For fixed $0 < n_1' \leq n_1 + n_2$, $0 \leq n_2' \leq n_2$ such that $n_1' + 2n_2' + k = n_1 + 2n_2$, we have

$$\begin{aligned}
\mathcal{B}^{((2^{n_2}1^{n_1},\emptyset),\alpha)}_{((1^{n_1'+n_2'},1^{n_2'}),\alpha')} &= \coprod_{\substack{\max\{0,k-n_1\}\leq k_2\leq \min\{k,2n_2\} \\ \max\{0,k_2-n_2\}\leq h<\frac{k_2}{2} \\ k_2-h=n_2-n_2'}} \mathcal{X}^7(k_2,h) \\
&= \coprod_{\substack{\max\{0,n_2-n_1'-n_2'\}\leq h \\ h\leq \min\{k-(n_2-n_2'),n_2-n_2'-1\}}} \mathcal{X}^7(h+n_2-n_2',h)
\end{aligned}$$

As in previous cases, if $\mathcal{X}^7(k_2,h) \neq \emptyset \neq \mathcal{X}^7(k_2+1,h+1)$, then $\mathcal{X}^7(k_2+1,h+1) \subset \overline{\mathcal{X}^7(k_2,h)}$. It follows that there is one irreducible piece of maximal dimension that contains all the other ones in its closure as positive codimension subvarieties hence $\mathcal{B}^{((2^{n_2}1^{n_1},\emptyset),\alpha)}_{((1^{n_1'+n_2'},1^{n_2'}),\alpha')}$ is irreducible. Notice that by (5.24) we have that the piece of maximal dimension is $\mathcal{X}(h+n_2-n_2',h)$ for minimal h.

If $n_2' = n_2 - k_2 + h > 0$, then all pieces are strictly lower dimensional than $d^{\alpha}_{(\mu,\nu)} - d^{\alpha'}_{(\mu',\nu')}$ and the shape of (μ',ν') is not contained in the shape of (μ,ν).

Otherwise, if $n_2' = 0$, we have that the minimal h is $\max\{0, n_2 - n_1'\}$, and this equals zero if and only if $n_1' \geq n_2$, if and only if the shape $(1^{n_1'},\emptyset)$ is obtained from $(2^{n_2}1^{n_1},\emptyset)$ by removing boxes with no two of them being in the same row.

In conclusion, $\mathcal{B}^{((2^{n_2}1^{n_1},\emptyset),\alpha)}_{((1^{n_1'+n_2'},1^{n_2'}),\alpha')}$ is an irreducible variety of dimension less or equal to $d^{\alpha}_{(\mu,\nu)} - d^{\alpha'}_{(\mu',\nu')}$ with equality if and only if the shape is obtained by removing boxes in the appropriate way.

Example 5.15. Here $n_1 = 2$, $n_2 = 3$. In the first example $k = 4$, $n_1' = 2$, $n_2' = 1 > 0$, so the diagrams are not nested. In the second example $k = 6$, and, while $n_2' = 0$, $n_1' = 2 < 3 = n_2$, so some boxes are removed from the same row, while in the third example $k = 4$, $n_2' = 0$ and $n_1' = 4 \geq 3 = n_2$, so the boxes are removed from different rows of the two tableaux.

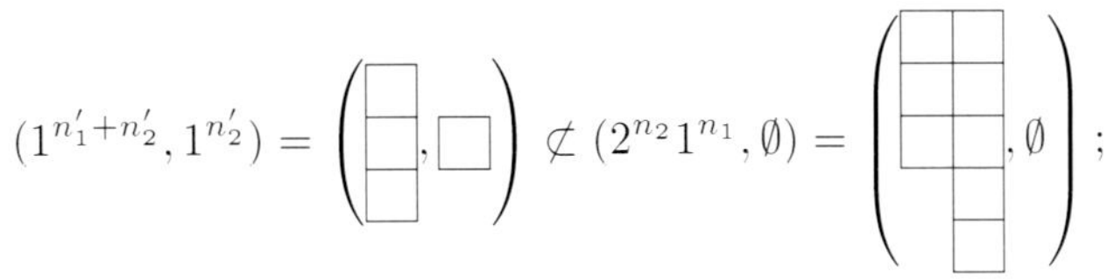

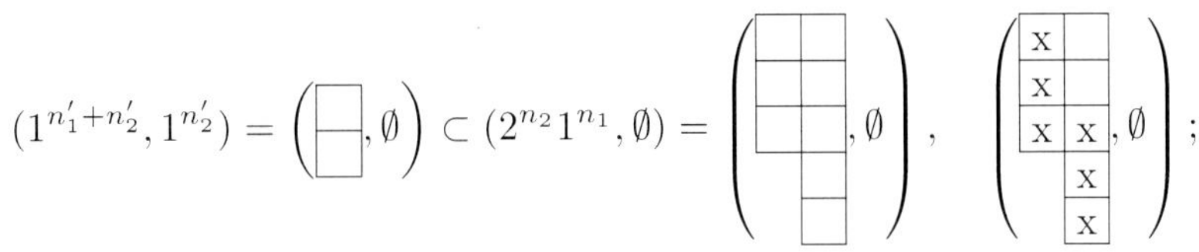

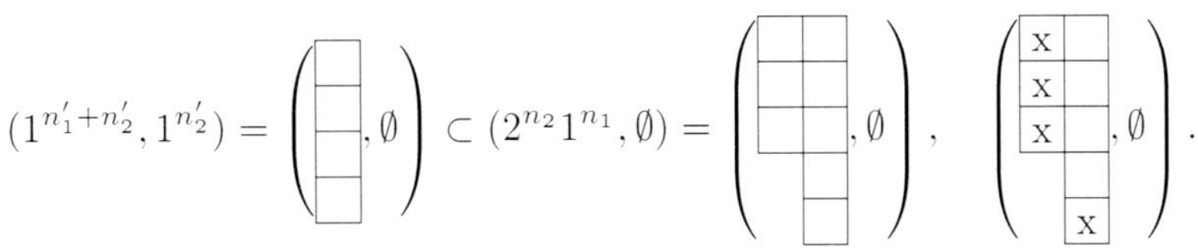

5.7.2. Suppose that $(\mu', \nu') = (2^{n_2'}1^{n_1'}, \emptyset) = (2^{n_2-k_2+h}1^{n_1-k_1+k_2-2h}, \emptyset)$, with $n_2' = n_2 - k_2 + h > 0$. Then by Section 4.2.6, we have $0 \neq xv \notin F_k$. We define the variety

$$\begin{aligned}\mathcal{X}^8(k_2, h) = \{F_k \in \mathrm{Gr}_k^{\perp}(\ker(x) \cap (\mathbb{C}[x]v)^{\perp}) \mid \dim(F_k \cap \mathrm{Im}(x)) = k_2,\\ \dim(F_k \cap x(F_k^{\perp})) = k_2 - 2h, xv \notin F_k\}.\end{aligned}$$

Notice that, since $xv \in x(F_k^{\perp}) = (F_k \cap \mathrm{Im}(x))^{\perp\!\!\perp}$, we have $F_k \cap \mathrm{Im}(x) \subset (\mathbb{C}xv)^{\perp\!\!\perp}$. Also the bilinear form $\langle\langle\ ,\ \rangle\rangle$ restricted to the $(2n_2-1)$-dimensional space $(\mathbb{C}xv)^{\perp\!\!\perp} \subset \mathrm{Im}(x)$ has rank $2n_2 - 2$ $(r = n_2 - 1)$. It follows that $F_k \cap x(F_k^{\perp}) \in Z := \{S \in \mathrm{Gr}_{k_2-2h}^{\perp\!\!\perp}((\mathbb{C}xv)^{\perp\!\!\perp}) \mid S \cap \mathbb{C}xv = 0\}$ which is an open dense subvariety of $\mathrm{Gr}_{k_2-2h}^{\perp\!\!\perp}((\mathbb{C}xv)^{\perp\!\!\perp})$, hence irreducible of dimension

$$\dim \mathrm{Gr}_{k_2-2h, 2n_2-1}^{\perp n_2-1} = (k_2 - 2h)(2n_2 - 1 - k_2 + 2h) - \tfrac{1}{2}(k_2 - 2h)(k_2 - 2h - 1).$$

Over every point $S \in Z$, we consider only the points $U \in (\pi_2^{k_2})^{-1}(S)$ such that $xv \notin U$, and $U \subset (\mathbb{C}xv)^{\perp\!\!\perp}$ i.e. we are choosing $U/S \in \mathrm{Gr}_{2h}((S^{\perp\!\!\perp} \cap (\mathbb{C}xv)^{\perp\!\!\perp})/S)$, such that $U/S \cap \mathbb{C}xv/S = 0$ (where $\mathbb{C}xv/S$ is the image of $\mathbb{C}xv$ under the projection $S^{\perp\!\!\perp} \cap (\mathbb{C}xv)^{\perp\!\!\perp} \to (S^{\perp\!\!\perp} \cap (\mathbb{C}xv)^{\perp\!\!\perp})/S)$ so each fibre, if nonempty, is an open subvariety of a Grassmannian, hence irreducible of dimension

$$\dim \mathrm{Gr}_{2h, 2n_2-2(k_2-2h)-1} = 2h(2n_2 - 2k_2 + 2h - 1).$$

This fibre is nonempty if and only if it is possible to choose U/S that does not contain $\mathbb{C}xv/S$, if and only if

$$\begin{aligned} 2h &< 2n_2 - 2(k_2 - 2h) - 1\\ 2h &< 2n_2 - 2k_2 + 4h - 1\\ 1 &< 2n_2 - 2k_2 + 2h\\ 1/2 &< n_2 - k_2 + h. \end{aligned}$$

which is equivalent to $n_2 - k_2 + h > 0$ since those are all integers.

Finally, over a point $U \in (\pi_2^{k_2})^{-1}(Z)$, (with $U \subset (\mathbb{C}xv)^{\perp\perp}$ and $xv \notin U$) we have that, when $n_1 \geq 1$, $(\pi_1^k)^{-1}(U)$ is isomorphic to an open dense subvariety of

$$\mathrm{Gr}_{k_1}^{\perp}(\ker(x) \cap (\mathbb{C}[x]v)^{\perp}/U) \simeq \mathrm{Gr}_{k_1, 2n_1+2n_2-1-k_2}^{\perp, n_1-1}$$

which is irreducible of dimension $k_1(2n_2 + 2n_1 - k_2 - k_1 - 1) - \frac{1}{2}k_1(k_1 - 1)$. If $n_1 = 0$, then necessarily $k_1 = 0$ and $(\pi_1^k)^{-1}(U)$ is a single point, so the formula for the dimension works in this case as well. In conclusion, $\mathcal{X}^8(k_2, h)$ is an irreducible variety of dimension

$$\begin{aligned}(5.26)\quad &(k_2 - 2h)(2n_2 - k_2 + 2h - 1) - \tfrac{1}{2}(k_2 - 2h)(k_2 - 2h - 1) \\ &+ 2h(2n_2 - 2k_2 + 2h - 1) + k_1(2n_2 + 2n_1 - k_2 - k_1 - 1) - \tfrac{1}{2}k_1(k_1 - 1).\end{aligned}$$

For $(\mu, \nu) = (2^{n_2}1^{n_1}, \emptyset)$, $(\mu', \nu') = (2^{n_2-k_2+h}1^{n_1-k_1+k_2-2h}, \emptyset)$ we compute the difference:

$$\begin{aligned}(5.27)\qquad & d^{\alpha}_{(\mu,\nu)} - d^{\alpha'}_{(\mu',\nu')} - \dim \mathcal{X}^8(k_2, h) \\ &= 2N(2^{n_2}1^{n_1}) - \frac{1}{2}\sum_{i\geq 1}(\alpha_i^2 - \alpha_i) - 2N(2^{n_2-k_2+h}1^{n_1-k_1+k_2-2h}) \\ &\quad + \frac{1}{2}\sum_{i\geq 2}(\alpha_i^2 - \alpha_i) - (5.26) \\ &= (n_1 + n_2)(n_1 + n_2 - 1) + n_2(n_2 - 1) - \tfrac{1}{2}k(k - 1) \\ &\quad - (n_1 + n_2 - k_1 - h)(n_1 + n_2 - k_1 - h - 1) \\ &\quad - (n_2 - k_2 + h)(n_2 - k_2 + h - 1) - (5.26) \\ &= h(2n_1 - 2k_1 + 1).\end{aligned}$$

Since $h \geq 0$ and $n_1 \geq k_1$ this difference is always greater or equal to zero and it equals zero if and only if $h = 0$.

For fixed $0 \leq n_1' \leq n_1 + n_2$, $0 < n_2' \leq n_2$ such that $n_1' + 2n_2' + k = n_1 + 2n_2$, we have

$$\begin{aligned}\mathcal{B}^{((2^{n_2}1^{n_1},\emptyset),\alpha)}_{((2^{n_2'}1^{n_1'},\emptyset),\alpha')} &= \coprod_{\substack{\max\{0,k-n_1\}\leq k_2\leq\min\{k,2n_2\} \\ \max\{0,k_2-n_2+1\}\leq h\leq\frac{k_2}{2} \\ k_2-h=n_2-n_2'}} \mathcal{X}^8(k_2, h) \\ &= \coprod_{\substack{\max\{0,n_2-n_1'-n_2'\}\leq h \\ h\leq\min\{k-(n_2-n_2'),n_2-n_2'-1\}}} \mathcal{X}^8(h + n_2 - n_2', h)\end{aligned}$$

As in previous cases, if $\mathcal{X}^8(k_2, h) \neq \emptyset \neq \mathcal{X}^8(k_2 + 1, h + 1)$, then $\mathcal{X}^8(k_2 + 1, h + 1) \subset \overline{\mathcal{X}^8(k_2, h)}$. It follows that there is one irreducible piece of maximal dimension that contains all the other ones in its closure as positive codimension subvarieties hence $\mathcal{B}^{((2^{n_2}1^{n_1},\emptyset),\alpha)}_{((2^{n_2'}1^{n_1'},\emptyset),\alpha')}$ is irreducible. By (5.27) we have that the piece of maximal dimension is $\mathcal{X}^8(h + n_2 - n_2', h)$ for minimal h i.e. $h = \max\{0, n_2 - n_1' - n_2'\}$. The minimal h is $h = 0$ if and only if $n_2 \leq n_1' + n_2'$ if and only if the shape $(2^{n_2'}1^{n_1'}, \emptyset)$ is obtained from $(2^{n_2}1^{n_1}, \emptyset)$ by removing boxes in a vertical strip. In conclusion, $\dim \mathcal{B}^{((\mu,\nu),\alpha)}_{((\mu',\nu'),\alpha')} = d^{\alpha}_{(\mu,\nu)} - d^{\alpha'}_{(\mu',\nu')}$ exactly when the shape is obtained by removing boxes from different rows and is strictly lower otherwise.

EXAMPLE 5.16. Here $n_1 = 2$, $n_2 = 3$, $k = 4$. In the first example $n'_1 + n'_2 = 0 + 2 < 3 = n_2$ so some boxes are removed from the same row, while in the second example $n'_1 + n'_2 = 2 + 1 = 3 \geq n_2$ so the boxes are removed from different rows.

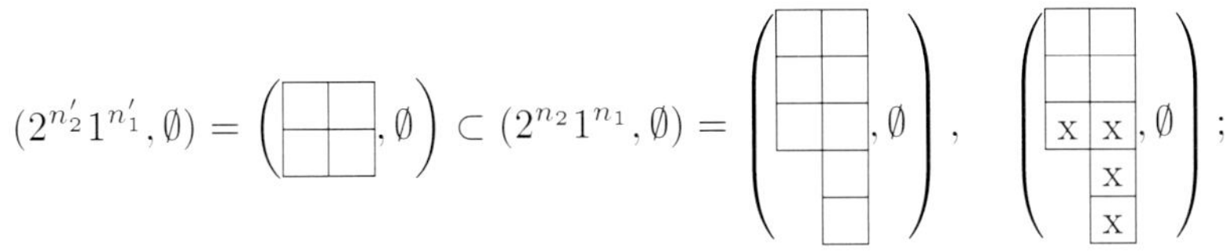

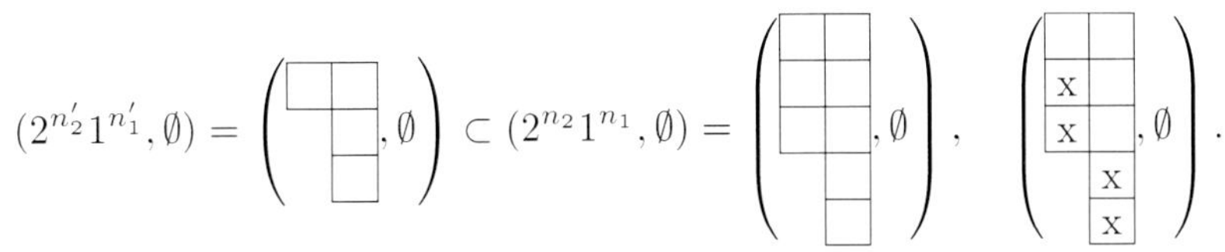

6. Extending to any $x^\ell = 0$

In this section we outline how our computations for the proof of the case of $x^2 = 0$ could be extended to any nilpotent x. We are not able to carry it out in this paper, but it should be the basis for future work.

Suppose that $x^\ell = 0$, $x^{\ell-1} \neq 0$ for some $\ell > 0$, then $\lambda = (\ell^{n_\ell} \cdots 2^{n_2} 1^{n_1})$ for some $n_j \geq 0$ and $J(x) = \left(\ell^{2n_\ell} \cdots 2^{2n_2} 1^{2n_1}\right)$. For any isotropic subspace $0 \subset F_k \subset F_k^\perp \subset V$ with $F_k \subset \ker(x)$, we define $k_j := \dim(F_k \cap \mathrm{Im}(x^{j-1})) - \dim(F_k \cap \mathrm{Im}(x^j))$ for $j = 1, \ldots, \ell$. Then by Lemma 3.2 we have that

$$J\left(x|_{V/F_k}\right) = J\left(x|_{F_k}\right) = \left(\ell^{2n_\ell - k_\ell} (\ell-1)^{2n_{\ell-1} - k_{\ell-1} + k_\ell} \cdots 2^{2n_2 - k_2 + k_3} 1^{2n_1 - k_1 + k_2}\right).$$

Now, notice that since $\mathrm{Im}(x) \subset F_k^\perp$, for all j we have that $\mathrm{Im}(x^j) \subset x^{j-1}(F_k)$. We have then a filtration of subspaces

$$\cdots \subset F_k \cap x^j(F_k^\perp) \subset F_k \cap \mathrm{Im}(x^j) \subset F_k \cap x^{j-1}(F_k^\perp) \subset F_k \cap \mathrm{Im}(x^{j-1}) \subset \cdots$$

By Lemma 4.3, $x^j(F_k^\perp) = \left(F_k \cap \mathrm{Im}(x^j)\right)^{\perp\!\!\perp_j}$, so $\left(F_k \cap \mathrm{Im}(x^j)\right) / \left(F_k \cap x^j(F_k^\perp)\right)$ is a symplectic space with non-degenerate skew-symmetric bilinear form given by the restriction of $\langle\langle\, ,\, \rangle\rangle_j$. We denote the dimension $\dim \left(F_k \cap \mathrm{Im}(x^j)\right) / \left(F_k \cap x^j(F_k^\perp)\right) = 2h_j$ (so the h we used in the previous sections would actually become h_1).

Then, we apply Lemma 3.7 and, since

$$a'_{j-1} - a'_j = \begin{cases} k_\ell - 2h_{\ell-1} & \text{if } j = \ell \\ k_j + 2h_j - 2h_{j-1} & \text{if } 1 < j < \ell\,, \\ k_1 + 2h_1 & \text{if j=1} \end{cases}$$

we obtain that

$$\begin{aligned} J\left(x|_{F_k^\perp/F_k}\right) = (&\ell^{2n_\ell - 2k_\ell + 2h_{\ell-1}} (\ell-1)^{2n_{\ell-1} - 2k_{\ell-1} + 2k_\ell - 4h_{\ell-1} + 2h_{\ell-2}} \\ &(\ell-2)^{2n_{\ell-2} - 2k_{\ell-2} + 2k_{\ell-1} + 2h_{\ell-3} - 4h_{\ell-2} + 2h_{\ell-1}} \cdots \\ &\cdots 2^{2n_2 - 2k_2 + 2k_3 + 2h_1 - 4h_2 + 2h_3} 1^{2n_1 - 2k_1 + 2k_2 - 4h_1 + 2h_2}) \end{aligned}$$

The difficulty then lies in computing $J\left(x|_{F_k^\perp/F_k + \mathbb{C}[x]v}\right)$ because there are many different possibilities in general for how v can fit within the various subspaces involved in the computation and we do not know how to treat it in general. The case-by-case approach we used for $\ell = 1, 2$ quickly becomes unwieldy for $\ell > 2$.

7. Exotic Robinson-Schensted-Knuth correspondence

The Robinson-Schensted correspondence is a combinatorial bijection between permutations and pairs of standard Young tableaux of the same shape, and it was given a geometric interpretation by Steinberg (see [**19**]) in terms of varieties of pairs of flags in Type A. In [**18**], one of the authors generalized the construction to pairs of partial flags, obtaining the Robinson-Schensted-Knuth (RSK) correspondence between matrices with nonnegative integer entries and pairs of semistandard Young tableaux of the same shape. We now outline how Conjecture 2.19 similarly implies an exotic Robinson-Schensted-Knuth correspondence that generalizes the one obtained for complete flags in [**16**].

DEFINITION 7.1 (Partial exotic Steinberg variety). Let α, β be compositions of n, we define the *exotic Steinberg variety of type* (α, β) to be

$$\mathfrak{Z}^{\alpha,\beta} := \{(F_\bullet, F'_\bullet, (v,x)) \in \mathcal{F}^\alpha(V) \times \mathcal{F}^\beta(V) \times \mathfrak{N} \mid F_\bullet \in \mathcal{C}^\alpha_{(v,x)},\ F'_\bullet \in \mathcal{C}^\beta_{(v,x)}\}.$$

Notice that for $\alpha = \beta = 1^n$, we obtain the exotic Steinberg variety of [**15**, Def. 6.1].

DEFINITION 7.2 (Relative position). Let $\alpha = (\alpha_1, \dots, \alpha_m) \models n$, $\beta = (\beta_1, \dots, \beta_\ell) \models n$, $F \in \mathcal{F}^\alpha(V)$, $F' \in \mathcal{F}^\beta(V)$. We define a matrix

$$A = A(F, F') = (a_{i,j})_{1\le i\le 2m, 1\le j\le 2\ell}$$

by

$$a_{i,j} = \dim((F_{\check{\alpha}_i} \cap F'_{\check{\beta}_j})/(F_{\check{\alpha}_{i-1}} \cap F'_{\check{\beta}_j} + F_{\check{\alpha}_i} \cap F'_{\check{\beta}_{j-1}}))$$

which we call the *relative position* of the partial flags F and F'.

REMARK 7.3. Notice that the row sums of $A(F, F')$ are
$\hat{\alpha} = (\alpha_1, \alpha_2, \dots, \alpha_m, \alpha_m, \dots, \alpha_1)$ and the column sums are
$\hat{\beta} = (\beta_1, \beta_2, \dots, \beta_\ell, \beta_\ell, \dots, \beta_1)$. Also, since $F^\perp_{\check{\alpha}_i} = F_{\check{\alpha}_{2m-i}}$ and $F'^\perp_{\check{\beta}_j} = F'_{\check{\beta}_{2\ell-j}}$, it follows that

$$a_{2m+1-i, 2\ell+1-j} = a_{i,j}. \tag{7.1}$$

For $\alpha, \beta \models n$, we define $M^{\alpha,\beta}$ to be the set of matrices with nonnegative integer entries, with row sums $\hat{\alpha}$, column sums $\hat{\beta}$ and satisfying (7.1).

REMARK 7.4. The set $M^{\alpha,\beta}$ parametrizes the orbits of the diagonal action of Sp_{2n} on $\mathcal{F}^\alpha(V) \times \mathcal{F}^\beta(V)$. Also, we can identify $M^{\alpha,\beta}$ with the double cosets $W_\alpha \backslash W / W_\beta$, where W is the Weyl group of Type C (signed permutations) and W_α, W_β are parabolic subgroups.

Let $\theta : \mathfrak{Z}^{\alpha,\beta} \to \mathcal{F}^\alpha(V) \times \mathcal{F}^\beta(V)$ defined by $\theta(F_\bullet, F'_\bullet, (v,x)) = (F, F')$, then for $A \in M^{\alpha,\beta}$ we define $\mathfrak{Z}^{\alpha,\beta}_A = \theta^{-1}(A)$.

On the other hand, for any bipartition of n, (μ, ν), and composition $\alpha \models n$, we let $\mathrm{SYB}^\alpha(\mu, \nu)$ be the set of all semistandard bitableaux of shape (μ, ν) and content α. Then for any $T \in \mathrm{SYB}^\alpha(\mu, \nu)$, $T' \in \mathrm{SYB}^\beta(\mu, \nu)$, we can consider the subvariety

$$\mathfrak{Z}^{\alpha,\beta}_{T,T'} := \{(F, F', (v,x)) \in \mathfrak{Z}^{\alpha,\beta} \mid \Phi(F) = T,\ \Phi(F') = T'\}.$$

CONJECTURE 7.5. *The variety* $\mathfrak{Z}^{\alpha,\beta}$ *(if nonempty) is pure dimensional of dimension* $2n^2 - \frac{1}{2}\sum_{i=1}^m(\alpha_i^2 - \alpha_i) - \frac{1}{2}\sum_{i=1}^\ell(\beta_i^2 - \beta_i)$ *and its irreducible components are*

$\{\overline{\mathfrak{Z}_A^{\alpha,\beta}} \mid A \in M^{\alpha,\beta}\}$. *The irreducible components of* $\mathfrak{Z}^{\alpha,\beta}$ *can also be parametrized as*

$$\{\overline{\mathfrak{Z}_{T,T'}^{\alpha,\beta}} \mid (\mu,\nu) \in \mathcal{Q}_n,\ T \in \mathrm{SYB}^\alpha(\mu,\nu),\ T' \in \mathrm{SYB}^\beta(\mu,\nu)\}.$$

This would follow from Conjecture 2.19 and give a geometrically defined bijection

$$M^{\alpha,\beta} \longleftrightarrow \coprod_{(\mu,\nu)\in\mathcal{Q}_n} \mathrm{SYB}^\alpha(\mu,\nu) \times \mathrm{SYB}^\beta(\mu,\nu)$$

given by $A \leftrightarrow (T,T')$ if and only if $\overline{\mathfrak{Z}_A^{\alpha,\beta}} = \overline{\mathfrak{Z}_{T,T'}^{\alpha,\beta}}$. We expect that its combinatorial description should be a 'semistandardization' of the algorithm of [**16**], analogously to what was done in [**18**] for the RSK correspondence in Type A.

EXAMPLE 7.6. Let $n = 2$, and $\alpha = \beta = (2)$, then there are three pairs of semistandard bitableaux of the same shape with contents α and β, which are

$$\left(\left(\boxed{1},\boxed{1}\right),\left(\boxed{1},\boxed{1}\right)\right),\quad \left(\left(\begin{array}{|c|}\hline 1\\\hline 1\\\hline\end{array},\varnothing\right),\left(\begin{array}{|c|}\hline 1\\\hline 1\\\hline\end{array},\varnothing\right)\right),\quad \left(\left(\varnothing,\begin{array}{|c|}\hline 1\\\hline 1\\\hline\end{array}\right),\left(\varnothing,\begin{array}{|c|}\hline 1\\\hline 1\\\hline\end{array}\right)\right).$$

It can then be checked that the geometric correspondence matches these respectively to the following matrices of relative positions (which comprise the set $M^{(2),(2)}$):

$$\begin{bmatrix} 2 & 0 \\ 0 & 2 \end{bmatrix} \quad \begin{bmatrix} 1 & 1 \\ 1 & 1 \end{bmatrix} \quad \begin{bmatrix} 0 & 2 \\ 2 & 0 \end{bmatrix}.$$

References

[1] Pramod N. Achar and Anthony Henderson, *Orbit closures in the enhanced nilpotent cone*, Adv. Math. **219** (2008), no. 1, 27–62, DOI 10.1016/j.aim.2008.04.008. MR2435419

[2] Jonathan Brundan and Victor Ostrik, *Cohomology of Spaltenstein varieties*, Transform. Groups **16** (2011), no. 3, 619–648, DOI 10.1007/s00031-011-9149-2. MR2827037

[3] Anders Skovsted Buch, Andrew Kresch, and Harry Tamvakis, *Quantum Pieri rules for isotropic Grassmannians*, Invent. Math. **178** (2009), no. 2, 345–405, DOI 10.1007/s00222-009-0201-y. MR2545685

[4] Corrado De Concini and Claudio Procesi, *Symmetric functions, conjugacy classes and the flag variety*, Invent. Math. **64** (1981), no. 2, 203–219, DOI 10.1007/BF01389168. MR629470

[5] J. Connor, S. Griffin and K. Purohit, *Irreducible components of two-column Δ-Springer fibers*, `arXiv:2411.17222`, 2024.

[6] Lucas Fresse, *Singular components of Springer fibers in the two-column case* (English, with English and French summaries), Ann. Inst. Fourier (Grenoble) **59** (2009), no. 6, 2429–2444, DOI 10.5802/aif.2495. MR2640925

[7] Lucas Fresse and Anna Melnikov, *Some characterizations of singular components of Springer fibers in the two-column case*, Algebr. Represent. Theory **14** (2011), no. 6, 1063–1086, DOI 10.1007/s10468-010-9227-5. MR2844756

[8] Lucas Fresse, *A unified approach on Springer fibers in the hook, two-row and two-column cases*, Transform. Groups **15** (2010), no. 2, 285–331, DOI 10.1007/s00031-010-9096-3. MR2657444

[9] Mee Seong Im, Chun-Ju Lai, and Arik Wilbert, *Irreducible components of two-row Springer fibers for all classical types*, Proc. Amer. Math. Soc. **150** (2022), no. 6, 2415–2432, DOI 10.1090/proc/15965. MR4399259

[10] Syu Kato, *An exotic Deligne-Langlands correspondence for symplectic groups*, Duke Math. J. **148** (2009), no. 2, 305–371, DOI 10.1215/00127094-2009-028. MR2524498

[11] Abel Lacabanne, Pedro Vaz, and Arik Wilbert, *Two-row Delta Springer varieties*, Algebr. Comb. **8** (2025), no. 4, 925–953, DOI 10.5802/alco.435. MR4952053

[12] Yiqiang Li, *Spaltenstein varieties of pure dimension*, Proc. Amer. Math. Soc. **148** (2020), no. 1, 133–144, DOI 10.1090/proc/14726. MR4042837

[13] Ronit Mansour, *Counting two-column Young tableaux corresponding to smooth components of Springer fibers*, J. Algebraic Combin. **61** (2025), no. 2, Paper No. 18, 9, DOI 10.1007/s10801-025-01390-x. MR4866596

[14] Ion Alexandru Mihai, *Odd symplectic flag manifolds*, Transform. Groups **12** (2007), no. 3, 573–599, DOI 10.1007/s00031-006-0053-0. MR2356323

[15] V. Nandakumar, D. Rosso, and N. Saunders, *Irreducible components of exotic Springer fibres*, J. Lond. Math. Soc. (2) **98** (2018), no. 3, 609–637, DOI 10.1112/jlms.12152. MR3893194

[16] Vinoth Nandakumar, Daniele Rosso, and Neil Saunders, *Irreducible components of exotic Springer fibres II: The exotic Robinson-Schensted algorithm*, Pacific J. Math. **310** (2021), no. 2, 447–485, DOI 10.2140/pjm.2021.310.447. MR4229244

[17] Nicolas Perrin and Evgeny Smirnov, *Springer fiber components in the two columns case for types A and D are normal*, Bull. Soc. Math. France **140** (2012), no. 3, 309–333, DOI 10.24033/bsmf.2629. MR3059118

[18] Daniele Rosso, *Classic and mirabolic Robinson-Schensted-Knuth correspondence for partial flags*, Canad. J. Math. **64** (2012), no. 5, 1090–1121, DOI 10.4153/CJM-2011-071-7. MR2979579

[19] Robert Steinberg, *An occurrence of the Robinson-Schensted correspondence*, J. Algebra **113** (1988), no. 2, 523–528, DOI 10.1016/0021-8693(88)90177-9. MR929778

[20] T. A. Springer, *A construction of representations of Weyl groups*, Invent. Math. **44** (1978), no. 3, 279–293, DOI 10.1007/BF01403165. MR491988

[21] N. Saunders and A. Wilbert, *Exotic Springer fibers for orbits corresponding to one-row bipartitions*, Transform. Groups **27** (2022), no. 3, 1111–1147, DOI 10.1007/s00031-020-09613-0. MR4475488

[22] N. Spaltenstein, *The fixed point set of a unipotent transformation on the flag manifold*, Indag. Math. **38** (1976), no. 5, 452–456. Nederl. Akad. Wetensch. Proc. Ser. A **79**. MR485901

[23] Roman Travkin, *Mirabolic Robinson-Schensted-Knuth correspondence*, Selecta Math. (N.S.) **14** (2009), no. 3-4, 727–758, DOI 10.1007/s00029-009-0508-y. MR2511197

Department of Mathematics and Actuarial Science, Indiana University Northwest, Gary, Indiana 46408

Email address: drosso@iu.edu

Department of Mathematics, School of Science and Technology, City St George's, University of London, Northampton Square, London EC1V 0HB, United Kingdom

Email address: neil.saunders.3@citystgeorges.ac.uk

Contemporary Mathematics
Volume **837**, 2026
https://doi.org/10.1090/conm/837/16791

Minimal semisimple Hessenberg schemes

Rebecca Goldin and Martha Precup

Abstract. We study a collection of Hessenberg varieties in the type A flag variety associated to a nonzero semisimple matrix whose conjugacy class has minimal dimension. We prove each such minimal semisimple Hessenberg variety is a union Richardson varieties and compute this set of Richardson varieties explicitly. Our methods leverage the notion of matrix Hessenberg schemes to answer questions about the geometry of minimal semisimple Hessenberg varieties using commutative algebra and known results on Schubert determinantal ideals. In particular, we show that all type A minimal semisimple matrix Hessenberg schemes are reduced.

1. Introduction

Let $G := GL_n(\mathbb{C})$ be the set of $n \times n$ invertible matrices and B be the subgroup of upper triangular matrices in G. A Hessenberg variety $Y_{\mathsf{x},h}$ is a subvariety of the flag variety G/B defined by a choice of matrix $\mathsf{x} \in \mathfrak{gl}_n(\mathbb{C}) = \mathrm{Lie}(G)$ and Hessenberg function h. In this paper, we consider Hessenberg varieties in the flag variety G/B for x a *minimal semisimple element*, that is, x is a non-scalar diagonalizable matrix such that the dimension of its conjugacy class in $\mathfrak{gl}_n(\mathbb{C})$ is as small as possible. In general, we say that any nonzero matrix x is *minimal* if its conjugacy class satisfies this property. If x is a minimal element, then we say that $Y_{\mathsf{x},h}$ is a *minimal Hessenberg variety.*

Consider the minimal nilpotent element $\mathsf{n} = E_{1n} \in \mathfrak{gl}_n(\mathbb{C})$, where E_{1n} is the elementary matrix with unique nonzero entry equal to 1 in row 1 and column n. Conjugation by B scales n so $Y_{\mathsf{n},h}$ is B-invariant, implying that $Y_{\mathsf{n},h}$ is a union of Schubert varieties. The minimal nilpotent Hessenberg varieties $Y_{\mathsf{n},h}$ were studied by Tymoczko [**Tym06**] and by Abe and Crooks [**AC16**], where they identify components of $Y_{\mathsf{n},h}$ explicitly. Here we characterize $Y_{\mathsf{x},h}$ in the minimal semisimple case, and show that they are unions of Richardson varieties determined by combinatorial data associated with the Hessenberg function h. We describe their defining ideals, allowing us to prove they are reduced as schemes. In a companion paper [**GP**] we prove that these Hessenberg schemes lie in flat families over lines in the union of all minimal orbits, and determine the (generally non-reduced) scheme structure on minimal nilpotent Hessenberg varieties as well.

The first author was supported by National Science Foundation grant #2152312 and the second author was supported by grant #1954001 and #2237057.

Throughout this paper, we identify G/B with the collection of complete flags in $\mathbb{C}^n$,

$$V_\bullet = 0 \subset V_1 \subset V_2 \subset \cdots \subset V_{n-1} \subset V_n = \mathbb{C}^n$$

with $\dim_{\mathbb{C}} V_i = i$ for all $1 \leq i \leq n$. Denote by B_- the subgroup of lower triangular matrices. Then $T := B \cap B_-$ is a maximal torus, consisting of diagonal matrices in G. We identify the Weyl group of G with the permutation group S_n on n letters. Given $w \in S_n$ we let $\dot{w}$ denote the permutation matrix with entry equal to 1 in row i and column j if $w(j) = i$. Our convention is that the *Schubert variety* $X_w = \overline{B\dot{w}B/B}$ is the closure of the left B-orbit of $\dot{w}B$ in G/B, while the *opposite Schubert variety* $X_w^{\mathrm{op}} = \overline{B_-\dot{w}B/B}$ is the closure the B_--orbit of $\dot{w}B$.

Hessenberg varieties are parameterized by a nondecreasing function

$$h : \{1, 2, \ldots, n\} \to \{1, 2, \ldots, n\}$$

satisfying $h(i) \geq i$, called a *Hessenberg function*, and an $n \times n$ matrix $\mathsf{x} \in \mathfrak{gl}_n(\mathbb{C})$. Explicitly,

$$Y_{\mathsf{x},h} := \{V_\bullet \mid \mathsf{x}(V_i) \subseteq V_{h(i)}\} \tag{1.1}$$

is the *Hessenberg variety* corresponding to the matrix x and Hessenberg function h. If x is a nilpotent (respectively, semisimple) element of $\mathfrak{gl}_n(\mathbb{C})$ then we say that $Y_{\mathsf{x},h}$ is a *nilpotent* (respectively, *semisimple*) *Hessenberg variety*.

We employ the term *variety* for $Y_{\mathsf{x},h}$ even though its defining ideal may not be reduced. The literature uses the term "Hessenberg variety" both to refer to the underlying reduced variety and to the scheme corresponding to the defining ideal of (1.1). Our results include a description of $Y_{\mathsf{x},h}$ as a scheme; we make this explicit in Section 3, where we lift $Y_{\mathsf{x},h}$ to a subscheme of $M_n(\mathbb{C})$, the affine scheme of $n \times n$ complex matrices.

In this article we characterize minimal semisimple Hessenberg varieties $Y_{\mathsf{x},h}$. We carry out this program by working "upstairs" in the matrix group G, realized as an open subscheme of $M_n(\mathbb{C})$. Following Insko–Tymoczko–Woo [**ITW20**] we introduce the *matrix Hessenberg scheme*, a right B-invariant subscheme $\mathcal{Y}_{\mathsf{x},h}$ of $M_n(\mathbb{C})$ equal to the closure of $\pi^{-1}(Y_{\mathsf{x},h}) \subseteq G$, where $\pi : G \to G/B$ is the quotient map.

We show in Sections 3 and 4 that, for all h, it suffices to consider the matrix Hessenberg scheme defined using the minimal semisimple element

$$\mathsf{s} := \mathrm{diag}(1, 0, \ldots, 0) \in \mathfrak{gl}_n(\mathbb{C}).$$

In Theorem 4.1, we identify all components of $\mathcal{Y}_{\mathsf{s},h}$ and prove that it is reduced for all Hessenberg functions h. We prove that the irreducible components of $\mathcal{Y}_{\mathsf{s},h}$ are in bijection with the *corners* $\mathcal{C}(h)$ of h, the set of $i \in [n]$ such that $h(i) > h(i-1)$ where, by convention, $h(0) = 0$. As a consequence, we obtain a corresponding decomposition of the Hessenberg variety $Y_{\mathsf{s},h}$ into a reduced union of Richardson varieties.

Theorem 1. For all Hessenberg functions h, the minimal semisimple Hessenberg variety $Y_{\mathsf{s},h}$ is equal to a reduced union of Richardson varieties,

$$Y_{\mathsf{s},h} = \bigcup_{i \in \mathcal{C}(h)} \left(X_{u[i]}^{\mathrm{op}} \cap X_{w_0 v[h(i)]} \right)$$

where $u[i] \in S_n$ is shortest permutation u such that $u(i) = 1$ and $v[j] \in S_n$ is the shortest permutation v such that $v(j) = n$.

While we show the minimal semisimple Hessenberg variety $Y_{\mathsf{s},h}$ is reduced for all h, it may not be equidimensional (see Corollary 4.5). The top-dimensional components are detected by its cohomology class in $H^*(G/B)$, which we describe as a function of h in Corollary 4.6. In a companion paper, we show that the minimal semisimple and nilpotent matrix Hessenberg schemes lie in a flat family [**GP**]. This provides a unified explanation for surprising similarities between minimal semisimple and minimal nilpotent Hessenberg varieties.

2. Background

This section contains the definitions and basic properties of Schubert determinantal ideals that will be used below. Many of the results in this section were originally proved by Fulton [**Ful92**] and also Knutson and Miller [**KM05**]; Chapter 15 of the text [**MS05**] is an excellent expository reference.

Let n be a positive integer and $[n] := \{1, 2, \dots, n\}$. As in the introduction, $G := GL_n(\mathbb{C})$ is the algebraic group of $n \times n$ invertible complex matrices with Lie algebra $\mathfrak{gl}_n(\mathbb{C})$. For any $n \times n$ matrix A, let $A_{p\times q}$ denote the submatrix of the top p rows and left-most q columns, i.e. the upper $p \times q$ northwest submatrix of A.

Let $w \in S_n$. For any p, q with $1 \leq p, q \leq n$, the rank of $\dot{w}_{p\times q}$ is given by

$$r_{pq}(w) := \mathrm{rank}(\dot{w}_{p\times q}) = |\{w(1), \dots, w(q)\} \cap \{1, \dots, p\}|.$$

Let $\mathbf{z} := \{z_{ij} \mid (i,j) \in [n] \times [n]\}$ denote n^2 variables, and $Z = (z_{ij})_{1\leq i,j\leq n} \in M_n(\mathbb{C}[\mathbf{z}])$ be the matrix with variable z_{ij} in the ijth entry of the matrix. The *Schubert determinantal ideal* $\mathcal{I}_w$ *corresponding to the permutation* w is the ideal in $\mathbb{C}[\mathbf{z}]$ generated by all minors of $Z_{p\times q}$ of size $1 + r_{pq}(w)$.[1]

For each $w \in S_n$, the *matrix Schubert variety* is given by $\mathcal{X}_w := \overline{B\dot{w}B}$ in $M_n(\mathbb{C}) = \mathrm{Spec}(\mathbb{C}[\mathbf{z}])$. Similarly, the *opposite matrix Schubert variety* is $\mathcal{X}_w^{\mathrm{op}} := \overline{B_-\dot{w}B}$.[2] We will frequently use the relationship that

$$w_0\mathcal{X}_w^{\mathrm{op}} = \mathcal{X}_{w_0w}, \tag{2.1}$$

where w_0 is the longest word in S_n. The following is well-known, see [**MS05**, Chap. 15].

PROPOSITION 2.1. *Let $w \in S_n$ and $\mathcal{I}_w$ denote the corresponding Schubert determinantal ideal. Recall that $\ell(w) := |\{(i,j) \mid 1 \leq i < j \leq n, w(i) > w(j)\}|$ is the* length *of w.*

1. *The opposite matrix Schubert variety $\mathcal{X}_w^{\mathrm{op}}$ is the reduced subscheme $\mathrm{Spec}(\mathbb{C}[\mathbf{z}]/\mathcal{I}_w)$; in particular, Schubert determinantal ideals are prime.*
2. *Let $\pi : G \to G/B$ be the natural projection. Then $\mathcal{X}_w^{\mathrm{op}} \cap G = \pi^{-1}(X_w^{\mathrm{op}})$.*
3. *The dimension of the opposite Schubert variety $\mathcal{X}_w^{\mathrm{op}}$ is $\dim \mathcal{X}_w^{\mathrm{op}} = n^2 - \ell(w)$.*

By Equation (2.1) and statement (1) above, we find that

$$\mathcal{X}_{w_0w} = \mathrm{Spec}(\mathbb{C}[\mathbf{z}]/w_0 \cdot \mathcal{I}_w) \tag{2.2}$$

where $w_0 \cdot \mathcal{I}_w$ is the image of the Schubert determinantal ideal $\mathcal{I}_w$ under the action of w_0 on $\mathbb{C}[\mathbf{z}]$ defined on variables z_{ij} by $w_0 \cdot z_{ij} = z_{w_0(i)j}$.

---

[1] [**MS05**] and [**KM05**] use the transpose matrix $\dot{w}^T$ rather than $\dot{w}$, resulting in a different ideal $\mathcal{J}_w$. Transposition on $M_n(\mathbb{C})$ induces an isomorphism on the coordinate ring that sends $\mathcal{J}_w$ to $\mathcal{I}_w$. Geometrically, transposition descends to an isomorphism from $B_-\backslash G$ to G/B.

[2] The varieties $\mathcal{X}_w^{\mathrm{op}}$ are called 'matrix Schubert varieties' by some authors.

Inclusions of matrix Schubert varieties (and also Schubert varieties) define a partial order $\leq_{\mathrm{Br}}$ on the set of permutations known as *Bruhat order*. That is, we have $v \leq_{\mathrm{Br}} w$ if and only if $\mathcal{X}_v \subseteq \mathcal{X}_w$. Consequently, $v \leq_{\mathrm{Br}} w$ if and only if $\mathcal{X}_v^{\mathrm{op}} \supseteq \mathcal{X}_w^{\mathrm{op}}$.

Particular Schubert and opposite Schubert varieties play a role in characterizing minimal semisimple Hessenberg schemes. For each $k \in [n]$, let

- $u[k]$ be the shortest permutation u in S_n such that $u(k) = 1$ and
- $v[k]$ be the shortest permutation v in S_n such that $v(k) = n$.

It is easy in practice to write down the one-line notation for each such permutation. Indeed, we obtain the one-line notation for $u[k]$ by placing 1 in the k-th position and filling all remaining entries with $2, \dots, n$ in increasing order from left to right. Similarly, we obtain the one-line notation for $v[k]$ by placing n in the k-th position and filling all remaining entries with $1, \dots, n-1$ in increasing order from left to right. For example, when $n = 5$, $u[2] = [2, 1, 3, 4, 5]$ and $v[3] = [1, 2, 5, 3, 4]$. Note that $u[1] = v[n] = e$ is the identity.

We now introduce the particular ideals of interest in this paper. For each $i \in [n-1]$, set

$$
\begin{aligned}
\mathcal{L}_i &:= \langle z_{11}, z_{12}, \dots, z_{1i} \rangle, \quad \text{and} \\
\mathcal{K}_i &:= \langle p_B \mid B \subseteq \{2, \dots n\}, |B| = i \rangle
\end{aligned} \tag{2.3}
$$

where p_B is the $i \times i$ minor of $Z = (z_{ij})_{1 \leq i,j \leq n}$ given by $\det(z_{k\ell})_{k \in B, 1 \leq \ell \leq i}$. Define $\mathcal{L}_0 := \langle 0 \rangle$ and $\mathcal{K}_n := \langle 0 \rangle$.

The following lemma is straightforward to verify using the definition of a matrix Schubert variety and the observation of Equation (2.2).

LEMMA 2.1. *For all $i \in [n-1]$, $\mathcal{L}_i$ is the Schubert determinantal ideal for the permutation $u[i+1]$ and $w_0 \cdot \mathcal{K}_i$ is the Schubert determinantal ideal for the permutation $v[i]$. Explicitly,*

$$\mathcal{X}_{u[i+1]}^{\mathrm{op}} = \mathrm{Spec}(\mathbb{C}[\mathbf{z}]/\mathcal{L}_i) \quad \textit{and} \quad \mathcal{X}_{w_0 v[i]} = \mathrm{Spec}(\mathbb{C}[\mathbf{z}]/\mathcal{K}_i).$$

Subvarieties of the flag variety equal to the intersection of a Schubert and opposite Schubert variety are called *Richardson varieties.* Let $u, v \in S_n$ and recall that the Richardson variety $X_u^{\mathrm{op}} \cap X_v$ is nonempty if and only if $u \leq v$ in Bruhat order. In that case, we say that the closure

$$\overline{\pi^{-1}(X_u^{\mathrm{op}} \cap X_v)} \subseteq M_n(\mathbb{C})$$

is a *matrix Richardson variety.* The next lemma shows that the sum of two of the Schubert ideals considered above defines a matrix Richardson variety.

LEMMA 2.2. *For all $i, j \in [n]$, the ideal $\mathcal{L}_i + \mathcal{K}_j$ is prime. It follow that, for all $1 \leq i < j \leq n$, the intersection*

$$\mathcal{X}_{u[i+1]}^{\mathrm{op}} \cap \mathcal{X}_{w_0 v[j]} = \mathrm{Spec}(\mathbb{C}[\mathbf{z}]/(\mathcal{L}_i + \mathcal{K}_j))$$

is a matrix Richardson variety.

PROOF. Observe that $\mathcal{L}_i = \langle z_{11}, z_{12}, \dots, z_{1i} \rangle$ and $\mathcal{K}_j \subseteq \mathbb{C}[\mathbf{z}']$ with $\mathbf{z}' = \{z_{ij} \mid 2 \leq i \leq n, 1 \leq j \leq n\}$ are prime ideals whose generators have no variables in common. Thus the sum $\mathcal{L}_i + \mathcal{K}_j$ is prime, and the affine subscheme $\mathcal{X}_{u[i+1]}^{\mathrm{op}} \cap \mathcal{X}_{w_0 v[j]}$ in $\mathbb{C}[\mathbf{z}]$ defined by $\mathcal{L}_i + \mathcal{K}_j$ is reduced. One may check that $u[i+1] \leq_{\mathrm{Br}} w_0 v[j]$ if and only if $i < j$, implying $X_{u[i+1]}^{\mathrm{op}} \cap X_{w_0 v[j]}$ nonempty exactly when $i < j$. Finally observe that $\pi^{-1}(X^{\mathrm{op}}_{u[i+1]} \cap X_{w_0v[j]}) = \pi^{-1}(X^{\mathrm{op}}_{u[i+1]}) \cap \pi^{-1}(X_{w_0v[j]}) \subseteq \mathcal{X}^{op}_{u[i+1]} \cap \mathcal{X}_{w_0v[j]}$, so the intersection being reduced implies

$$\overline{\pi^{-1}(X^{\mathrm{op}}_{u[i+1]} \cap X_{w_0v[j]})} = \mathcal{X}^{\mathrm{op}}_{u[i+1]} \cap \mathcal{X}_{w_0v[j]}. \qquad \square$$

Toward the proof of Theorem 1, we highlight several needed properties of the ideals $\mathcal{L}_i$ and $\mathcal{K}_j$ in the following lemmas.

LEMMA 2.3. *The following statements all hold for $i, j, k, \ell \in [n-1]$.*

(1) *Suppose $i \leq j$. Then $\mathcal{L}_i \subseteq \mathcal{L}_j$ and $\mathcal{K}_j \subseteq \mathcal{K}_i$.*
(2) *Suppose $i < j$ and $k < \ell$. Then $\mathcal{L}_i + \mathcal{K}_j \subseteq \mathcal{L}_k + \mathcal{K}_\ell$ if and only if $i \leq k$ and $j \geq \ell$.*

PROOF. The assertions of (1) follow easily from the definitions.

To prove (2), observe that $i \leq k$ and $j \geq \ell$ immediately implies $\mathcal{L}_i + \mathcal{K}_j \subseteq \mathcal{L}_k + \mathcal{K}_\ell$ by (1). Suppose now that $\mathcal{L}_i + \mathcal{K}_j \subseteq \mathcal{L}_k + \mathcal{K}_\ell$. By Lemma 2.2, there an inclusion of Richardson varieties,

$$X^{\mathrm{op}}_{u[k+1]} \cap X_{w_0v[\ell]} \subseteq X^{\mathrm{op}}_{u[i+1]} \cap X_{w_0v[j]}.$$

The inclusion implies $u[i+1] \leq_{\mathrm{Br}} u[k+1]$, which in turn implies $i \leq k$. Similarly, the inclusion implies $w_0v[\ell] \leq_{\mathrm{Br}} w_0v[j]$, or, equivalently, $v[\ell] \geq_{\mathrm{Br}} v[j]$. A quick check shows $j \geq \ell$. $\square$

Next, we argue that the product of ideals $\mathcal{L}_i \cdot \mathcal{K}_j$ is radical.

LEMMA 2.4. *For all $i, j \in [n-1]$, the product $\mathcal{L}_i \cdot \mathcal{K}_j$ is radical. In particular, $\mathcal{L}_i \cdot \mathcal{K}_j = \mathcal{L}_i \cap \mathcal{K}_j$.*

PROOF. Let $\preceq$ be a monomial order on $\mathbb{C}[\mathbf{z}]$ and recall that if the monomial ideal generated by leading terms of polynomials in an ideal $\mathcal{I}$ with respect to $\preceq$ is square free, then $\mathcal{I}$ is radical [**HH11**, Prop. 3.3]. This fact can be used to prove that $\mathcal{L}_i \cdot \mathcal{K}_j$ is radical using a simple Gröbner basis argument (see [**CLO15**] for background on Gröbner bases); we provide an outline below for the interested reader.

Let $\preceq$ be any diagonal term order on $\mathbb{C}[\mathbf{z}]$. It is well known (see e.g. [**KM05**, Thm. B]) that

$$\{p_B \mid B \subseteq \{2, \ldots, n\}, |B| = j\}$$

is a Gröbner basis for $\mathcal{K}_j$ with respect to this ordering and it's obvious that $\{z_{11}, z_{12}, \ldots, z_{1i}\}$ is a Gröbner basis for $\mathcal{L}_i$. Since the leading monomial of any p_B and any z_{1k} are relatively prime, it is straightforward to check that

$$\{z_{1k}p_B \mid B \subseteq \{2, \ldots, n\}, |B| = j, 1 \leq k \leq i\}$$

is a Gröbner basis for $\mathcal{L}_i \cdot \mathcal{K}_j$. The leading term of every polynomial in this Gröbner basis is square free, so $\mathcal{L}_i \cdot \mathcal{K}_j$ is radical. Thus

$$\mathcal{L}_i \cdot \mathcal{K}_j = \mathrm{rad}(\mathcal{L}_i \cdot \mathcal{K}_j) = \mathrm{rad}(\mathcal{L}_i \cap \mathcal{K}_j) = \mathrm{rad}(\mathcal{L}_i) \cap \mathrm{rad}(\mathcal{K}_j) = \mathcal{L}_i \cap \mathcal{K}_j$$

as desired. $\square$

3. Matrix Hessenberg schemes

Recall from the Introduction that a function $h : [n] \to [n]$ is a *Hessenberg function* if $h(i-1) \leq h(i)$ and $h(i) \geq i$ for all i. We denote a Hessenberg function $h : [n] \to [n]$ by listing its values, e.g. $h = (h(1), \ldots, h(n))$. Although 0 is not technically in the domain of h, it will be convenient for us to assume $h(0) = 0$.

There is a bijection from the set of all Hessenberg functions $h : [n] \to [n]$ to the set of Dyck paths of length $2n$, as follows. Each Hessenberg function h may be uniquely identified with a lattice path from the upper left corner of an $n \times n$ grid to the lower right corner obtained by requiring that the horizontal step in column i occurs $h(i)$ rows from the top. We call the resulting image *diagram of h*. The diagrams of the functions $h_1 = (2, 3, 3, 4)$ and $h_2 = (2, 2, 4, 4)$ are displayed below.

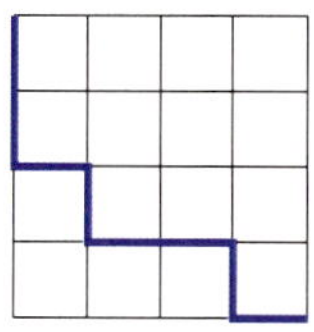

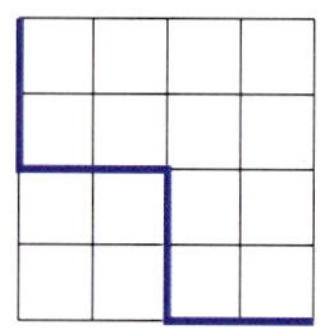

The diagrams help visualize examples presented in later sections.

For $j \in [n]$, let v_j denote the jth column vector of the matrix $Z = (z_{ij})$, i.e.

$$v_j := \sum_{i=1}^{n} z_{ij} e_i. \tag{3.1}$$

Let $\mathsf{x} \in \mathfrak{gl}_n(\mathbb{C})$. Define the ideal $\mathcal{I}_{\mathsf{x},h,i} \subseteq \mathbb{C}[\mathbf{z}]$ as follows. If $h(i) = n$, then $\mathcal{I}_{\mathsf{x},h,i} := 0$. For each $i \in [n]$ such that $h(i) < n$, define $\mathcal{I}_{\mathsf{x},h,i} \subseteq \mathbb{C}[\mathbf{z}]$ to be the ideal in $\mathbb{C}[\mathbf{z}]$ generated by the rank condition:

$$\operatorname{rk} \begin{bmatrix} | & | & & | & | & | & & | \\ \mathsf{x}v_1 & \mathsf{x}v_2 & \cdots & \mathsf{x}v_i & v_1 & v_2 & \cdots & v_{h(i)} \\ | & | & & | & | & | & & | \end{bmatrix} \leq h(i). \tag{3.2}$$

That is, $\mathcal{I}_{\mathsf{x},h,i}$ is the ideal generated by all $(h(i)+1) \times (h(i)+1)$-minors of the $n \times (h(i)+i)$-matrix above.

REMARK 3.1. When $\mathsf{s} = \operatorname{diag}(1, 0, \ldots, 0)$, the ideal $\mathcal{I}_{\mathsf{s},h,i}$ is determined by the rank condition:

$$\operatorname{rk} \begin{bmatrix} z_{11} & z_{12} & \cdots & z_{1i} & z_{11} & z_{12} & \cdots & z_{1h(i)} \\ 0 & 0 & \cdots & 0 & z_{21} & z_{22} & \cdots & z_{2h(i)} \\ \vdots & \vdots & \ddots & \vdots & \vdots & \vdots & \ddots & \vdots \\ 0 & 0 & \cdots & 0 & z_{n1} & z_{n2} & \cdots & z_{nh(i)} \end{bmatrix} \leq h(i).$$

In particular, $\mathcal{I}_{\mathsf{s},h,i} = \mathcal{L}_i \cdot \mathcal{K}_{h(i)}$, with $\mathcal{L}_i$ and $\mathcal{K}_j$ defined in (2.3).

We define

$$\mathcal{I}_{\mathsf{x},h} := \sum_{i \in [n]} \mathcal{I}_{\mathsf{x},h,i} \subseteq \mathbb{C}[\mathbf{z}]. \tag{3.3}$$

Consider the action of B on $M_n(\mathbb{C})$ by right multiplication. This induces an action of B on $\mathbb{C}[\mathbf{z}]$ defined on each variable z_{ij} by

$$z_{ij} \cdot b = \sum_k z_{ik} b_{kj} = \sum_{k \leq j} b_{kj} z_{ik}$$

for each upper triangular matrix $b = (b_{ij}) \in B$. This action of B amounts to rescaling each column vector v_j by a nonzero value, and adding any multiple of 'earlier' columns v_ℓ with $\ell < j$. Note that the ideal $\mathcal{I}_{\mathsf{x},h}$ is obviously invariant under this action of B on $\mathbb{C}[\mathbf{z}]$ since the rank condition (3.2) is invariant under such column operations.

Recall that $G = GL_n(\mathbb{C})$ is a principal open subset of the affine scheme $M_n(\mathbb{C}) = \mathrm{Spec}(\mathbb{C}[\mathbf{z}])$ of $n \times n$ matrices. As is typical, we identify $GL_n(\mathbb{C})$ as an affine subscheme of $\mathrm{Spec}(\mathbb{C}[\mathbf{z}, y])$ with coordinate ring $\mathbb{C}[\mathbf{z}, d^{-1}] := \mathbb{C}[\mathbf{z}, y]/\langle yd - 1\rangle$ obtained via localization, where $d \in \mathbb{C}[\mathbf{z}]$ denotes the determinant function $\det(Z)$. Let

$$\iota : \mathbb{C}[\mathbf{z}] \hookrightarrow \mathbb{C}[\mathbf{z}, d^{-1}] \tag{3.4}$$

denote the canonical injection. Given an ideal $\mathcal{I} \subseteq \mathbb{C}[\mathbf{z}]$, we write $\widetilde{\mathcal{I}}$ to denote the preimage of the ideal generated by $\iota(\mathcal{I})$.

DEFINITION 3.2. Given a matrix $\mathsf{x} \in \mathfrak{gl}_n(\mathbb{C})$ and Hessenberg function $h : [n] \to [n]$, the associated *matrix Hessenberg scheme* is the affine scheme

$$\mathcal{Y}_{\mathsf{x},h} := \mathrm{Spec}(\mathbb{C}[\mathbf{z}]/\widetilde{\mathcal{I}}_{\mathsf{x},h}).$$

Recall that $\pi : G \to G/B$ defines a correspondence between (right) B-invariant subschemes of G and subschemes of G/B. Thus, given a Hessenberg scheme $Y_{\mathsf{x},h}$ in the flag variety, we may instead consider the (right) B-invariant subscheme $\pi^{-1}(Y_{\mathsf{x},h})$ of $GL_n(\mathbb{C})$. By construction, the matrix Hessenberg scheme $\mathcal{Y}_{\mathsf{x},h}$ is precisely the Zariski closure of $\pi^{-1}(Y_{\mathsf{x},h})$ in $M_n(\mathbb{C}) = \mathrm{Spec}(\mathbb{C}[\mathbf{z}])$. Furthermore, though the inclusion $\mathcal{I}_{\mathsf{x},h} \subset \widetilde{\mathcal{I}}_{\mathsf{x},h}$ may be strict, $\mathcal{Y}_{\mathsf{x},h}$ is precisely the closure of the open set of prime ideals in the scheme $\mathrm{Spec}(\mathbb{C}[\mathbf{z}]/\mathcal{I}_{\mathsf{x},h})$ that do not contain the determinant function d. We summarize these observations in the following remark.

REMARK 3.3. Given the matrix $\mathsf{x} \in \mathfrak{gl}_n(\mathbb{C})$ and Hessenberg function $h : [n] \to [n]$, let $Y_{\mathsf{x},h}$ denote corresponding the Hessenberg scheme in the flag variety $GL_n(\mathbb{C})/B$ (also refered to as the Hessenberg variety). Then

(1) $\pi^{-1}(Y_{\mathsf{x},h}) = GL_n(\mathbb{C}) \cap \mathrm{Spec}(\mathbb{C}[\mathbf{z}]/\mathcal{I}_{\mathsf{x},h})$, and
(2) $\mathcal{Y}_{\mathsf{x},h} = \overline{\pi^{-1}(Y_{\mathsf{x},h})}$.

A natural question is to identify suitable generators of $\widetilde{\mathcal{I}}_{\mathsf{x},h}$. Here "suitable" means explicit generators that can be used for computation using commutative algebra, such as a Gröbner basis. In general, the ideals $\mathcal{I}_{\mathsf{x},h}$ and $\widetilde{\mathcal{I}}_{\mathsf{x},h}$ may be distinct, demonstrated by Example 3.4. However, when the ideals *are* equal, we can leverage tools from commutative algebra on $\mathbb{C}[\mathbf{z}]$ to study matrix Hessenberg schemes via the generators of $\mathcal{I}_{\mathsf{x},h}$. This is exactly the approach of our arguments below for x a minimal semisimple element.

While the equations defining $\mathcal{I}_{\mathsf{x},h}$ are given by rank conditions, it isn't obvious how to write down equations defining $\widetilde{\mathcal{I}}_{\mathsf{x},h}$. For example, Berget and Fink [**BF18**] study ideals defining certain torus orbits in the set of $r \times n$ matrices over $\mathbb{C}$, and find that even in this case the answer is not straightforward.

EXAMPLE 3.4. Let $n = 5$ and $h = (3, 3, 5, 5, 5)$. Let E_{ij} denote the elementary matrix with 1 in the (i, j)-th entry and all other entries equal to 0. Let $\mathsf{x} = E_{12} + E_{23} + E_{34} + E_{45}$; in this case $Y_{\mathsf{x},h}$ is called a regular nilpotent Hessenberg variety, and $Y_{\mathsf{x},h}$ is known to be irreducible and reduced (see [**ADGH18**, **AFZ20**]).

It is easy to confirm that $\mathcal{I}_{\mathsf{x},h} = \mathcal{I}_{\mathsf{x},h,2}$ (see Lemma 3.8 below). Thus, $\mathcal{I}_{\mathsf{x},h}$ is generated by all 4×4 minors of

$$\begin{bmatrix} | & | & | & | & | \\ \mathsf{x}v_1 & \mathsf{x}v_2 & v_1 & v_2 & v_3 \\ | & | & | & | & | \end{bmatrix} = \begin{bmatrix} z_{21} & z_{22} & z_{11} & z_{12} & z_{13} \\ z_{31} & z_{32} & z_{21} & z_{22} & z_{23} \\ z_{41} & z_{42} & z_{31} & z_{32} & z_{33} \\ z_{51} & z_{52} & z_{41} & z_{42} & z_{43} \\ 0 & 0 & z_{51} & z_{52} & z_{53} \end{bmatrix}.$$

Since $Y_{\mathsf{x},h}$ is irreducible, the matrix Hessenberg scheme $\mathcal{Y}_{\mathsf{x},h}$ is also irreducible. However, Macaulay2 can be used to verify that $\mathcal{I}_{\mathsf{x},h}$ is not prime, but the intersection of two prime ideals. One of these ideals defines an affine subscheme that does not intersect $GL_5(\mathbb{C})$. In particular, $\mathcal{I}_{\mathsf{x},h} \subsetneq \widetilde{\mathcal{I}}_{\mathsf{x},h}$.

REMARK 3.5. Let h be a fixed Hessenberg function. Matrix Hessenberg schemes for h corresponding to conjugate matrices are isomorphic. Indeed, suppose $\mathsf{x}, \mathsf{x}' \in \mathfrak{gl}_n(\mathbb{C})$ such that $\mathsf{x} = g\mathsf{x}'g^{-1}$ for some $g = (g_{ij})_{1\leq i,j\leq n} \in G$. The action of g on the matrix $Z \in M_n(\mathbb{C}[\mathbf{z}])$ by left multiplication induces an isomorphism $\mathbb{C}[\mathbf{z}] \to \mathbb{C}[\mathbf{z}]$ defined by $z_{ij} \mapsto g_{i1}z_{1j} + g_{i2}z_{2j} + \cdots + g_{in}z_{nj}$ for each $1 \leq i, j \leq n$. This isomorphism maps $\mathcal{I}_{\mathsf{x},h}$ onto $\mathcal{I}_{\mathsf{x}',h}$ and thus results in an isomorphism of schemes.

The definition of the ideal $\mathcal{I}_{\mathsf{x},h}$ given in (3.3) is overdetermined by the sum. It is sufficient to restrict to 'corners' of the Hessenberg function h.

DEFINITION 3.6. Let $h : [n] \to [n]$ be a Hessenberg function. We say $i \in [n]$ is a *corner of* h whenever $h(i) > h(i-1)$. Denote the set of all corners by $\mathcal{C}(h)$.

Our convention that $h(0) = 0$ implies 1 is always a corner of h. The corners of h are an indexing set for the columns in the diagram of h containing corners. More specifically, $i \in \mathcal{C}(h)$ implies that $(h(i), i)$ is a corner box in the diagram of h. For each $i \in [n]$, let

$$i^* := \max\{k \in [n] \mid h(k) = h(i)\}.$$

Notice that if i is a corner of h, then $i^* + 1 \in \mathcal{C}(h)$. Furthermore, there is no corner k with $i < k < i^* + 1$ in $\mathcal{C}(h)$.

EXAMPLE 3.7. The Hessenberg function $h = (4, 4, 4, 6, 6, 6)$ has corners 1 and 4 with $1^* = 3$ and $4^* = 6$. The function $h = (3, 3, 4, 5, 6, 6)$ has corners 1, 3, 4, and 5 with $1^* = 2$, $3^* = 3$, $4^* = 4$, and $5^* = 6$. The diagrams for these Hessenberg functions are displayed below, with corner boxes $(h(i), i)$ highlighted.

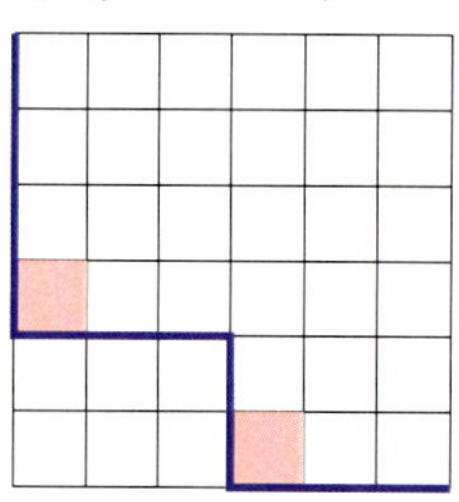

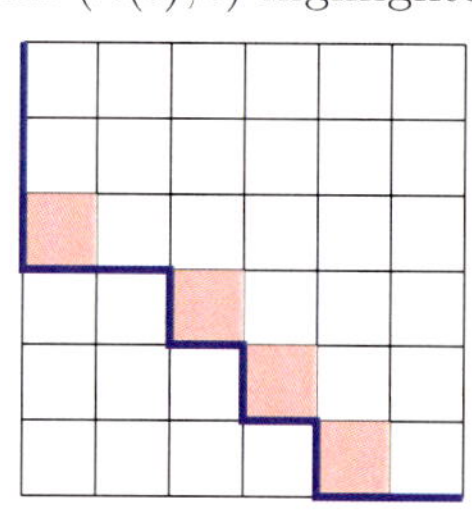

LEMMA 3.8. *For all Hessenberg functions* h *and matrices* $\mathsf{x} \in \mathfrak{gl}_n(\mathbb{C})$,

$$\mathcal{I}_{\mathsf{x},h} = \sum_{i \in \mathcal{C}(h)} \mathcal{I}_{\mathsf{x},h,i^*}.$$

Recall that $\mathcal{I}_{\mathsf{x},h,n} = 0$*, so the corner* $i \in \mathcal{C}(h)$ *such that* $h(i) = n$ *does not contribute to the sum.*

PROOF. The desired formula follows from that fact that if $i, j \in [n]$ such that $i < j$ and $h(i) = h(j)$, then $\mathcal{I}_{\mathsf{x},h,i} \subseteq \mathcal{I}_{\mathsf{x},h,j}$. Indeed, our assumptions imply that the matrix

$$\begin{bmatrix} | & | & & | & | & | & & | \\ \mathsf{x}v_1 & \mathsf{x}v_2 & \cdots & \mathsf{x}v_i & v_1 & v_2 & \cdots & v_{h(i)} \\ | & | & & | & | & | & & | \end{bmatrix} \tag{3.5}$$

can be obtained from

$$\begin{bmatrix} | & | & & | & | & | & & | \\ \mathsf{x}v_1 & \mathsf{x}v_2 & \cdots & \mathsf{x}v_j & v_1 & v_2 & \cdots & v_{h(j)} \\ | & | & & | & | & | & & | \end{bmatrix} \tag{3.6}$$

by deleting the columns $\mathsf{x}v_{i+1}, \ldots, \mathsf{x}v_j$. Furthermore, since $h(i) = h(j)$, the collection of all $(h(i)+1) \times (h(i)+1)$-minors of (3.5) is a subset of all $(h(j)+1) \times (h(j)+1)$-minors of (3.6). Thus $\mathcal{I}_{\mathsf{x},h,i} \subseteq \mathcal{I}_{\mathsf{x},h,j}$ as desired. □

EXAMPLE 3.9. Let $n = 4$ and $h = (2, 4, 4, 4)$. The corner set of h is $\mathcal{C}(h) = \{1, 2\}$, and we have $1^* = 1$ and $2^* = 4$. The diagram for h is displayed below, with the corner boxes $(h(i), i)$ highlighted.

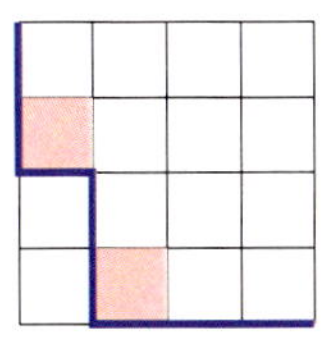

Since $\mathcal{I}_{\mathsf{x},h,4} = 0$, Lemma 3.8 implies $\mathcal{I}_{\mathsf{x},h} = \mathcal{I}_{\mathsf{x},h,1}$, which is generated by the 3×3 determinantal conditions derived from

$$\operatorname{rk} \begin{bmatrix} | & | & | \\ \mathsf{x}v_1 & v_1 & v_2 \\ | & | & | \end{bmatrix} \leq 2. \tag{3.7}$$

EXAMPLE 3.10. We continue Example 3.9, with $h = (2, 4, 4, 4)$ and $\mathsf{x} = \mathsf{s} = \operatorname{diag}(1, 0, 0, 0)$. The rank conditions of (3.7) are

$$\operatorname{rk} \begin{bmatrix} z_{11} & z_{11} & z_{12} \\ 0 & z_{21} & z_{22} \\ 0 & z_{31} & z_{32} \\ 0 & z_{41} & z_{42} \end{bmatrix} \leq 2.$$

The rank conditions are satisfied exactly when all 3×3 minors vanish, so

$$\mathcal{I}_{\mathsf{s},h} = \langle z_{11}(z_{21}z_{32} - z_{22}z_{31}), z_{11}(z_{21}z_{42} - z_{22}z_{41}), z_{11}(z_{31}z_{42} - z_{32}z_{41}) \rangle$$

Macaulay2 confirms that $\mathcal{I}_{\mathsf{s},h}$ is a reduced ideal, with associated primes

$$\mathcal{K}_2 = \langle z_{32}z_{41} - z_{31}z_{42}, z_{22}z_{41} - z_{21}z_{42}, z_{22}z_{31} - z_{21}z_{32} \rangle \text{ and } \mathcal{L}_1 = \langle z_{11} \rangle.$$

We identify these as Schubert determinantal ideals using Lemma 2.1. Thus $\mathcal{Y}_{\mathsf{s},h}$ is a union of the corresponding matrix Schubert varieties:

$$\mathcal{Y}_{\mathsf{s},h} = \mathcal{X}_{w_0[1423]} \cup \mathcal{X}^{\mathrm{op}}_{[2134]} = \mathcal{X}_{[4132]} \cup \mathcal{X}^{\mathrm{op}}_{[2134]}.$$

4. The main theorem

Suppose $\mathsf{x} \in \mathfrak{gl}_n(\mathbb{C})$ is a minimal semisimple element. It is straightforward to show that any such matrix is conjugate to $\mathsf{s}' = \mathrm{diag}(c_1, c_2, \dots, c_2)$ for some $c_1, c_2 \in \mathbb{C}$ such that $c_1 \neq c_2$. The rank inequalities of (3.2) are satisfied for s' if and only if they are satisfied for $\mathsf{s}'' = \mathrm{diag}(c_1 - c_2, 0, \dots, 0)$ which are themselves satisfied if and only if they are satisfied for $\mathsf{s} = \mathrm{diag}(1, 0, \dots, 0)$. Thus $\mathcal{I}_{\mathsf{s}',h} = \mathcal{I}_{\mathsf{s},h}$ for any fixed choice of Hessenberg function h. Using Remark 3.5, we conclude that any minimal semisimple Hessenberg scheme $\mathcal{Y}_{\mathsf{x},h}$ is isomorphic to $\mathcal{Y}_{\mathsf{s},h}$. For this reason, it is sufficient to restrict our attention to the geometry of $\mathcal{Y}_{\mathsf{s},h}$.

THEOREM 4.1. *Let* $\mathsf{s} = \mathrm{diag}(1, 0, \dots, 0)$ *and let* $h : [n] \to [n]$ *be a Hessenberg function. The matrix Hessenberg scheme* $\mathcal{Y}_{\mathsf{s},h}$ *is a reduced union of matrix Richardson schemes:*

$$\mathcal{Y}_{\mathsf{s},h} = \bigcup_{i \in \mathcal{C}(h)} \left(\mathcal{X}^{\mathrm{op}}_{u[i]} \cap \mathcal{X}_{w_0 v[h(i)]} \right).$$

Moreover, the ideal defining $\mathcal{Y}_{\mathsf{s},h}$ *satisfies*

$$\widetilde{\mathcal{I}}_{\mathsf{s},h} = \mathcal{I}_{\mathsf{s},h} = \bigcap_{i \in \mathcal{C}(h)} \left(\mathcal{L}_{i-1} + \mathcal{K}_{h(i)} \right). \tag{4.1}$$

PROOF. The statement about the structure of the scheme follows from the claim about the ideals by Lemma 2.2. Recall the definition of $\mathcal{L}_i$ and $\mathcal{K}_j$ in Equation (2.3). By Remark 3.1, $\mathcal{I}_{\mathsf{s},h,i} = \mathcal{L}_i \cdot \mathcal{K}_{h(i)}$. Then

$$\mathcal{I}_{\mathsf{s},h} = \sum_{i \in \mathcal{C}(h)} \mathcal{I}_{\mathsf{s},h,i} = \sum_{i \in \mathcal{C}(h)} \mathcal{L}_{i^*} \cdot \mathcal{K}_{h(i)} = \sum_{i \in \mathcal{C}(h)} \mathcal{L}_{i^*} \cap \mathcal{K}_{h(i)} \tag{4.2}$$

where the first equality follows from Lemma 3.8 and the last equality follows from Lemma 2.4. We aim to prove the second equality of (4.1) using induction on the size of $\mathcal{C}(h)$.

If h has one corner, then $h = (n, n, \dots, n)$. It follows that $1^* = h(1) = n$, and (4.2) becomes

$$\mathcal{I}_{\mathsf{s},h} = \mathcal{L}_{1^*} \cap \mathcal{K}_{h(1)} = \mathcal{L}_n \cap \mathcal{K}_n = \langle 0 \rangle$$

since $\mathcal{K}_n = \langle 0 \rangle$. On the other hand, $\bigcap_{i \in \mathcal{C}(h)} \left(\mathcal{L}_{i-1} + \mathcal{K}_{h(i)} \right) = \mathcal{L}_0 + \mathcal{K}_n = 0$, since $\mathcal{K}_n = \mathcal{L}_0 = \langle 0 \rangle$. This proves the second equality of (4.1) in this case.

Now suppose the equality holds for any Hessenberg function whose corner set has size k, and suppose $|\mathcal{C}(h)| = k+1$. Observe that $\{i \in \mathcal{C}(h) \mid h(i) < n\}$ is nonempty, since $\mathcal{C}(h)$ has at least two elements. Let $m = \max\{i \in \mathcal{C}(h) \mid h(i) < n\}$. Then $m^* + 1$ is the maximum element of $\mathcal{C}(h)$, and $h(m^* + 1) = n$.

Define $h' : [n] \to [n]$ to be the Hessenberg function such that $h'(i) = h(i)$ for all $i < m$ and $h'(i) = n$ for all $i \geq m$. By construction, $\mathcal{C}(h') = \mathcal{C}(h) \setminus \{m^* + 1\}$, m is the largest corner of h', and $\mathcal{K}_{h(i)} = \mathcal{K}_{h'(i)}$ for all $i < m$. Applying (4.2) and the inductive assumption to h' and the fact that $\mathcal{K}_n = \langle 0 \rangle$, we obtain

$$\begin{aligned}\sum_{i \in \mathcal{C}(h),\, i<m} \mathcal{L}_{i^*} \cap \mathcal{K}_{h(i)} &= \sum_{i \in \mathcal{C}(h')} \mathcal{L}_{i^*} \cap \mathcal{K}_{h'(i)} = \mathcal{I}_{\mathsf{s},h'} = \bigcap_{i \in \mathcal{C}(h')} \left(\mathcal{L}_{i-1} + \mathcal{K}_{h'(i)} \right) \\ &= \left(\bigcap_{i \in \mathcal{C}(h),\, i<m} \left(\mathcal{L}_{i-1} + \mathcal{K}_{h(i)} \right) \right) \cap \mathcal{L}_{m-1}.\end{aligned}$$

Let $\mathcal{J} = \bigcap_{i\in\mathcal{C}(h),\, i<m} \big(\mathcal{L}_{i-1} + \mathcal{K}_{h(i)}\big)$ for notational simplicity. By the previous calculation,

$$\mathcal{I}_{\mathsf{s},h} = \left(\sum_{i\in\mathcal{C}(h), i<m} \mathcal{L}_{i^*} \cap \mathcal{K}_{h(i)}\right) + \mathcal{L}_{m^*} \cap \mathcal{K}_{h(m)} = \mathcal{J} \cap \mathcal{L}_{m-1} + \mathcal{L}_{m^*} \cap \mathcal{K}_{h(m)}$$

where the first equality follows from (4.2) and $\mathcal{K}_{h(m^*+1)} = \mathcal{K}_n = \langle 0\rangle$. Note that $h(i) \leq h(m)$ for all $i < m$, so that $\mathcal{K}_{h(m)} \subseteq \mathcal{L}_{i-1} + \mathcal{K}_{h(i)}$ and thus $\mathcal{K}_{h(m)} \subseteq \mathcal{J}$. Therefore,

$$\begin{aligned}
\mathcal{I}_{\mathsf{s},h} &= (\mathcal{J} \cap \mathcal{L}_{m-1}) + (\mathcal{L}_{m^*} \cap \mathcal{K}_{h(m)}) \\
&= (\mathcal{L}_{m^*} \cap \mathcal{J} \cap \mathcal{L}_{m-1}) + (\mathcal{L}_{m^*} \cap \mathcal{K}_{h(m)}) && \text{since } \mathcal{L}_{m-1} \subseteq \mathcal{L}_{m^*}, \text{ as } m-1<m\leq m^* \\
&= \mathcal{L}_{m^*} \cap \big[(\mathcal{J} \cap \mathcal{L}_{m-1}) + \mathcal{K}_{h(m)}\big] && \text{since } \mathcal{J} \cap \mathcal{L}_{m-1} \subseteq \mathcal{L}_{m^*} \\
&= \mathcal{L}_{m^*} \cap \big[(\mathcal{J} \cap \mathcal{L}_{m-1}) + (\mathcal{J} \cap \mathcal{K}_{h(m)})\big] && \text{since } \mathcal{K}_{h(m)} \subseteq \mathcal{J} \\
&= \mathcal{J} \cap \big(\mathcal{L}_{m-1} + \mathcal{K}_{h(m)}\big) \cap \mathcal{L}_{m^*} \\
&= \bigcap_{i\in\mathcal{C}(h)} \big(\mathcal{L}_{i-1} + \mathcal{K}_{h(i)}\big)
\end{aligned}$$

where the last equality follows from the definition of $\mathcal{J}$ and the fact that $\mathcal{L}_{m^*} + \mathcal{K}_{h(m^*+1)} = \mathcal{L}_{m^*} + \mathcal{K}_n = \mathcal{L}_{m^*}$. This proves the second equality of (4.1).

Each ideal $\mathcal{L}_{i-1}+\mathcal{K}_{h(i)}$ in (4.1) is prime by Lemma 2.2, so $\mathcal{I}_{\mathsf{s},h}$ is an intersection of prime ideals and therefore radical. By Lemma 2.3, we have $\mathcal{L}_{i-1} + \mathcal{K}_{h(i)} \subseteq \mathcal{L}_{j-1} + \mathcal{K}_{h(j)}$ if and only if $i \leq j$ and $h(i) \geq h(j)$. However, for distinct corners $i, j \in \mathcal{C}(h)$, the inequality $i \leq j$ implies $h(i) < h(j)$. It follows that the primes in the set $\big\{\mathcal{L}_{i-1} + \mathcal{K}_{h(i)} \mid i \in \mathcal{C}(h)\big\}$ are distinct and none is contained in any of the others. Consequently, $\big\{\mathcal{L}_{i-1} + \mathcal{K}_{h(i)} \mid i \in \mathcal{C}(h)\big\}$ is the set of associated primes of $\mathcal{I}_{\mathsf{s},h}$.

Recall that $d \in \mathbb{C}[\mathbf{z}]$ denotes the determinant function of the generic $n \times n$ matrix $Z = (z_{ij})_{1\leq i,j\leq n}$. If $d \in \mathcal{L}_{i-1}+\mathcal{K}_{h(i)}$ for some i then $\mathrm{GL}_n(\mathbb{C})\cap\mathrm{Spec}(\mathbb{C}[\mathbf{z}]/(\mathcal{I}_{i-1}+\mathcal{K}_{h(i)})) = \emptyset$, contradicting Lemma 2.2. This proves $d \notin \mathcal{L}_{i-1} + \mathcal{K}_{h(i)}$ for all i, so

$$\widetilde{\mathcal{L}_{i-1} + \mathcal{K}_{h(i)}} = \mathcal{L}_{i-1} + \mathcal{K}_{h(i)}.$$

Since the localization map is injective, it preserves intersections of ideals, and the previous observation implies $\widetilde{\mathcal{I}}_{\mathsf{s},h} = \mathcal{I}_{\mathsf{s},h}$, proving the first equality of (4.1). □

COROLLARY 4.2. *For any Hessenberg function $h : [n] \to [n]$, the minimal semisimple Hessenberg variety $Y_{\mathsf{s},h}$ is isomorphic to a reduced union of Richardson varieties,*

$$Y_{\mathsf{s},h} = \bigcup_{i\in\mathcal{C}(h)} \left(X^{\mathrm{op}}_{u[i]} \cap X_{w_0 v[h(i)]}\right).$$

Theorem 4.1 shows that the structure of the semisimple Hessenberg variety $Y_{\mathsf{s},h}$ can be read directly from the data of the Hessenberg function. We recover the results of Example 3.10; another example appears below.

EXAMPLE 4.3. Let $\mathsf{s} = \mathrm{diag}(1,0,0,0)$. First, we consider $h = (1,2,3,4)$. The corner set of h is $\mathcal{C}(h) = \{1,2,3,4\}$ and the corresponding diagram is the following.

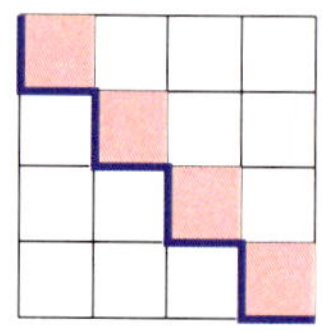

In this case, the variety $Y_{\mathsf{s},(1,2,3,4)} = \pi(GL_4(\mathbb{C}) \cap \mathcal{Y}_{\mathsf{s},(1,2,3,4)})$ is the Grothedieck–Springer fiber over s. We have

$$u[1] = e,\ u[2] = [2134],\ u[3] = [2314],\ u[4] = [2341],$$

and

$$v[1] = [4123],\ v[2] = [1423],\ v[3] = [1243],\ v[4] = e.$$

Thus, by Theorem 4.1, $\mathcal{Y}_{\mathsf{s},(1,2,3,4)}$ is the union of four irreducible components

$$\begin{aligned}\mathcal{Y}_{\mathsf{s},(1,2,3,4)} &= \mathcal{X}_{w_0v[1]} \cup \left(\mathcal{X}^{\mathrm{op}}_{u[2]} \cap \mathcal{X}_{w_0v[2]}\right) \cup \left(\mathcal{X}^{\mathrm{op}}_{u[3]} \cap \mathcal{X}_{w_0v[3]}\right) \cup \mathcal{X}^{\mathrm{op}}_{u[4]} \\ &= \mathcal{X}_{[1432]} \cup \left(\mathcal{X}^{\mathrm{op}}_{[2134]} \cap \mathcal{X}_{[4132]}\right) \cup \left(\mathcal{X}^{\mathrm{op}}_{[2314]} \cap \mathcal{X}_{[4312]}\right) \cup \mathcal{X}^{\mathrm{op}}_{[2341]},\end{aligned}$$

each of which is a matrix Richardson variety.

Next, consider $h = (2,2,4,4)$. The corner set of h is $\mathcal{C}(h) = \{1,3\}$ and the corresponding diagram is the following.

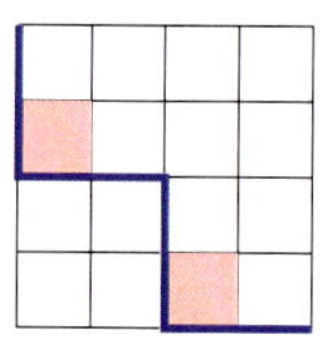

In this case Theorem 4.1 tells us $\mathcal{Y}_{\mathsf{s},(2,2,4,4)}$ is the union of two irreducible components,

$$\mathcal{Y}_{\mathsf{s},(2,2,4,4)} = \mathcal{X}_{w_0v[2]} \cup \mathcal{X}^{\mathrm{op}}_{u[3]} = \mathcal{X}_{[4132]} \cup \mathcal{X}^{\mathrm{op}}_{[2314]}.$$

COROLLARY 4.4. *Let $h : [n] \to [n]$ be a Hessenberg function. The dimension of the minimal semisimple matrix Hessenberg scheme $\mathcal{Y}_{\mathsf{s},h}$ is*

$$\dim \mathcal{Y}_{\mathsf{s},h} = \frac{n(n+1)}{2} + \frac{(n-1)(n-2)}{2} + \max_{i\in\mathcal{C}(h)} \{h(i) - i\}.$$

PROOF. By Proposition 2.1(3), the dimension of each Richardson variety $X^{\mathrm{op}}_{u[i]} \cap X_{w_0v[h(i)]}$ is

$$\begin{aligned}\ell(w_0v[h(i)]) - \ell(u[i]) &= \left(\frac{n(n-1)}{2} - (n-h(i))\right) - (i-1) \\ &= \frac{(n-1)(n-2)}{2} + (h(i)-i).\end{aligned}$$

Since

$$\begin{aligned}\dim \mathcal{X}^{\mathrm{op}}_{u[i]} \cap \mathcal{X}_{w_0v[h(i)]} &= \dim B + \dim\left(X^{\mathrm{op}}_{u[i]} \cap X_{w_0v[h(i)]}\right) \\ &= \frac{n(n+1)}{2} + \frac{(n-1)(n-2)}{2} + (h(i)-i)\end{aligned} \tag{4.3}$$

the result now follows immediately from Theorem 4.1. □

It is now straightforward to determine when the minimal semisimple Hessenberg scheme is equidimensional.

COROLLARY 4.5. *Let $h : [n] \to [n]$ be a Hessenberg function. The minimal semisimple matrix Hessenberg scheme $\mathcal{Y}_{\mathsf{s},h}$ is equidimensional if and only if all of the corners of the associated diagram of h lie on the same subdiagonal. In other words, $\mathcal{Y}_{\mathsf{s},h}$ is equidimensional if and only if $h(i) - i = h(j) - j$ for all distinct corners i and j of h.*

PROOF. Given distinct corners i and j of the Hessenberg function h, the formula (4.3) from the proof of Corollary 4.4 implies that the corresponding irreducible components of $\mathcal{Y}_{\mathsf{s},h}$ have the same dimension if and only if $h(i)-i = h(j)-j$. Since the subdiagonal containing the corner box $(h(i), i)$ is the set of all (a, b) such that $a - b = h(i) - i$, the result is proved. □

Notice that minimal semisimple Hessenberg schemes appearing in Example 4.3 are equidimensional, which is obvious from Corollary 4.5 by looking at the diagrams for the corresponding Hessenberg functions. Similarly, Example 3.10 considers a case that is not equidimensional, which is again obvious in light of Corollary 4.5 since the two corners appear in different subdiagonals of the diagram for h.

Next, we compute the cohomology class of the minimal semisimple Hessenberg variety $Y_{\mathsf{s},h}$. Recall Borel's description of the integral cohomology ring $H^*(GL_n(\mathbb{C})/B, \mathbb{Z})$ as the ring of coinvariants, that is,

$$H^*(GL_n/B, \mathbb{Z}) \simeq \mathbb{Z}[x_1, \dots, x_n]/I,$$

where $I \subset \mathbb{Z}[x_1, \dots, x_n]$ is the ideal generated by the symmetric polynomials without a constant term. The Schubert polynomial $\mathfrak{S}_w \in \mathbb{Z}[x_1, \dots, x_n]$ is a polynomial representative for the class $[X_w^{\mathrm{op}}]$ of the opposite Schubert variety X_w^{op}. It is well-known that the product of Schubert polynomials $\mathfrak{S}_u \cdot \mathfrak{S}_v$ is a polynomial representative for the cohomology class $[X_u^{\mathrm{op}} \cap X_{w_0 v}]$ of the Richardson variety $X_u^{\mathrm{op}} \cap X_{w_0 v}$. For a more detailed definition of Schubert polynomials see [**Man01**].

With this notation in place, Corollary 4.2 allows us to compute a representative for $[Y_{\mathsf{s},h}]$ as a sum of Schubert polynomials.

COROLLARY 4.6. *Suppose $n \geq 3$, and let h be a Hessenberg function and $d_h = \max\{h(i) - i \mid i \in [n]\}$. The cohomology class of the Hessenberg variety $Y_{\mathsf{s},h}$ is*

$$[Y_{\mathsf{s},h}] = \begin{cases} \displaystyle\sum_{i=1}^{n-1} 2\ \mathfrak{S}_{w[i+1,i]} & \text{if } h = (1, 2, \dots, n) \\ \displaystyle\sum_{\substack{i \in \mathcal{C}(h) \\ d_h = h(i)-i}} \mathfrak{S}_{w[i,h(i)]} & \text{if } h \neq (1, 2, \dots, n) \end{cases}$$

where $w[i, j]$ is the shortest permutation w in S_n with $w(i) = 1$ and $w(j) = n$,

PROOF. By Corollary 4.2,

$$Y_{\mathsf{s},h} = \bigcup_{i \in \mathcal{C}(h)} \left(X_{u[i]}^{\mathrm{op}} \cap X_{w_0 v[h(i)]} \right).$$

The largest dimensional terms occur when $h(i) - i = d_h$. The cohomology class is therefore given by

$$[Y_{\mathsf{s},h}] = \sum_{\substack{i \in \mathcal{C}(h) \\ d_h = h(i) - i}} \mathfrak{S}_{u[i]} \cdot \mathfrak{S}_{v[h(i)]}.$$

Consider first the case that $h \neq (1, 2, \ldots, n)$. Then $d_h > 0$ and thus $i < h(i)$ for each i in the sum. Since $i < h(i)$, $u[i]$ and $v[h(i)]$ are in commuting subgroups of the permutation group, implying $\mathfrak{S}_{u[i]}\mathfrak{S}_{v[h(i)]} = \mathfrak{S}_{u[i]v[h(i)]} = \mathfrak{S}_{w[i,h(i)]}$. The second equality follows.

When $h = (1, 2, \ldots, n)$, observe that $\mathcal{C}(h) = [n]$ and $d_h = 0$ is achieved for all $i \in [n]$. Since $u[1] = v[n] = e$, the expression simplifies to

$$[Y_{\mathsf{s},h}] = \sum_{i=1}^{n} \mathfrak{S}_{u[i]} \cdot \mathfrak{S}_{v[i]} = \mathfrak{S}_{v[1]} + \left(\sum_{i=2}^{n-1} \mathfrak{S}_{u[i]} \cdot \mathfrak{S}_{v[i]} \right) + \mathfrak{S}_{u[n]}.$$

The terms in the middle sum can be evaluated using the fact that $u[i] = s_1 s_2 \cdots s_{i-1}$ (here s_i denotes the simple transposition swapping i and $i+1$) and the Pieri rule, which dictates that

$$\mathfrak{S}_{u[i]} \cdot \mathfrak{S}_{v[i]} = \mathfrak{S}_{w[i,i-1]} + \mathfrak{S}_{w[i+1,i]}, \quad \text{for } 1 < i < n.$$

Since $v[1] = w[2,1]$ and $u[n] = w[n, n-1]$, we obtain

$$[Y_{\mathsf{s},h}] = \mathfrak{S}_{w[2,1]} + \left(\sum_{i=2}^{n-1} \left(\mathfrak{S}_{w[i,i-1]} + \mathfrak{S}_{w[i+1,i]} \right) \right) + \mathfrak{S}_{w[n,n-1]} = \sum_{i=1}^{n-1} 2\, \mathfrak{S}_{w[i+1,i]}.$$

This concludes the proof. □

We show in [**GP**] that $\mathcal{Y}_{\mathsf{s},h}$ admits a flat degeneration to the minimal nilpotent matrix Hessenberg scheme $\mathcal{Y}_{\mathsf{n},h}$. Since the underlying variety of $\mathcal{Y}_{\mathsf{n},h}$ is a union of matrix Schubert varieties, the factor of 2 occurring when $h = (1, 2, \ldots, n)$ implies that $\mathcal{Y}_{\mathsf{n},h}$ is not a reduced scheme.

Acknowledgment

We thank Rahul Singh for several helpful conversations.

References

[AC16] Hiraku Abe and Peter Crooks, *Hessenberg varieties for the minimal nilpotent orbit*, Pure Appl. Math. Q. **12** (2016), no. 2, 183–223, DOI 10.4310/PAMQ.2016.v12.n2.a1. MR3767215

[ADGH18] Hiraku Abe, Lauren DeDieu, Federico Galetto, and Megumi Harada, *Geometry of Hessenberg varieties with applications to Newton-Okounkov bodies*, Selecta Math. (N.S.) **24** (2018), no. 3, 2129–2163, DOI 10.1007/s00029-018-0405-3. MR3816501

[AFZ20] Hiraku Abe, Naoki Fujita, and Haozhi Zeng, *Geometry of regular Hessenberg varieties*, Transform. Groups **25** (2020), no. 2, 305–333, DOI 10.1007/s00031-020-09554-8. MR4098882

[BF18] Andrew Berget and Alex Fink, *Matrix orbit closures*, Beitr. Algebra Geom. **59** (2018), no. 3, 397–430, DOI 10.1007/s13366-018-0402-x. MR3844635

[CLO15] David A. Cox, John Little, and Donal O'Shea, *Ideals, varieties, and algorithms: An introduction to computational algebraic geometry and commutative algebra*, 4th ed., Undergraduate Texts in Mathematics, Springer, Cham, 2015, DOI 10.1007/978-3-319-16721-3. MR3330490

[Ful92] William Fulton, *Flags, Schubert polynomials, degeneracy loci, and determinantal formulas*, Duke Math. J. **65** (1992), no. 3, 381–420, DOI 10.1215/S0012-7094-92-06516-1. MR1154177

[GP] Rebecca Goldin and Martha Precup, *Matrix Hessenberg schemes over the minimal sheet*, In preparation.

[HH11] Jürgen Herzog and Takayuki Hibi, *Monomial ideals*, Graduate Texts in Mathematics, vol. 260, Springer-Verlag London, Ltd., London, 2011, DOI 10.1007/978-0-85729-106-6. MR2724673

[ITW20] Erik Insko, Julianna Tymoczko, and Alexander Woo, *A formula for the cohomology and K-class of a regular Hessenberg variety*, J. Pure Appl. Algebra **224** (2020), no. 5, 106230, 14, DOI 10.1016/j.jpaa.2019.106230. MR4046236

[KM05] Allen Knutson and Ezra Miller, *Gröbner geometry of Schubert polynomials*, Ann. of Math. (2) **161** (2005), no. 3, 1245–1318, DOI 10.4007/annals.2005.161.1245. MR2180402

[Man01] Laurent Manivel, *Symmetric functions, Schubert polynomials and degeneracy loci*, SMF/AMS Texts and Monographs, vol. 6, American Mathematical Society, Providence, RI; Société Mathématique de France, Paris, 2001. Translated from the 1998 French original by John R. Swallow; Cours Spécialisés [Specialized Courses], 3. MR1852463

[MS05] Ezra Miller and Bernd Sturmfels, *Combinatorial commutative algebra*, Graduate Texts in Mathematics, vol. 227, Springer-Verlag, New York, 2005. MR2110098

[Tym06] Julianna S. Tymoczko, *Hessenberg varieties are not pure dimensional*, Pure Appl. Math. Q. **2** (2006), no. 3, Special Issue: In honor of Robert D. MacPherson., 779–794, DOI 10.4310/PAMQ.2006.v2.n3.a8. MR2252117

Department of Mathematical Sciences, George Mason University, 4400 University Drive, Fairfax, Virginia 22030

Email address: rgoldin@gmu.edu

Department of Mathematics, Washington University in St. Louis, One Brookings Drive, St. Louis, Missouri 63130

Email address: martha.precup@wustl.edu

Contemporary Mathematics
Volume **837**, 2026
https://doi.org/10.1090/conm/837/16792

On the pure dimensionality of Spaltenstein varieties: A family of counterexamples

Drew Meyer

ABSTRACT. We investigate the pure dimensionality of Spaltenstein varieties coming from complex classical groups. It is known that these objects are not pure dimensional in general. Spaltenstein himself presented a counterexample associated with the partition $(1, 2^2, 3)$. However, there is a recent paper by Y. Li that establishes the pure dimensionality of such Spaltenstein varieties when the associated nilpotent element is of a given type. Assume that the nilpotent element has m_i Jordan blocks of size i. We have pure dimensionality if $m_i m_{i+1} = 0$ for all $i \geq 1$. It is now natural to consider Spaltenstein varieties where a single condition $m_i m_{i+1} = 0$ fails to hold. In this paper, we show the existence of a family of Spaltenstein varieties that is not pure dimensional with $m_1 m_2 \neq 0$.

Introduction

Let G be a complex reductive group and P be a parabolic subgroup of G. We have the associated flag variety G/P. Let x be a nilpotent element of Lie(G) and $\mathfrak{n}$ be the nilpotent radical of Lie(P). We can define the associated Spaltenstein variety as $X_x^P = \{gP \in G/P \,|\, g^{-1}xg \in \mathfrak{n}\}$. Consider the case $G = GL(m, \mathbb{C})$ and $P = B$ is Borel. We can realize G/P as the set of flags in an m-dimensional complex vector space. By the term flag, we mean a chain of subspaces $0 = F_0 \subsetneq F_1 \subsetneq \cdots \subsetneq F_{m-1} \subsetneq F_m = \mathbb{C}^m$ with $\dim F_i = i\,(0 \leq i \leq m)$. In turn, when we fix a nilpotent matrix $N \in \mathfrak{gl}(m, \mathbb{C})$, the Spaltenstein variety is then realized as the subset of flags subject to the condition $N \cdot F_i \subseteq F_{i-1}\,(1 \leq i \leq m)$. This object can be referred to as a Springer fiber (of type A) and is the initial object studied by Spaltenstein. In his paper (see [**4**]) published in 1976, we see his landmark result on the pure dimensionality of this object. Alternatively, we can see an independent proof of this result in the paper (see [**6**]) published by Steinberg in 1976.

Relaxing the condition on P, there is a natural generalization of realizing G/P as the set of partial flags in an m-dimensional complex vector space. By a partial flag, we mean a chain of subspaces $0 = F_0 \subsetneq F_1 \subsetneq \cdots \subsetneq F_{l-1} \subsetneq F_l = \mathbb{C}^m$ with $\dim(F_i) = m_1 + \cdots + m_i\,(0 \leq i \leq l)$, where $(m_1, \ldots, m_l)$ is a composition of m. The Spaltenstein variety (of type A) is analogously realized and turns out to also be pure dimensional. This result can be seen in the paper (see [**3**]) by Nakajima published in 1994 as an application of quiver varieties and symplectic geometry. We can also see a proof of this result via crystals in the paper (see [**7**]) by Malkin published in 2001. Additionally, the reader may reference the paper (see [**1**]) by

Ostrik and Brundan published in 2011 to see a clear and comprehensive proof of this result that is similar in style to Spaltenstein's original proof.

Relaxing the condition on G instead, we have the case of the general Springer fiber. This too is also pure dimensional. The paper (see [**5**]) published by Spaltenstein in 1977 contains a proof of this result. In short, the pure dimensionality of the Spaltenstein variety at the extremities is well known, but for a time little was known beyond the aforementioned results. However, in Y. Li's paper (see [**2**]) published in 2020, substantial work was done in the interior of this problem.

Restrict G to be a complex classical group (type A, B, C, or D). Write the Jordan type of x as the partition $(1^{m_1}, 2^{m_2}, \dots)$. That is, the number of Jordan blocks in x of dimension i is m_i. As a result of Li's work, we know that these Spaltenstein varieties are pure dimensional if $m_i m_{i+1} = 0$ for all $i \geq 1$. Now it is known that these objects are not pure dimensional in general. In particular, there is a Spaltenstein variety (of type D) associated with the partition $(1, 2^2, 3)$ that is not pure dimensional. Refer to [**2**] to see Spaltenstein's well-known counterexample. However, it is now natural to study the vacuum left by Li's result and Spaltenstein's counterexample by considering the objects that fail a single condition $m_i m_{i+1} = 0$. In this paper, we investigate the pure dimensionality of these Spaltenstein varieties by showing the following result:

THEOREM. *Let $n \in \mathbb{Z}$ with $n \geq 3$. There exists a Spaltenstein variety of type B associated to the composition $(1, \dots, 1, 2, 1, 2, 1, \dots, 1)$ and partition $(1^{2n-3}, 2^2)$ that is not pure dimensional.*

Note that the above theorem is a bit sharper of a result than Spaltenstein's counterexample since only the condition $m_1 m_2 = 0$ fails to hold for this family of counterexamples. Moreover, we see that Li's result is inhibited from extending in general to the aforementioned type of Spaltenstein varieties.

Preliminaries

Let V be a complex finite-dimensional vector space and $\beta : V \times V \to \mathbb{C}$ be a nondegenerate symmetric bilinear form. Given the form β, we have the set of invertible linear operators on V that preserve this form. This set is called the orthogonal group and is denoted by $O(m, \mathbb{C})$. Precisely, $A \in O(m, \mathbb{C})$ is an invertible linear operator on V that satisfies $\beta(Av, Aw) = \beta(v, w)$ for all $v, w \in V$. We then have the associated Lie algebra $\mathfrak{o}(m, \mathbb{C})$ of the orthogonal group $O(m, \mathbb{C})$. Specifically, $T \in \mathfrak{o}(m, \mathbb{C})$ is a linear operator on V that satisfies $\beta(Tv, w) = -\beta(v, Tw)$ for all $v, w \in V$. Also, given a subspace W of V, we define the isotropic complement $W^\perp = \{v \in V \mid \beta(v, w) = 0\,(w \in W)\}$. Note that $W^\perp$ is also a subspace of V. Recall that we have $\dim V = \dim W + \dim W^\perp$. We say that W is isotropic if $W \subseteq W^\perp$. That is, for all $w_1, w_2 \in W$, we have $\beta(w_1, w_2) = 0$.

Set $m = \dim V$. Let $\overrightarrow{m} = (m_1, \dots, m_l)$ be a composition of m. Hence $m_i\,(1 \leq i \leq l)$ is a positive integer and $m_1 + \dots + m_l = m$. We consider chains of subspaces $0 = F_0 \subseteq F_1 \subseteq \dots \subseteq F_{l-1} \subseteq F_l = \mathbb{C}^m$ such that $\dim F_i = m_1 + \dots + m_i\ (0 \leq i \leq l)$ and $F_i^\perp = F_{l-i}\,(0 \leq i \leq l)$. In other words, the first "half" is a chain of isotropic subspaces and the second "half" is the chain of their isotropic complements in reverse order. Note that we must have $m_i = m_{l+1-i}\,(1 \leq i \leq l)$ in order for the above definition to be well-defined. Let $X^{\overrightarrow{m}}$ denote the set of all isotropic flags of type $(m_1, \dots, m_l)$. Now there is a natural action of $G = O(m, \mathbb{C})$ on $X^{\overrightarrow{m}}$. Given any $A \in G$, we can send the flag $0 = F_0 \subseteq F_1 \subseteq \cdots \subseteq F_{l-1} \subseteq F_l = \mathbb{C}^m$ to the flag $0 = G_0 \subseteq G_1 \subseteq \cdots \subseteq G_{l-1} \subseteq G_l = \mathbb{C}^m$ with $G_i = A \cdot F_i\,(0 \le i \le l)$. Note that this action is transitive. Fixing a flag in $X^{\vec{m}}$ and letting P be the stabilizer of this flag, we have the identification $G/P \cong X^{\vec{m}}$. It follows that we can think of $X^{\vec{m}}$ as a projective homogeneous G-space. We shall let $\psi : G \to X^{\vec{m}}$ denote the canonical morphism. In the particular case when m is odd, we could restrict the above action to $SO(m, \mathbb{C})$ and it would remain transitive. It then follows that $X^{\vec{m}}$ is also irreducible and smooth. We only consider $X^{\vec{m}}$ with m odd henceforth.

Set $n = \lfloor \frac{m}{2} \rfloor$. Fix an integer k with $1 \le k \le n$. Consider the flags corresponding to the composition $\vec{m} = (k, m-k)$. That is, flags of the form $0 \subseteq W \subseteq W^\perp \subseteq \mathbb{C}^m$ with W an isotropic subspace of dimension k. Note that these flags are determined by just the subspace W. Denote this flag variety by $OGr(k, 2n+1)$. This object is called the orthogonal Grassmanian (at least when $m = 2n+1$ is odd). Since $OGr(k, 2n+1)$ is a particular instance of $X^{\vec{m}}$, it is a smooth irreducible projective homogeneous G-space. In what follows, we treat an element $W \in OGr(k, 2n+1)$ as just an isotropic subspace of dimension k.

Finally, let $N \in \mathfrak{o}(m, \mathbb{C})$ be nilpotent. Define $X_N^{\vec{m}} = \{F \in X^{\vec{m}} \mid N \cdot F_i \subseteq F_{i-1}\,(1 \le i \le l)\}$. Note that $X_N^{\vec{m}}$ is a closed subvariety of $X^{\vec{m}}$. This is the associated Spaltenstein variety.

Description of the counterexample

Fix $n \in \mathbb{Z}$ with $n \ge 3$. Let $V = \mathbb{C}^{2n+1}$ and $e_1, \ldots, e_{2n+1}$ be the standard basis. Define $\beta : V \times V \to \mathbb{C}$ by setting $\beta(e_i, e_j) = \delta_{i,2n+2-j}$. Pick $A \in O(2n+1, \mathbb{C})$ with $A = (a_{i,j})_{i,j=1}^{2n+1}$. The invertible matrix A is characterized by the following conditions:

$$\sum_{k=1}^{2n+1} a_{k,i}\, a_{2n+2-k,2n+2-j} = \delta_{i,j} \text{ for all } 1 \le i, j \le 2n+1.$$

Pick $T \in \mathfrak{o}(2n+1, \mathbb{C})$ with $T = (t_{i,j})_{i,j=1}^{2n+1}$. The matrix T is characterized by the following conditions:

$$t_{2n+2-j,2n+2-i} = -t_{i,j} \text{ for all } 1 \le i, j \le 2n+1.$$

In particular, the entries of the anti-diagonal are 0. We consider the flag variety $X^{\vec{m}}$ corresponding to $\vec{m} = (1, \ldots, 1, 2, 1, 2, 1, \ldots, 1)$. Precisely, $F \in X^{\vec{m}}$ is of the form $F = (F_0, \ldots, F_{2n-1})$ with $F_i^\perp = F_{2n-1-i}\,(0 \le i \le 2n-1)$, $\dim F_i = i$ $(0 \le i \le n-2)$, and $\dim F_{n-1} = n$. Choose P to be the stabilizer of the flag $0 \subseteq \langle e_1 \rangle \subseteq \cdots \subseteq \langle e_1, \ldots, e_{n-2} \rangle \subseteq \langle e_1, \ldots e_n \rangle \subseteq \langle e_1, \ldots, e_n \rangle^\perp \subseteq \cdots \subseteq \langle e_1 \rangle^\perp \subseteq \mathbb{C}^{2n+1}$. We consider N given by the partition $(1^{2n-3}, 2^2)$. In particular, we may choose the following matrix:

$$N = \begin{bmatrix} 0 & 1 & & & & & \\ 0 & 0 & & & & & \\ & & 0 & & & & \\ & & & \ddots & & & \\ & & & & 0 & & \\ & & & & & 0 & -1 \\ & & & & & 0 & 0 \end{bmatrix} \in \mathfrak{o}(2n+1, \mathbb{C})$$

We then have the Spaltenstein variety $X_N^{\vec{m}} = \{F \in X^{\vec{m}} \mid N \cdot F_i \subseteq F_{i-1}\,(1 \leq i \leq 2n-1)\}$. For simplicity, let $X = X^{\vec{m}}$ and $X_N = X_N^{\vec{m}}$ in what follows. We now show that X_N is not pure dimensional.

Proof of theorem

As expected, we show that X_N has two irreducible components that do not have the same dimension. Define $E_1 = \{F \in X_N \mid F_1 \subseteq \langle e_1, e_{2n}\rangle\}$. We start by determining the relevant properties of E_1.

LEMMA 1. *E_1 is an irreducible closed subvariety of X_N of dimension $(n-1)^2$.*

PROOF. It is straightforward to show that E_1 is a closed subvariety of X_N. Pick $A \in O(2n+1, \mathbb{C})$. Denote the i-th column of A by A_i. Now the condition $\langle A_1\rangle \subseteq \langle e_1, e_{2n}\rangle$ is characterized by the vanishing of entry $i\,(i \neq 1, 2n)$ in the first column of A. This is of course a closed condition on matrices in $O(2n+1, \mathbb{C})$ and implies that E_1 is closed. Now we have the morphism $\varphi_1 : X \to OGr(1, 2n+1)$ defined by $F \mapsto F_1$. Set $Y = \{\langle \alpha e_1 + \beta e_{2n}\rangle \mid |\alpha| + |\beta| \neq 0\}$. Note that Y is a closed subset of $OGr(1, 2n+1)$ by the same argument as above. Also, note that $\varphi_1(E_1) = Y$. Therefore we have the restriction $E_1 \to Y$. Now Y is just the projective space $\mathbb{P}^1$. It thus follows that the image of $E_1 \to Y$ is irreducible of dimension 1. Pick any $\langle \alpha e_1 + \beta e_{2n}\rangle \in Y$. We next describe the fiber $\varphi_1^{-1}(\langle \alpha e_1 + \beta e_{2n}\rangle) \cap E_1$.

Let X' be the set of isotropic flags in $\mathbb{C}^{2n-1}$ associated to the composition $(1, \ldots, 1, 2, 1, 2, 1, \ldots, 1)$. We have the identification $O(2n-1, \mathbb{C})/P' \cong X'$ with P' chosen in the same way as P. That is, P' is the stabilizer of the flag $0 \subseteq \langle e_1\rangle \subseteq \cdots \subseteq \langle e_1, \ldots, e_{n-3}\rangle \subseteq \langle e_1, \ldots, e_{n-1}\rangle \subseteq \langle e_1, \ldots, e_{n-1}\rangle^{\perp} \subseteq \cdots \subseteq \langle e_1\rangle^{\perp} \subseteq \mathbb{C}^{2n-1}$. Consider the map $O(2n-1, \mathbb{C}) \to O(2n+1, \mathbb{C}) \to O(2n+1, \mathbb{C})/P$ defined by the following:

$$x \longmapsto \begin{bmatrix} 1 & 0 & 0 \\ 0 & x & 0 \\ 0 & 0 & 1 \end{bmatrix} P$$

It is clear that this map is a morphism of varieties and sends P' into P. We therefore have a map $O(2n-1, \mathbb{C})/P' \to O(2n+1, \mathbb{C})/P$. Note that this map is injective. Now fix any $g \in O(2n+1, \mathbb{C})$ such that $g \cdot \langle e_1\rangle = \langle \alpha e_1 + \beta e_{2n}\rangle$. This matrix induces an isomorphism $O(2n+1, \mathbb{C})/P \to O(2n+1, \mathbb{C})/P$ by left multiplication. We now consider the composition $O(2n-1, \mathbb{C})/P' \to O(2n+1, \mathbb{C})/P \to O(2n+1, \mathbb{C})/P$. The image of this map is exactly $\varphi_1^{-1}(\langle \alpha e_1 + \beta e_{2n}\rangle) \cap E_1$. An important observation is to notice that $F \in X$ with $F_1 \subseteq \langle e_1, e_{2n}\rangle$ is automatically an element of X_N. Indeed, we have $\langle \alpha e_1 + \beta e_{2n}\rangle^{\perp} = \langle e_1, e_3, \ldots, e_{2n}, \alpha e_2 - \beta e_{2n+1}\rangle$. Hence we have $N \cdot \mathbb{C}^{2n+1} \subseteq \langle \alpha e_1 + \beta e_{2n}\rangle^{\perp}$, $N \cdot \langle \alpha e_1 + \beta e_{2n}\rangle^{\perp} \subseteq \langle \alpha e_1 + \beta e_{2n}\rangle$, and $N \cdot \langle \alpha e_1 + \beta e_{2n}\rangle \subseteq 0$. Finally, since $O(2n-1, \mathbb{C})/P'$ is projective, we know that the above map is closed and thus a homeomorphism onto its image. It follows immediately that $\varphi_1^{-1}(\langle \alpha e_1 + \beta e_{2n}\rangle) \cap E_1$ is irreducible. Now we must compute the dimension of this fiber. Consider the subalgebra $\mathrm{Lie}(P')$ of the Lie algebra $\mathfrak{o}(2n-1, \mathbb{C})$. Now the elements of $\mathrm{Lie}(P')$ are the matrices in $\mathfrak{o}(2n-1, \mathbb{C})$ that are block upper triangular with block sizes given by the composition $(1, \ldots, 1, 2, 1, 2, 1, \ldots, 1)$. In particular, these matrices will have the following form above the zero anti-diagonal:

$$\begin{bmatrix}
* & * & \cdots & * & * & * & * & * & * & \cdots & * & * \\
0 & * & \cdots & * & * & * & * & * & * & \cdots & * & \\
\vdots & \ddots & \ddots & \vdots & \vdots & \vdots & \vdots & \vdots & \vdots & \iddots & & \\
0 & \cdots & 0 & * & * & * & * & * & * & & & \\
0 & \cdots & 0 & 0 & * & * & * & * & & & & \\
0 & \cdots & 0 & 0 & * & * & * & & & & & \\
0 & \cdots & 0 & 0 & 0 & 0 & & & & & & \\
0 & \cdots & 0 & 0 & 0 & & & & & & & \\
0 & \cdots & 0 & 0 & & & & & & & & \\
\vdots & \iddots & \iddots & & & & & & & & & \\
0 & 0 & & & & & & & & & & \\
0 & & & & & & & & & & & \\
 & & & & & & & & & & &
\end{bmatrix}$$

Now the number of zeros shown in the above matrix will give the dimension of our fiber. Therefore the dimension of $\varphi_1^{-1}(\langle \alpha e_1 + \beta e_{2n}\rangle) \cap E_1$ is $\frac{(2n-1)^2-2(2n-1)+1}{4} - 1 = (n-1)^2 - 1$.

In summary, Y is irreducible of dimension 1 and the fibers of the map $E_1 \to Y$ are irreducible of dimension $(n-1)^2-1$. It follows that E_1 is irreducible of dimension $1 + (n-1)^2 - 1 = (n-1)^2$. □

Now define $E_{n-1} = \{F \in X_N \mid \langle e_1, e_{2n}\rangle \subseteq F_{n-1}\}$. We next determine the relevant properties of E_{n-1}.

LEMMA 2. *E_{n-1} is an irreducible closed subvariety of X_N of dimension $(n-1)^2 - 1$.*

PROOF. It is straightforward to show that E_{n-1} is a closed subvariety of X_N. Pick $A \in O(2n+1, \mathbb{C})$. Denote the i-th column of A by A_i. Now the condition $\langle e_1, e_{2n}\rangle \subseteq \langle A_1, \dots A_n\rangle$ is equivalent to the condition $\langle A_1, \dots, A_{n+1}\rangle \subseteq \langle e_i\rangle_{i \neq 2, 2n+1}$. The latter of which is characterized by the vanishing of entry 2 and $2n+1$ in the front $n+1$ columns of A. This is of course a closed condition on matrices in $O(2n+1, \mathbb{C})$ and implies that E_{n-1} is closed. Now we have the morphism $\varphi_{n-1} : X \to OGr(n, 2n+1)$ defined by $F \mapsto F_{n-1}$. Set $Z = \{W \in OGr(n, 2n+1) \mid \langle e_1, e_{2n}\rangle \subseteq W\}$. Note that Z is a closed subset of $OGr(n, 2n+1)$ by the same argument as above. Also, note that $\varphi_{n-1}(E_{n-1}) = Z$. Therefore we have the restriction $E_{n-1} \to Z$.

Let us now describe Z. We make the identification

$$O(2n+1, \mathbb{C})/P_1 \cong OGr(n, 2n+1)$$

with P_1 the stabilizer of $\langle e_1, \dots, e_n\rangle$. This implies that P_1 consists of the matrices in $O(2n+1, \mathbb{C})$ that are block upper triangular with block sizes given by the composition $(n, n+1)$. Also, we make the identification $O(2n-3, \mathbb{C})/P_2 \cong OGr(n-2, 2n-3)$ with P_2 the stabilizer of $\langle e_1, \dots, e_{n-2}\rangle$. This implies that P_2 consists of the matrices in $O(2n-3, \mathbb{C})$ that are block upper triangular with block sizes given by the composition $(n-2, n-1)$. Consider the map $O(2n-3, \mathbb{C}) \to O(2n+1, \mathbb{C}) \to O(2n+1, \mathbb{C})/P_1$

defined by the following:

$$x \longmapsto \begin{bmatrix} 1 & 0 & 0 & 0 & 0 \\ 0 & 0 & 0 & 1 & 0 \\ 0 & 0 & x & 0 & 0 \\ 0 & 1 & 0 & 0 & 0 \\ 0 & 0 & 0 & 0 & 1 \end{bmatrix} P_1$$

It is clear that this is a morphism of varieties. Note that it is constant on P_2-cosets. We therefore have a morphism $O(2n-3,\mathbb{C})/P_2 \to O(2n+1,\mathbb{C})/P_1$. Note that this map is injective and has image equal to Z. Moreover, since $O(2n-3,\mathbb{C})/P_2$ is projective, we know that the map is closed and thus a homeomorphism onto its image. It follows that Z is irreducible and has dimension equal to $(n-2)+(n-3)+\cdots+1 = \frac{n^2-3n+2}{2}$. The dimension calculation follows from the fact that $\mathrm{Lie}(P_2)$ is the subalgebra of $\mathfrak{o}(2n-3,\mathbb{C})$ consisting of matrices that above the zero anti-diagonal have the following form:

$$\begin{bmatrix} * & \cdots & * & * & \cdots & * & \\ \vdots & & \vdots & \vdots & \iddots & & \\ * & \cdots & * & * & & & \\ 0 & \cdots & 0 & & & & \\ \vdots & \iddots & & & & & \\ 0 & & & & & & \\ & & & & & & \end{bmatrix}$$

Let us now consider the fibers of $E_{n-1} \to Z$. Define X'' to be the set of partial flags in $\mathbb{C}^n$ associated with the composition $(1,\dots,1,2)$. We have the identification $GL(n,\mathbb{C})/P'' \cong X''$. Note that we can choose P'' to be the stabilizer of the flag $0 \subseteq \langle e_1\rangle \subseteq \cdots \subseteq \langle e_1,\dots,e_{n-2}\rangle \subseteq \mathbb{C}^n$. This implies that P'' consists of the matrices in $GL(n,\mathbb{C})$ that are block upper triangular with block sizes given by the composition $(1,\dots,1,2)$. Given a square matrix $B = (b_{i,j})_{i,j=1}^m$, define $B^\tau = (b_{m+1-j,m+1-i})_{i,j=1}^m$. Note that this is just reflecting the matrix across the anti-diagonal. It is straightforward to check that $(BD)^\tau = D^\tau\, B^\tau$ and $(B^{-1})^\tau = (B^\tau)^{-1}$. Now consider the map $GL(n,\mathbb{C}) \to O(2n+1,\mathbb{C}) \to O(2n+1,\mathbb{C})/P$ defined by the following:

$$x \longmapsto \begin{bmatrix} x & 0 & 0 \\ 0 & 1 & 0 \\ 0 & 0 & (x^{-1})^\tau \end{bmatrix}$$

This map is polynomial in x_{ij}, $\frac{1}{\det(x)}$ and thus is a morphism of varieties. Also, it maps P'' into P. Hence we have a morphism $GL(n,\mathbb{C})/P'' \to O(2n+1,\mathbb{C})/P$. Note that this morphism is injective. Fix $W \in Z$. Pick any $g \in O(2n+1,\mathbb{C})$ with $g\cdot\langle e_1,\dots,e_n\rangle = W$. This induces an isomorphism $O(2n+1,\mathbb{C})/P \to O(2n+1,\mathbb{C})/P$ by left multiplication. We now consider the composition $GL(n,\mathbb{C})/P'' \to O(2n+1,\mathbb{C})/P \to O(2n+1,\mathbb{C})/P$. The image of this morphism is precisely $\varphi_{n-1}^{-1}(W) \cap E_{n-1}$. An important observation is to notice that $F \in X$ with $\langle e_1, e_{2n}\rangle \subseteq F_{n-1}$ is automatically an element of X_N. Indeed, the fact that $\langle e_1, e_{2n}\rangle \subseteq F_{n-1}$ implies that $F_{n-1}^\perp \subseteq \langle e_i\rangle_{i\neq 2,2n+1} = \mathrm{Ker}(N)$. Hence we have $N\cdot\mathbb{C}^{2n+1} \subseteq F_{n-1}^\perp$ and $N\cdot F_{n-1}^\perp \subseteq 0$. Finally, since $GL(n,\mathbb{C})/P''$ is projective, we know that the above map is closed and thus a homeomorphism onto its image. It follows that $\varphi_{n-1}^{-1}(W)\cap E_{n-1}$ is irreducible and has dimension $\frac{n^2-(n+2)}{2}$. The dimension calculation comes from the fact that Lie(P'') is the subalgebra of $\mathfrak{gl}(n,\mathbb{C})$ of matrices of the following form:

$$\begin{bmatrix} * & * & \cdots & * & * & * \\ 0 & * & \cdots & * & * & * \\ \vdots & \ddots & \ddots & \vdots & \vdots & \vdots \\ 0 & \cdots & 0 & * & * & * \\ 0 & \cdots & 0 & 0 & * & * \\ 0 & \cdots & 0 & 0 & * & * \end{bmatrix}$$

In summary, Z is irreducible of dimension $\frac{n^2-3n+2}{2}$ and the fibers of the map $E_{n-1} \to Z$ are irreducible of dimension $\frac{n^2-n-2}{2}$. It follows that E_{n-1} is irreducible of dimension $\frac{n^2-3n+2}{2} + \frac{n^2-n-2}{2} = (n-1)^2 - 1$. □

Lastly, it is not too difficult to show that E_{n-1} is maximal. We can then quickly have the lemma below.

LEMMA 3. *E_{n-1} is an irreducible component of X_N.*

PROOF. Define $D = \{F \in X \mid F_{n-2} \cap \langle e_1, e_{2n}\rangle \neq 0\}$. Consider the subset $\psi^{-1}(D)$ of $O(2n+1,\mathbb{C})$. Given $A \in O(2n+1,\mathbb{C})$, we can show that A in $\psi^{-1}(D)$ is completely characterized by having the determinant of each $n \times n$ minor of the matrix with columns $e_1, e_{2n}, A_1, \dots, A_{n-2}$ equal 0. This implies that D is closed. Now set $D_N = X_N \cap D$. We show that $X_N \setminus D_N \subseteq E_{n-1}$. Assume that we have $F \in X_N \setminus D_N$. The fact that $F_{n-2} \cap \langle e_1, e_{2n}\rangle = 0$ implies that $F_{n-1} \subseteq \langle e_i\rangle_{i\neq 2,2n+1} = \mathrm{Ker}(N)$. It is then clear that $F_{n-1} + \langle e_1\rangle$ is an isotropic subspace. Noting that F_{n-1} is a maximal isotropic subspace, it follows that $e_1 \in F_{n-1}$. Similarly, we have $e_{2n} \in F_{n-1}$. Hence $F \in E_{n-1}$. Now we have the union $X_N = D_N \cup E_{n-1}$. Given an irreducible component C' of X_N, we must have $C' \subseteq D_N$ or $C' \subseteq E_{n-1}$ since both D_N and E_{n-1} are closed. Let C be the irreducible component containing E_{n-1}. Suppose that $C \subseteq D_N$. This quickly implies that $X_N \setminus D_N \subseteq D_N$ and so $X_N \setminus D_N = \emptyset$. However, the isotropic flag $0 \subseteq \langle e_3\rangle \subseteq \langle e_3, e_4\rangle \subseteq \cdots \subseteq \langle e_3, \dots, e_n\rangle \subseteq \langle e_1, e_3, \dots, e_n, e_{2n}\rangle \subseteq \langle e_1, e_3, \dots, e_n, e_{2n}\rangle^\perp \subseteq \cdots \subseteq \langle e_3, e_4\rangle^\perp \subseteq \langle e_3\rangle^\perp \subseteq \mathbb{C}^{2n+1}$ gives a contradiction. It follows that $C \subseteq E_{n-1}$ and thus E_{n-1} is an irreducible component of X_N. □

In conclusion, we know that E_1 is an irreducible closed subvariety of X_N of dimension $(n-1)^2$ and E_{n-1} is an irreducible component of X_N of dimension $(n-1)^2 - 1$. It follows immediately that X_N is not pure dimensional.

References

[1] Jonathan Brundan and Victor Ostrik, *Cohomology of Spaltenstein varieties*, Transform. Groups **16** (2011), no. 3, 619–648, DOI 10.1007/s00031-011-9149-2. MR2827037

[2] Yiqiang Li, *Spaltenstein varieties of pure dimension*, Proc. Amer. Math. Soc. **148** (2020), no. 1, 133–144, DOI 10.1090/proc/14726. MR4042837

[3] Hiraku Nakajima, *Instantons on ALE spaces, quiver varieties, and Kac-Moody algebras*, Duke Math. J. **76** (1994), no. 2, 365–416, DOI 10.1215/S0012-7094-94-07613-8. MR1302318

[4] Nicolas Spaltenstein, *The fixed point set of a unipotent transformation on the flag manifold*, Indag. Math. **38** (1976), no. 5, 452–456. Nederl. Akad. Wetensch. Proc. Ser. A **79**. MR485901

[5] Nicolas Spaltenstein, *On the fixed point set of a unipotent element on the variety of Borel subgroups*, Topology **16** (1977), no. 2, 203–204, DOI 10.1016/0040-9383(77)90022-2. MR447423

[6] Robert Steinberg, *On the desingularization of the unipotent variety*, Invent. Math. **36** (1976), 209–224, DOI 10.1007/BF01390010. MR430094
[7] Anton Malkin, *Tensor product varieties and crystals: GL case*, Trans. Amer. Math. Soc. **354** (2002), no. 2, 675–704, DOI 10.1090/S0002-9947-01-02899-9. MR1862563

Department of Mathematics, University at Buffalo, Buffalo, New York 14260
Email address: drewmeye@buffalo.edu

Contemporary Mathematics
Volume **837**, 2026
https://doi.org/10.1090/conm/837/16793

Connections on the projective line whose differential Galois groups are as large as possible

Daniel S. Sage

ABSTRACT. It is a well-known classical result that a generic nth order polynomial equation with rational coefficients has Galois group as large as possible, namely equal to S_n. However, it is challenging to exhibit explicit examples of equations whose Galois group is S_n. In this paper, we discuss the analogue of this problem for differential Galois groups of linear differential equations with coefficients in the field of complex rational functions. The differential Galois group of an nth order linear differential equation is the symmetry group of its solutions; it is an algebraic subgroup of $\mathrm{GL}_n(\mathbb{C})$. More generally, if G is a complex connected reductive algebraic group, the differential Galois group of a G-connection is an algebraic subgroup of G. For G simple, we generalize earlier work of Katz in type A to give a criterion for the differential Galois group of a meromorphic G-connection to be "large": usually all of G, but certain Dynkin diagrams allow one or two more possibilities depending on the symmetry properties of the connection. We will then apply this result to Coxeter G-connections on $\mathbb{G}_m$, a class of connections that has appeared in recent work on the geometric Langlands program and includes both Airy and Kloosterman equations in type A. In particular, we will exhibit many explicit G-connections with differential Galois group G.

Contents

1. Introduction

It is a basic fact of Galois theory that the Galois group of an nth degree polynomial acts faithfully on the set of roots and is thus a subgroup of S_n, the symmetric

2020 *Mathematics Subject Classification.* Primary 14D24, 20G25, 22E50, 22E67.
Key words and phrases. Differential Galois group, Reductive groups, Frenkel–Gross connection, Kloosterman connections, Airy connections, Coxeter connections.
The author received support from Simons Collaboration Grant 637367.

group on n elements. This leads to a natural question that goes back to the nineteenth century: does there exists a degree n polynomial with rational coefficients whose Galois group is the entire symmetric group? An affirmative, but nonconstructive answer may be given using Hilbert's Irreducibility Theorem. Let $E_1, \dots, E_n$ be the elementary symmetric polynomials in the indeterminates $X_1, \dots, X_n$. It is well-known that the field of rational functions $\mathbb{Q}(X_1, \dots, X_n)$ is a Galois extension of $\mathbb{Q}(E_1, \dots, E_n)$ with Galois group S_n; moreover, this is the Galois group of the polynomial $Y^n - E_1 Y + E_2 Y^2 + \cdots + (-1)^n E_n$. Since the elementary symmetric polynomials are algebraically independent, $\mathbb{Q}(X_1, \dots, X_n) \cong \mathbb{Q}(E_1, \dots, E_n)$, and it follows that there is a degree n polynomial $p(X_1, \dots, X_n, T) \in \mathbb{Q}[X_1, \dots, X_n, T]$ whose Galois group over $\mathbb{Q}(X_1, \dots, X_n)$ is S_n. Hilbert's irreducibility theorem now implies that one can specialize the X_i's to rational numbers a_i for which $p(a_1, \dots, a_n, T) \in \mathbb{Q}[T]$ is irreducible, and the Galois group of this polynomial is S_n.

In fact, Hilbert's Irreducibility Theorem allows one to conclude that almost all rational polynomials have Galois group S_n. More precisely, consider the $(2H+1)^n$ monic degree n polynomials with integer coefficients of absolute value less than the positive integer H. Let $E_n(H)$ be the number of such polynomials whose Galois group is not S_n. Hilbert's Theorem can be used to show that $E_n(H)$ is $o(H^n)$ as $H \to \infty$. In 1936, van der Waerden further conjectured that $E_n(H)$ is $O(H^{n-1})$ [**vdW36**], and this was proved recently by Bhargava [**Bha21**].

Although a generic degree n polynomial has Galois group S_n, it is challenging to write down explicit examples of such polynomials. Perhaps the best known example is the polynomial $x^n - x - 1$. Selmer proved that this polynomial is irreducible for all n in 1956 [**Sel56**], and Nart and Vila showed that the Galois group is S_n almost a quarter century later [**NV79**]. The problem of extending this result to other trinomials, even assuming the trinomial is irreducible, is quite subtle, and there have been several other papers by Osada, Cohen, Movahedi, Salnier among others exploring this question [**Osa87a, Osa87b, CMS97, CMS99**]. For example, the latter three authors have given easily verifiable conditions for an irreducible trinomial to have a "large" Galois group, in the sense that it contains $A_n = [S_n, S_n]$.

In this paper, we will discuss an analogous problem for *differential Galois groups*. The differential Galois group of an order n linear differential equation, say with coefficients in the field of rational functions $\mathbb{C}(z)$, is an algebraic subgroup of $\mathrm{GL}_n(\mathbb{C})$. It is now natural to ask for explicit differential equations whose differential Galois group is the entire general linear group. This question was addressed in a fundamental paper by Katz in 1987 [**Kat87**]. In this paper, Katz considered two important classes of differential equations, *Airy equations* and *Kloosterman equations*, and showed that under mild conditions, their differential Galois groups are "large", in a sense that will be made precise below.

First, we consider Airy differential equations. Let $P_n, Q_m \in \mathbb{C}[z]$ be polynomials of degrees $n \geq 2$ and $m \geq 1$ respectively, and write $P_n(z) = a_n z^n + a_{n-1} z^{n-1} + \cdots + a_0$. Assume further that n and m are relatively prime. The differential equation $(P(\frac{d}{dz}) + Q(z))f = 0$ is called an Airy differential equation of bidegree (n, m). This equation is regular on $\mathbb{A}^1$ and has an irregular singular point at infinity. The standard nth order Airy equation is the special case arising from the operator $\frac{d^n}{dz^n} - z$. If n is odd, then the differential Galois group of any Airy differential equation contains $\mathrm{SL}_n(\mathbb{C})$: it is $\mathrm{SL}_n(\mathbb{C})$ if $a_{n-1} = 0$ and is $\mathrm{GL}_n(\mathbb{C})$ otherwise. The situation is somewhat more complicated when n is even. Here, $a_{n-1} = 0$ implies that the differential Galois group is $\mathrm{SL}_n(\mathbb{C})$ or $\mathrm{Sp}_n(\mathbb{C})$, depending on whether the differential equation satisfies a certain self-duality property. Similarly, if $a_{n-1} \neq 0$, then the differential Galois group is either $\mathrm{GL}_n(\mathbb{C})$ or $\mathrm{GSp}_n(\mathbb{C})$.

Kloosterman differential equations are also defined in terms of two polynomials P_n and Q_m as above, but with the additional restriction that $Q(0) = 0$. One then defines the Kloosterman differential equations of bidegree (n, m) via $(P(z\partial) + Q(z))f = 0$. These equations are regular on $\mathbb{G}_m$ with an irregular singular point at ∞ and a regular singular point at 0. In [**Kat87**], Katz only computes the differential Galois group of Kloosterman differential equations under the hypothesis that all roots of P_n lie in $\mathbb{Z}$. Assuming this, he shows that that the differential Galois group is $\mathrm{SL}_n(\mathbb{C})$ if n is odd and $\mathrm{Sp}_n(\mathbb{C})$ is n is even.

Katz computes the differential Galois groups of these very explicit differential equations using a much more general criterion that is the main theorem of [**Kat87**].

THEOREM 1.1 (Katz). *Suppose ∇ is a rank n algebraic connection with trivial determinant on a smooth connected complex algebraic curve. Assume*

(i) *the Zariski closure of the image of the monodromy map of ∇, considered as a holomorphic connection on the corresponding Riemann surface, is connected; and*
(ii) *the connection ∇ has a singular point where the slope*[1] *is r/n, with $\gcd(r, n) = 1$.*

Then the differential Galois group of ∇ is either $\mathrm{SL}_n(\mathbb{C})$ or, if n is even and ∇ is "self-dual", $\mathrm{Sp}_n(\mathbb{C})$.

To apply this theorem, one writes an order n differential equation as a first-order system $X' + A(z)X = 0$ with $A(z)$ an $n \times n$ matrix whose entries are rational functions. This is equivalent to the rank n trivial vector bundle on $\mathbb{P}^1$ endowed with the meromorphic connection $d + A(z)\, dz$.

The goal of this paper is to describe some recent work of Kamgarpour and the author that generalizes these results of Katz to G-connections, where G is a simple complex algebraic group [**KS23**]. We have proved an analogue of Katz' theorem which gives a criterion for the differential Galois group of a meromorphic G-connection to be "large": usually all of G, but certain Dynkin diagrams allow one or two more possibilities depending on the symmetry properties of the connection. We will then apply this result to *Coxeter G-connections* on $\mathbb{G}_m$, a class of connections that includes both Airy and Kloosterman equations in type A. In particular, we will exhibit many explicit G-connections with differential Galois group G.

We remark that the fact that the differential Galois group of a connection on $\mathbb{P}^1$ is large is important for determining whether the connection is (cohomologically) rigid, i.e., the moduli space of connections with the specified behavior at the singularities is discrete. In particular, one can use the computations of the differential Galois group in this paper to determine which Coxeter connections are rigid. We will not pursue this matter here, but details can be found in [**KS21**, **KLM$^+$22**, **KS23**].

2. Preliminaries

Let G be a simple complex algebraic group with Lie algebra $\mathfrak{g}$, and let $\mathrm{Rep}(G)$ be the category of finite-dimensional complex representations of G. We will fix a

[1] See 2.2 for the definition.

Borel subgroup B with unipotent radical U and a maximal torus $T \subset B$. The corresponding Lie subalgebras will be denoted $\mathfrak{b}$, $\mathfrak{u}$, and $\mathfrak{t}$. Let Φ denote the set of roots and $\Delta \subset \Phi$ the set of simple roots determined by T and B. We let $\mathfrak{g}_\alpha \subset \mathfrak{g}$ denote the root space corresponding to the root α. The character group of T will be denoted by $X^*(T)$.

2.1. ***G*-connections.** Let $\mathcal{E}$ be a principal G-bundle over a smooth projective complex curve X. By Tannakian formalism, $\mathcal{E}$ can be viewed as a tensor functor from $\mathrm{Rep}(G)$ to the category of vector bundles over X via $V \mapsto V_{\mathcal{E}} = \mathcal{E} \times_G V$. A *meromorphic G-connection* on X is a tensor functor $(\mathcal{E}, \nabla)$ from $\mathrm{Rep}(G)$ to the category of vector bundles endowed with a meromorphic connection. We can be much more explicit if $X = \mathbb{P}^1$ and we assume that $\mathcal{E}$ is trivializable. Upon fixing a trivialization ϕ, we can write ∇ explicitly in this trivialization as $\nabla = d + [\nabla]_\phi$, where $[\nabla]_\phi \in \mathfrak{g}(\mathbb{C}(z))\frac{dz}{z}$. The corresponding tensor functor takes the representation (V, ρ) to the flat vector bundle $d + \rho([\nabla]_\phi)$.

Formal G-connections are defined similarly, where X is replaced by the formal punctured disc $\mathrm{Spec}(\mathbb{C}(\!(z)\!))$. A meromorphic G connection ∇ on X induces a formal connection $\widehat{\nabla}_p$ at each $p \in X$. We view this formal connection as determining the local behavior of ∇ at p.

The global connection ∇ has a singular point at p if the connection matrix (defined in terms of some local trivialization around p) has a singular point. Explicitly, this means that the corresponding formal connection $\widehat{\nabla}_p$ has a pole at p. Note that the order of the singularity is not an invariant of a formal connection $\widehat{\nabla}$. Two formal connections are isomorphic if they are gauge-equivalent, i.e., if their connection matrices are related by a formal change of trivialization of the underlying formal G-bundle. Gauge-change by $g \in G(\!(z)\!)$ takes $d + A(z)dz$ to $d + (\mathrm{Ad}(g)(A(z)) - g'g^{-1})dz$, and this clearly does not preserve the order of the singularity of $A(z)$.

We say that a formal connection is *regular singular* if it has a simple pole in some trivialization; if it has a higher order pole in every trivialization, it is called *irregular singular*.

2.2. Slope. The *slope* of a formal G-connection $\widehat{\nabla}$ is a nonnegative rational number that provides one measure of how irregular the connection is. In particular, the slope is positive if and only if $\widehat{\nabla}$ is irregular singular. Here, we give two equivalent ways of defining the slope.

One approach uses the Jordan form for formal G-connections discovered by Babbitt and Varadarajan. This is a generalization of the well-known Levelt–Turrittin normal form [**Tur55**, **Lev75**] for rank n formal connections.

THEOREM 2.1 ([**BV83**]). *Any formal G-connection $\widehat{\nabla}$ is isomorphic to a connection of the form*

$$d + (D_r z^{-r/b} + D_{r-1} z^{-(r-1)/b} + \cdots + D_1 z^{-1} + D_0 + N)\frac{dz}{z}, \tag{2.1}$$

where $D_i \in \mathfrak{t}$, $D_r \neq 0$, $N \in \mathfrak{u}$ commutes with each D_i, and $b \in \mathbb{N}$.

The rational number r/b is an invariant of the connection, namely the slope of $\widehat{\nabla}$.

It is not easy to use this result to compute the slope as it is often difficult to compute the Jordan form. A more computable approach uses the theory of fundamental strata for formal connections due to Bremer and Sage [**BS13, BS18, Sag17**]. Let $\widehat{\nabla} = d + (A_{-s}z^{-s} + A_{1-s}z^{1-s} + \dots)\frac{dz}{z}$ be a formal connection, with $A_i \in \mathfrak{g}$ for all i and $A_{-s} \neq 0$. Neither the integer s nor the naive leading term of the connection $A_{-s}z^{-s}\frac{dz}{z}$ is very useful unless A_{-s} is non-nilpotent.

A better notion of the "leading term" of a connection is provided by the geometric theory of fundamental strata [**BS18**]. Recall that the Bruhat-Tits building $\mathcal{B}$ for $G(\mathbb{C}((z)))$ is a certain simplicial complex reflecting the structure of the loop group. Moy and Prasad have associated to every $x \in \mathcal{B}$ a decreasing $\mathbb{R}$-filtration $\{\mathfrak{g}(\mathbb{C}((z)))_{x,r}\}_{r\in\mathbb{R}}$ of $\mathfrak{g}(\mathbb{C}((z)))$ by $\mathbb{C}[[z]]$-lattices [**MP94**]. This filtration is 1-periodic, in the sense that $\mathfrak{g}(\mathbb{C}((z)))_{x,r+1} = z\,\mathfrak{g}(\mathbb{C}((z)))_{x,r}$. Moreover, if we set $\mathfrak{g}(\mathbb{C}((z)))_{x,r+} = \bigcup_{s>r}\mathfrak{g}(\mathbb{C}((z)))_{x,s}$, then the set of r where $\mathfrak{g}(\mathbb{C}((z)))_{x,r} \neq \mathfrak{g}(\mathbb{C}((z)))_{x,r+}$ is discrete.

We can now consider the leading terms of $\widehat{\nabla} = d + A(z)\frac{dz}{z}$ with respect to these filtrations. A G-stratum is a triple (x, r, β) with $x \in \mathcal{B}$, $r \geq 0$, and $\beta \in \mathfrak{g}(\mathbb{C}((z)))_{x,-r}/\,\mathfrak{g}(\mathbb{C}((z)))_{x,-r+}$. The stratum is called *fundamental* if every coset representative of β is non-nilpotent. Note that this implies that $\mathfrak{g}(\mathbb{C}((z)))_{x,-r} \neq \mathfrak{g}(\mathbb{C}((z)))_{x,-r+}$. We say that $\widehat{\nabla}$ contains this stratum if $A(z) \in \mathfrak{g}(\mathbb{C}((z)))_{x,-r}$ and is a coset representative of β.[2]

It turns out that every formal connection contains a fundamental stratum, and the depth of any such stratum is the slope.

THEOREM 2.2 ([**BS18**, Theorem 2.14]). *Every formal G-connection $\widehat{\nabla}$ contains a fundamental stratum (x, r, β), where x is in the fundamental alcove and $r \in \mathbb{Q}$; the depth r is positive if and only if $\widehat{\nabla}$ is irregular singular. Moreover,*

(1) *If $\widehat{\nabla}$ contains the stratum (y, r', β'), then $r' \geq r$.*
(2) *If $\widehat{\nabla}$ is irregular singular, a stratum (y, r', β') contained in $\widehat{\nabla}$ is fundamental if and only if $r' = r$.*

We conclude this section with one final remark about the slope. It can be shown that the slope of a formal G-connection can be written in the form a/b, where b is a fundamental degree of G and $a \in \mathbb{N}$ [**CK17**]. Since the Coxeter number h is the largest fundamental degree, it is the largest possible denominator of the slope in reduced form.

3. Formal Coxeter connections

Formal G-connections are particularly well-behaved when they contain a stratum (x, r, β) which is *regular* [**BS13, BS20, Sag17, Sag24**]. This means that every coset representative of $\beta/\frac{dz}{z}$ is regular semisimple. In this case, the centralizers of these representatives are conjugate maximal tori in the loop group $G(\mathbb{C}((z)))$, and if S is one such maximal torus, we abuse notation to say that the connection is S-toral.

Not all classes of maximal tori give rise to S-toral connections; in fact, this condition is quite restrictive. Recall that there is a bijection between the conjugacy classes of maximal tori in the loop group $G(\mathbb{C}((z)))$ and conjugacy classes in the Weyl group W [**KL88**]. A Weyl group element is called *regular* if its action on the reflection representation has a regular eigenvector—an eigenvector which does not lie on a root hyperplane. It turns out that there exists an S-toral connection if and only if the class of S corresponds to a regular conjugacy class in W; moreover,

[2]This is the definition for $r > 0$; see [**BS13, BS18**] for the general definition.

there is an S-toral connection of slope $r \geq 0$ precisely when $e^{2\pi i r}$ is the eigenvalue corresponding to a regular eigenvalue of the class.

The most familiar examples of toral connections are those whose naive leading term is regular semisimple. Here, the maximal tori are the split maximal tori, i.e., those conjugate to $T(\mathbb{C}((z)))$, and such toral connections have integral slope. In this paper, we will be interested in what may be viewed as the opposite extreme, the *formal Coxeter connections.*

The Coxeter class in the Weyl group is a regular class. We call the corresponding class of maximal tori the Coxeter maximal tori. We can construct a representative of this class by means of a choice of "affine pinning" of $\mathfrak{g}$. Choose generators $X_\alpha \in \mathfrak{g}_\alpha$ for each negative simple root $\alpha \in -\Delta$ and a generator $E \in \mathfrak{g}_\theta$, where θ is the highest root. Now set $\omega_{-1} = z^{-1}E + \sum_{\alpha \in -\Delta} X_\alpha$. It can be shown that ω_{-1} is regular semisimple [**KS21**], and we denote its connected centralizer by $\mathcal{C}$. This maximal torus is a Coxeter maximal torus, and we will refer to it (resp. its Lie algebra $\mathfrak{c}$) as the Coxeter maximal torus (resp. Coxeter Cartan subalgebra) determined by our fixed affine pinning. The Coxeter torus is *elliptic*, meaning that it has no rational cocharacters.

The Coxeter Cartan subalgebra $\mathfrak{c}$ is endowed with a unique Moy-Prasad filtration. This filtration is inherited from the *standard Iwahori filtration* on $\mathfrak{g}(\mathbb{C}((z)))$, the Moy-Prasad filtration determined by x_I, the barycenter of the fundamental alcove.[3] This standard Iwahori filtration is determined by the following $\frac{1}{h}\mathbb{Z}$-grading, given in terms of the heights of the roots of $\mathfrak{g}$:

(3.1)

$$\mathfrak{g}_{x_I}(m+\tfrac{s}{h}) = z^m \left[\left(\bigoplus_{\mathrm{ht}(\alpha)=s} \mathfrak{g}_\alpha \right) \oplus \left(z \bigoplus_{\mathrm{ht}(\alpha)=s-h} \mathfrak{g}_\alpha \right) \right] \quad \text{for } m \in \mathbb{Z} \text{ and } 0 \leq s < h.$$

The Moy-Prasad filtration on $\mathfrak{c}$ now comes from the grading $\mathfrak{c}(r/h) = \mathfrak{g}_{x_I}(r/h) \cap \mathfrak{c}$. This graded piece contains a regular semisimple element if and only if $(r, h) = 1$, in which case $\mathfrak{c}(r/h)$ is one-dimensional and each nonzero element is regular semisimple. The ellipticity of $\mathcal{C}$ implies that $\mathfrak{c}(0) = \{0\}$.

We can now write down rational canonical forms for all formal connections whose slope has maximal denominator in lowest form. If $r > 0$ is relatively prime to h, then let $\mathcal{A}(\mathcal{C}, r/h)$ be the subset of $\left(\bigoplus_{i=1}^{r} \mathfrak{c}(-i/h)\right)\frac{dz}{z}$ with nonzero component in degree $-r/h$; we call it the set of $\mathcal{C}$-formal types of depth r/h.

THEOREM 3.1 ([**KS23**, Theorem 10]). *Let $r > 0$ be relatively prime to h. Then a formal connection has slope r/h if and only if it is gauge equivalent to an element of $\mathcal{A}(\mathcal{C}, r/h)$. In particular, any connection of slope r/h is $\mathcal{C}$-toral.*

The rational canonical form given in the theorem is unique up to multiplication by an element of μ_h.

It turns out that $\mathcal{A}(\mathcal{C}, 1/h) \cong C^*$, so every formal Coxeter connections of slope $1/h$ is isomorphic to $d + a\omega_{-1}\frac{dz}{z}$ for some $a \in \mathbb{C}^*$. These connections are called *formal Frenkel–Gross connections.* In fact, the formal Frenkel–Gross connections are precisely the formal G-connections with the lowest possible nonzero slope [**KS19**].

---

[3] The terminology is explained as follows. Each cell in the Bruhat-Tits building corresponds to a "parahoric subgroup", with the alcoves corresponding to "Iwahori subgroups", which are the minimal parahoric subgroups. The fundamental alcove corresponds to the "standard Iwahori subgroup" I, so the Moy-Prasad filtration at its barycenter is called the standard Iwahori filtration.

G	K
$A_{2n-1}, n \geq 2$	C_n
B_3	G_2
D_4	B_3, G_2
$D_n, n \geq 5$	B_{n-1}
E_6	F_4

TABLE 1. Proper reductive subgroups $K \subsetneq G$ of maximal degree.

4. Global differential Galois groups

The differential Galois group of a G-connection is an algebraic subgroup of G. It turns out that one can always find a meromorphic G-connection on $\mathbb{P}^1$ whose differential Galois group equals G. In this section, we will state our main theorem on global G-connections whose differential Galois group is as large as possible. In the following section, we will apply this theorem to give explicit examples of such connections.

Our theorem actually shows that if a meromorphic G-connection satisfies certain geometric conditions, then its differential Galois group is a subgroup of G satisfying a certain group-theoretic property. It turns out that this property is very restrictive and leaves 1, 2, or 3 possibilities for the differential Galois group, depending on the root system of G. We will then show how all of these possibilities may be realized explicitly.

4.1. Reductive subgroups of maximal degree. Let $K \subset G$ be a (closed, connected) reductive subgroup. We say that K has *maximal degree* if one of its fundamental degrees equals h. Such subgroups of simple complex groups were classified in [**KS23**]. Up to conjugacy, there are at most 3 such distinct subgroups, depending on the root system of G.

Recall that a pinning of G (for fixed B and T) consists of a choice of generator for each simple root space. A pinned automorphism of G (or $\mathfrak{g}$) is an automorphism which preserves the pinning; this means that it stabilizes B and T while permuting the fixed generators. The set Σ of pinned automorphisms with respect to a fixed pinning is isomorphic to the automorphism group of the Dynkin diagram of G.

We can now state the classification theorem for reductive subgroups of maximal degree.

THEOREM 4.1 ([**KS23**, Theorem 3]). *Let $K \subseteq G$ be a reductive subgroup of maximal degree. Then*

(1) *K is simple.*
(2) *If $L \subseteq G$ is another reductive subgroup of maximal degree which is isomorphic to K (as algebraic groups), then L is G-conjugate to K except for the case where G has type D_4 and K has type B_3.*
(3) *In all types, except $G_2 \hookrightarrow B_3$, there exists $\sigma \in \Sigma$ such that $K = G^{\sigma,\circ}$. For the embedding $G_2 \hookrightarrow B_3$, we have $K = H^{\sigma,\circ}$, where H has type D_4 and σ is an automorphism of D_4.*
(4) *The possibilities for the pairs (G, K) with $K \neq G$ are given in Table 1.*

It is possible to choose the embeddings above so that they induce embedding of maximal tori and Cartan subalgebras. We will always assume below that this is the case, i.e., $\mathfrak{t}_K = \mathfrak{t} \cap \mathrm{Lie}(K)$ and $\mathfrak{t}_{G_2} \subset \mathfrak{t}_{B_3} \subset \mathfrak{t}_{D_4}$.

4.2. The main theorems. In [**FG09**], Frenkel and Gross introduced a meromorphic G-connection on $\mathbb{P}^1$ which was a generalization of a particular Kloosterman connection in type A. This connection has two singular points: an irregular singular point at ∞ with slope $1/h$ and a regular singular point with principal unipotent monodromy at 0. Among other things, they showed that the differential Galois group of this connection is usually, but not always G. Their proof depended heavily on the behavior of the connection at both singular points. In particular, the fact that the regular singular point has principal unipotent monodromy was crucial for their argument.

It turns out that the details of the singular point at 0 are not necessary to compute the differential Galois group. All one needs is the fact that the connection has a formal Coxeter singularity and that the geometric monodromy group is connected. (For the Frenkel–Gross connection, the latter condition follows from the unipotent monodromy at 0. This is explained in Section 5.) We thus have the following generalization of Katz' Theorem to G-connections.

THEOREM 4.2 ([**KS23**, Theorem 5]). *Let X be a smooth connected projective complex algebraic curve and S a finite subset of X. Let ∇ be an algebraic G-connection on $X - S$. Suppose*

(i) *the slope of ∇ at some $s \in S$ equals r/h where $\gcd(r, h) = 1$; and*
(ii) *the geometric monodromy group of ∇ is connected.*

Then ∇ is irreducible and the global differential Galois group $G_\nabla \subseteq G$ is a (connected) reductive subgroup of maximal degree.

In order to determine G_∇ exactly when G is one of the types in Table 1, one needs to examine the effect of nontrivial pinned automorphisms on ∇.

THEOREM 4.3. *Let G and ∇ satisfy the assumptions of 4.2.*

(i) *If G is of a type that does not appear in Table 1, then $G_\nabla = G$.*
(ii) *Suppose G is of type A_{2n-1} for $n \geq 2$, D_n for $n \geq 5$, or E_6. If there exists a nontrivial pinned automorphism σ such that $\sigma(\nabla)$ is gauge-equivalent to ∇, then $G_\nabla = G^{\sigma,\circ}$. Otherwise, $G_\nabla = G$. The same statement holds in type B_3 if we allow σ to be an order 3 pinned automorphism of $D_4 \supseteq B_3$.*
(iii) *Suppose G is of type D_4. If ∇ is gauge-equivalent to $\sigma(\nabla)$ for some order 3 pinned automorphism σ, then $G_\nabla = G^{\sigma,\circ} = G_2$. If not, but there is an order 2 pinned automorphism τ such that $\tau(\nabla)$ is gauge-equivalent to ∇, then $G_\nabla = G^{\tau,\circ} = B_3$. If neither of these cases holds, then $G_\nabla = G$.*

We refer the reader to [**KS23**] for the proofs.

5. Explicit connections on $\mathbb{G}_m$ with large differential Galois groups

We now return to our original problem of writing down explicit G-connections whose differential Galois group is G. We will do so by first exhibiting a class of connections on $\mathbb{G}_m$ called *framable Coxeter connections* and showing that such connections usually have differential Galois groups that are reductive of maximal degree. We will then consider two subclasses, generalizing Kloosterman and Airy connections respectively, and study their differential Galois groups.

5.1. Framable Coxeter connections. In order to apply the main theorem, we need our global G-connection to have at least one irregular singular point, which must be a formal Coxeter singularity. From now on, we will restrict ourselves to connections on $\mathbb{P}^1$ with exactly one irregular singular point of slope r/h with $(r,h) = 1$ and possibly one additional regular singular point. Without loss of generality, we will consider connections on $\mathbb{G}_m$ with the formal Coxeter singularity at ∞. Such connections are called *Coxeter connections*; they have appeared in recent work on geometric Langlands, cf. [**FG09, HNY13, KS21, Zhu17, KLM$^+$22, LSN23, JKY23**].

To be as explicit as possible, we will actually work with connections in which the singular point at ∞ is manifestly a formal Coxeter singularity. Consider the "opposite Iwahori grading", which is the $\frac{1}{h}\mathbb{Z}$-grading on $\mathfrak{g}[z, z^{-1}]$ defined by

$$\mathfrak{g}_{x_{\bar{I}}}(m + i/h) = \Big(\bigoplus_{\mathrm{ht}(\alpha)=h-i} z^{m+1}\,\mathfrak{g}_\alpha \Big) \oplus \Big(\bigoplus_{\mathrm{ht}(\alpha)=-i} z^m\,\mathfrak{g}_\alpha \Big). \tag{5.1}$$

Let ∇ be a G-connection on a trivial G-bundle over $\mathbb{G}_m$; it is given by $\nabla = d + A\frac{dz}{z}$, where $A \in \mathfrak{g}(\mathbb{C}[z, z^{-1}])$. We say that ∇ is a *framable*[4] *Coxeter connection* if

$$A \in \mathfrak{b} \oplus \bigoplus_{i=1}^{r} \mathfrak{g}_{x_{\bar{I}}}(i/h) \subset \mathfrak{g}(\mathbb{C}[z]), \tag{5.2}$$

and the projection of A onto $\mathfrak{g}_{x_{\bar{I}}}(r/h)$ is regular semisimple. It follows from the theory of fundamental and regular strata that ∇ has an irregular singular point of slope r/h at ∞ and possibly a regular singular point at 0.[5]

EXAMPLE 5.1. Let

$$\bar{\omega}_3 = \begin{pmatrix} 0 & 0 & z \\ 1 & 0 & 0 \\ 0 & 1 & 0 \end{pmatrix} \in \mathfrak{sl}_3(\mathbb{C}(\!(z)\!)).$$

If r is a positive integer not divisible by 3, then the SL_3-connection $\nabla_r = d + \bar{\omega}_3^r$ is a "homogeneous" Coxeter SL_3-connection with slope $r/3$ at ∞. If $r = 1$ (resp. $r = 4$), one obtains the Frenkel–Gross (resp. the standard Airy) connection for SL_3. One can write down explicit examples of Coxeter SL_n-connections in a similar way.

In order to apply Theorem 4.2 to framable Coxeter connections, we first need to determine when their geometric monodromy groups are connected. The global monodromy of any connection ∇ on $\mathbb{G}_m$ is a homomorphism $\rho_\nabla : \mathbb{Z} \to G$, which is the same as the local monodromy map around 0, and the geometric monodromy group is the Zariski closure of its image.

Before continuing, we recall the criterion for the connectedness of the subgroup $\overline{\langle \exp(2\pi i Y)\rangle}$ for $Y \in \mathfrak{g}$. Let Y_s be the semisimple part of Y, and choose any maximal torus T' whose Lie algebra $\mathfrak{t}'$ contains Y_s. Let $X^*(Y_s) = \{d\lambda(Y_s) \mid \lambda \in X^*(T')\} \subset \mathbb{C}$; this set is independent of the choice of T'. The algebraic group $\overline{\langle \exp(2\pi i Y)\rangle}$ is

[4] Framable connections are a generalization of Fuchsian connections for which irregular singularities are allowed. For details, see [**Sag24**].

[5] The analogous construction with the standard Iwahori grading would reverse the roles of 0 and ∞. We have used the opposite Iwahori grading so that our definition of the Frenkel–Gross connection agrees with the one given in [**FG09**].

connected if and only if $X^*(Y_s)\cap(\mathbb{Q}\setminus\mathbb{Z})=\varnothing$. (For a proof, see [**KS23**, Proposition 5].)

Let $\nabla = d + A\frac{dz}{z}$ be a framable Coxeter connection. The residue of ∇ is $A(0)$, and one might expect that $\rho_\nabla(1) = \exp(2\pi i A(0))$. This is almost always true—indeed it holds as long as $A(0)$ is *nonresonant*, meaning that $\mathrm{ad}(A(0))$ has no nonzero integral eigenvalues. However, it is always the case that the semisimple parts of $\rho_\nabla(1)$ and $A(0)$ coincide. Combining Theorem 4.2 with the above connectedness criterion, we now obtain the following theorem.

THEOREM 5.2. *Let $\nabla = d + A\frac{dz}{z}$ be a framable Coxeter G-connection. If $X^*(A(0)_s)\cap(\mathbb{Q}\setminus\mathbb{Z})=\varnothing$, then G_∇ is a reductive subgroup of maximal degree. In particular, this holds if ∇ has unipotent monodromy.*

The last statement holds because if ∇ has unipotent monodromy, then one can take $A(0)$ to be nilpotent, so that $X^*(A(0)_s) = X^*(0) = \{0\}$. More directly, one can simply use the fact that the algebraic group generated by a unipotent element is connected.

In order to compute the differential Galois group precisely in the ambiguous cases, one needs to investigate the behavior of ∇ under pinned automorphisms as in Theorem 4.3. In the next two sections, we will apply Theorem 4.3 to two important types of Coxeter connections: generalized Frenkel–Gross connections and Airy connections. This will allow us to exhibit explicit G-connections whose differential Galois group is any reductive subgroup of maximal degree.

5.2. Generalized Frenkel–Gross connections. Generalized Frenkel–Gross connections are the Coxeter connections with the smallest possible slope at the irregular singular point, namely $1/h$. These are the G-connection analogues of Kloosterman connections.

Generalized Frenkel–Gross connections are *cohomologically rigid*, meaning that they are isolated points in the moduli space of connections with the given formal types at 0 and ∞ [**KS23**, Proposition 44]. More recently, it has been shown that they are *physically rigid*, i.e. uniquely determined by these formal types [**HJ24**, **Yi24**].

It is a consequence of physical rigidity that generalized Frenkel–Gross connection are framable. In fact, it is shown in [**KS23**] that every framable generalized Frenkel–Gross connection is equivalent to a connection of the form

$$\nabla = d + a(N + X + Ez)\frac{dz}{z}, \qquad a\in\mathbb{C}^*, \quad X\in\mathfrak{t}. \tag{5.3}$$

The residue at 0 of this connection is $N+X$, so in the notation of the previous section, $A(0) = N+X$. Moreover, the semisimple part of $N+X$ is G-conjugate to X. It follows that the condition on $A(0)_s$ in Theorem 5.2 can be replaced by the analogous condition on X.

We now state our main result on differential Galois groups of generalized Frenkel–Gross connections. It shows explicitly how to realize all possible reductive subgroups of maximal degree as differential Galois groups of such connections by appropriate choices of X. Moreover, when G is simply connected, it gives the differential Galois group of every generalized Frenkel–Gross connection satisfying very mild conditions on X.

Let $\widetilde{W} = W \ltimes X_*(T)$ denote the extended affine Weyl group of G.

THEOREM 5.3. Let $\nabla = d + a(N + X + Ez)\frac{dz}{z}$ be a generalized Frenkel–Gross connection. Suppose $X^*(X) \cap \mathbb{Q}/\mathbb{Z} = \varnothing$. Then $G_\nabla \subseteq G$ is a reductive subgroup of maximal degree. More specifically:

(i) If G is of a type that does not appear in Table 1, then $G_\nabla = G$.

(ii) Suppose G is of type A_{2n-1} for $n \geq 2$, D_n for $n \geq 5$, or E_6. Let σ be a nontrivial pinned automorphism of G. If X and $\sigma(X)$ are not $\widetilde{W}$-conjugate, then $G_\nabla = G$; otherwise, $G_\nabla = G^{\sigma,\circ}$. (In particular, if $X \in \mathfrak{t}^\sigma$, then $G_\nabla = G^{\sigma,\circ}$.) The same statement holds in type B_3 if we allow σ to be an order 3 automorphism of $D_4 \supseteq B_3$.

(iii) Suppose G is of type D_4. If X and $\sigma(X)$ are $\widetilde{W}$-conjugate for some order 3 pinned automorphism σ, then $G_\nabla = G^{\sigma,\circ} = G_2$. If not, but there is an order 2 pinned automorphism τ such that X and $\tau(X)$ are $\widetilde{W}$-conjugate, then $G_\nabla = G^{\tau,\circ} = B_3$. If neither of these cases holds, then $G_\nabla = G$. (In particular, if $X \in \mathfrak{t}^\sigma$, then $G_\nabla = G^{\sigma,\circ} = G_2$. If $X \in \mathfrak{t}^\tau$, and X and $\sigma(X)$ are not $\widetilde{W}$-conjugate for any order 3 pinned automorphism, then $G_\nabla = G^{\tau,\circ} = B_3$.)

REMARK 5.4. If $X = 0$, we obtain the original Frenkel–Gross connection. The theorem applies trivially in this case, and we recover the computation of Frenkel and Gross of the differential Galois group of this connection.

In the classical case, we can give even more explicit criteria for generalized Frenkel–Gross G-connections with differential Galois group G. The criteria will involve the eigenvalues of $X \in \mathfrak{t}$ viewed as a matrix in the standard representation. In type D_n (resp. B_n), we will choose the orthogonal form so that $\mathfrak{t}$ consists of diagonal matrices of the form $\operatorname{diag}(\lambda_1, \dots, \lambda_n, -\lambda_n, \dots, -\lambda_1)$ (resp. $\operatorname{diag}(\lambda_1, \dots, \lambda_n, 0, -\lambda_n, \dots, -\lambda_1)$. For convenience, we will restrict to the simply connected case, so the extended Weyl group is the affine Weyl group W_{aff}. Recall that the closure of the open fundamental alcove $\mathcal{A} \subset \mathfrak{t}_\mathbb{R}$ is a fundamental domain for the affine Weyl group.

THEOREM 5.5. Let G be a simple, simply-connected classical group for which there exists a proper reductive subgroup of maximal degree. Let $\nabla = d + a(N + X + Ez)\frac{dz}{z}$ be a generalized Frenkel–Gross connection. Assume that $X \in \mathcal{A} + i\mathfrak{t}_\mathbb{R}$ and that no $\mathbb{Z}$-linear combination of the eigenvalues of X lies in $\mathbb{Q} \setminus \mathbb{Z}$. In each of the following cases, if the eigenvalues of X satisfy the given condition, then $G_\nabla = G$.

(1) *Type A_{2n-1}*: There exists an eigenvalue λ for which the multiplicities of λ and $-\lambda$ are different;

(2) *Type D_n, $n \geq 4$*: The kernel of X is trivial, i.e., 0 is not an eigenvalue of X;

(3) *Type B_3*: There are no triples of eigenvalues of X which sum to 0.

PROOF. The statement about $\mathbb{Z}$-linear combinations of the eigenvalues is the matrix version of the condition $X^*(X) \cap \mathbb{Q}/\mathbb{Z} = \varnothing$. We can thus apply the previous result. Moreover, two elements of $\mathcal{A} + i\mathfrak{t}_\mathbb{R}$ are W_{aff}-conjugate if and only if they are equal. Since pinned automorphisms in these types preserve the set $\mathcal{A} + i\mathfrak{t}_\mathbb{R}$, the conditions in Theorem 5.3 reduce to checking that X is not fixed by any nontrivial pinned automorphisms.

First, take $\mathfrak{g} = \mathfrak{sl}_{2n}(\mathbb{C})$. If $X = \operatorname{diag}(\lambda_1, \dots, \lambda_{2n})$, then $\sigma(X) = \operatorname{diag}(-\lambda_{2n}, \dots, -\lambda_1)$. It follows that these two matrices in $\mathcal{A} + i\mathfrak{t}_\mathbb{R}$ coincide if and only if the eigenvalues appear in $\pm$-pairs. Thus, $\sigma(X) \neq X$ precisely when there exists an eigenvalue λ whose multiplicity differs from $-\lambda$.

Next, consider $\mathfrak{g} = \mathfrak{so}_{2n}(\mathbb{C})$. One can choose σ so that it operates by switching the sign of λ_n. If the λ_i are all nonzero, then $\sigma(X) \neq X$.

Finally, the order three automorphism σ of $\mathfrak{so}_8(\mathbb{C})$ can be chosen so that the eigenvalues of $\sigma(X)$ are of the form $1/2 \sum_{i=1}^4 \pm\lambda_i$. If $X \in \mathfrak{so}_7(\mathbb{C}) \subset \mathfrak{so}_8(\mathbb{C})$, then $\lambda_4 = 0$. If $\sigma(X) = X$, then one of the eigenvalues of $\sigma(X)$ must be 0, so we must have $\sum_{i=1}^3 \pm\lambda_i = 0$ for some choice of signs. Since the eigenvalues of X appear in $\pm$ pairs, this means that there is a triple of eigenvalues of X which sum to 0, contradicting the assumption. $\square$

A similar argument to the B_3 case in the theorem shows how to find explicit generalized Frenkel–Gross connections in type D_4 with differential Galois group of type B_3. Take $X \in (\mathcal{A} + i\mathfrak{t}_{\mathbb{R}}) \cap \mathfrak{t}_{\mathfrak{so}_7(\mathbb{C})} \subset \mathfrak{t}_{\mathfrak{so}_8(\mathbb{C})}$ (so $\lambda_4 = 0$ in the notation above) with the additional mild genericity assumption from the statement of the theorem. If there are no triples of eigenvalues which sum to 0, then the differential Galois group of the corresponding generalized Frenkel–Gross connection has type B_3.

5.3. Airy connections. The order n Airy differential equations considered by Katz have an irregular singular point at ∞ of slope $1 + 1/n$. It is thus natural to define Airy G-connections to be connections on $\mathbb{A}^1$ whose sole singular point has slope $1 + 1/h$. Airy connections are thus a particular class of Coxeter connections. They were first introduced in [**KS21**].

As in the case of generalized Frenkel–Gross connections, Airy connections are physically rigid [**HJ24**], and this can be used to show that they are framable. In fact, Yi has shown that any Airy connection is equivalent to a connection of the form

$$\nabla = d + a(N + X + zE)dz, \tag{5.4}$$

where $a \in \mathbb{C}^*$, and $X \in \mathfrak{b}$ [**Yi25**].

Since the monodromy of (5.4) is trivial, we see that the differential Galois group of an Airy connection is reductive of maximal degree.

We conclude the paper by showing how to compute the differential Galois group exactly for certain Airy connections. Recall from Theorem 3.1 that the formal connection at ∞ of (5.4) has a formal type in $\mathcal{A}(\mathcal{C}, \frac{h+1}{h})$. This formal type is unique if we take the leading term to be $a(N + zE)$; we denote this by $\mathrm{ft}(\nabla_\infty)$. The formal type can be computed explicitly by an algorithm given in [**BS13, BS20**].

THEOREM 5.6. *Let ∇ be the Airy connection $\nabla = d + a(N + X + zE)dz$. Then $G_\nabla \subseteq G$ is a reductive subgroup of maximal degree. More specifically:*

(i) *If G is of a type that does not appear in Table 1, then $G_\nabla = G$.*
(ii) *Suppose G is of type A_{2n-1} for $n \geq 2$, D_n for $n \geq 5$, or E_6. Let σ be the nontrivial pinned automorphism of G (with respect to the pinning underlying our fixed affine pinning). If $\mathrm{ft}(\sigma(\nabla_\infty)) \neq \mathrm{ft}(\nabla_\infty)$, then $G_\nabla = G$. If $\sigma(X) = X$, then $G_\nabla = G^{\sigma,\circ}$. The same statement holds in type B_3 if we allow σ to be a order 3 automorphism of $D_4 \supseteq B_3$.*
(iii) *Suppose G is of type D_4. If $\mathrm{ft}(\sigma(\nabla_\infty)) \neq \mathrm{ft}(\nabla_\infty)$ for all nontrivial pinned automorphisms, then $G_\nabla = G$. If $\sigma(X) = X$ for a pinned automorphism of order 3, then G_∇ has type G_2. If $\tau(X) = X$ for a pinned involution τ,*

but $\mathrm{ft}(\sigma(\nabla_\infty)) \neq \mathrm{ft}(\nabla_\infty)$ for a pinned automorphism of order 3, then G_∇ has type B_3.

As in the Frenkel–Gross case, one can realize all possible reductive subgroups of maximal degree as differential Galois groups of Airy connections. We refer the reader to [**KS23**] for the details.

References

[Bha21] M. Bhargava, *Galois groups of random integer polynomials and van der Waerden's conjecture*, Ann. of Math. (2) **201** (2025), no. 2, 339–377, DOI 10.4007/annals.2025.201.2.1. MR4878222

[BS13] C. L. Bremer and D. S. Sage, *Moduli spaces of irregular singular connections*, Int. Math. Res. Not. IMRN **8** (2013), 1800–1872, DOI 10.1093/imrn/rns103. MR3047490

[BS18] C. Bremer and D. S. Sage, *A theory of minimal K-types for flat G-bundles*, Int. Math. Res. Not. IMRN (2018), 3507–3555.

[BS20] C. Bremer and D. S. Sage, *Flat G-bundles and regular strata for reductive groups*, arXiv:1309.6060, 2020.

[BV83] D. G. Babbitt and V. S. Varadarajan, *Formal reduction theory of meromorphic differential equations: a group theoretic view*, Pacific J. Math. **109** (1983), no. 1, 1–80. MR716289

[CK17] T.-H. Chen and M. Kamgarpour, *Preservation of depth in the local geometric Langlands correspondence*, Trans. Amer. Math. Soc. **369** (2017), no. 2, 1345–1364, DOI 10.1090/tran/6794. MR3572276

[CMS97] S. D. Cohen, A. Movahhedi, and A. Salinier, *Double transitivity of Galois groups of trinomials*, Acta Arith. **82** (1997), no. 1, 1–15, DOI 10.4064/aa-82-1-1-15. MR1475762

[CMS99] S. D. Cohen, A. Movahhedi, and A. Salinier, *Galois groups of trinomials*, J. Algebra **222** (1999), no. 2, 561–573, DOI 10.1006/jabr.1999.8033. MR1734229

[FG09] E. Frenkel and B. Gross, *A rigid irregular connection on the projective line*, Ann. of Math. (2) **170** (2009), no. 3, 1469–1512, DOI 10.4007/annals.2009.170.1469. MR2600880

[HNY13] J. Heinloth, B.-C. Ngô, and Z. Yun, *Kloosterman sheaves for reductive groups*, Ann. of Math. (2) **177** (2013), no. 1, 241–310, DOI 10.4007/annals.2013.177.1.5. MR2999041

[HJ24] A. Hohl and K. Jakob, *Stokes phenomenon of Kloosterman and Airy connections*, arXiv:2404.09582, 2024.

[JKY23] K. Jakob, M. Kamgarpour, and L. Yi, *Airy sheaves for reductive groups*, Proc. Lond. Math. Soc. (3) **126** (2023), no. 1, 390–428, DOI 10.1112/plms.12494. MR4535023

[Kat87] N. M. Katz, *On the calculation of some differential Galois groups*, Invent. Math. **87** (1987), no. 1, 13–61, DOI 10.1007/BF01389152. MR862711

[KL88] D. Kazhdan and G. Lusztig, *Fixed point varieties on affine flag manifolds*, Israel J. Math. **62** (1988), no. 2, 129–168, DOI 10.1007/BF02787119. MR947819

[KLM+22] M. C. Kulkarni, N. Livesay, J. P. Matherne, B. Nguyen, and D. S. Sage, *The Deligne-Simpson problem for connections on $\mathbb{G}_m$ with a maximally ramified singularity*. part B, Adv. Math. **408** (2022), no. part B, Paper No. 108596, 28, DOI 10.1016/j.aim.2022.108596. MR4458913

[KS19] M. Kamgarpour and D. S. Sage, *A geometric analogue of a conjecture of Gross and Reeder*, Amer. J. Math. **141** (2019), no. 5, 1457–1476, DOI 10.1353/ajm.2019.0038. MR4011806

[KS21] M. Kamgarpour and D. S. Sage, *Rigid connections on $\mathbb{P}^1$ via the Bruhat-Tits building*, Proc. Lond. Math. Soc. (3) **122** (2021), no. 3, 359–376, DOI 10.1112/plms.12346. MR4230058

[KS23] M. Kamgarpour and D. S. Sage, *Connections whose differential Galois groups are reductive of maximal degree*, arXiv:2309.11742, 2023.

[Lev75] A. H. M. Levelt, *Jordan decomposition for a class of singular differential operators*, Ark. Mat. **13** (1975), 1–27, DOI 10.1007/BF02386195. MR500294

[LSN23] N. Livesay, D. S. Sage, and B. Nguyen, *Explicit constructions of connections on the projective line with a maximally ramified irregular singularity*, arXiv:2303.06581, 2023.

[MP94] A. Moy and G. Prasad, *Unrefined minimal K-types for p-adic groups*, Invent. Math. **116** (1994), no. 1-3, 393–408, DOI 10.1007/BF01231566. MR1253198

[NV79] E. Nart and N. Vila, *Equations of the type $x^n + aX + b$ with absolute Galois group S_n*, Proceedings of the sixth conference of Portuguese and Spanish mathematicians, Part II (Santander, 1979), Rev. Univ. Santander **2** (1979), 821–825. MR754788

[Osa87a] H. Osada, *The Galois groups of the polynomials $X^n + aX^l + b$*, J. Number Theory **25** (1987), no. 2, 230–238, DOI 10.1016/0022-314X(87)90029-1. MR873881

[Osa87b] H. Osada, *The Galois groups of the polynomials $x^n + ax^s + b$. II*, Tohoku Math. J. (2) **39** (1987), no. 3, 437–445, DOI 10.2748/tmj/1178228289. MR902581

[Sag17] D. S. Sage, *Regular strata and moduli spaces of irregular singular connections*, New trends in analysis and interdisciplinary applications, Trends Math. Res. Perspect., Birkhäuser/Springer, Cham, 2017, pp. 69–75, DOI 10.1007/978-3-319-48812-7_10. MR3695627

[Sag24] D. S. Sage, *Meromorphic connections on the projective line with specified local behavior*, A glimpse into geometric representation theory, Contemp. Math., vol. 804, Amer. Math. Soc., [Providence], RI, [2024] ©2024, pp. 171–203, DOI 10.1090/conm/804/16114. MR4780755

[Sel56] E. S. Selmer, *On the irreducibility of certain trinomials*, Math. Scand. **4** (1956), 287–302, DOI 10.7146/math.scand.a-10478. MR85223

[Tur55] H. L. Turrittin, *Convergent solutions of ordinary linear homogeneous differential equations in the neighborhood of an irregular singular point*, Acta Math. **93** (1955), 27–66, DOI 10.1007/BF02392519. MR68689

[vdW36] B. L. van der Waerden, *Die Seltenheit der reduziblen Gleichungen und der Gleichungen mit Affekt* (German), Monatsh. Math. Phys. **43** (1936), no. 1, 133–147, DOI 10.1007/BF01707594. MR1550517

[Yi24] L. Yi, *On the physical rigidity of Frenkel-Gross connection*, Selecta Math. (N.S.) **30** (2024), no. 3, Paper No. 41, 23, DOI 10.1007/s00029-024-00931-9. MR4727771

[Yi25] L. Yi, *An explicit local geometric Langlands for supercuspidal representations: the toral case*, `arXiv:2506.13179`, 2025.

[Zhu17] X. Zhu, *Frenkel-Gross' irregular connection and Heinloth-Ngô-Yun's are the same*, Selecta Math. (N.S.) **23** (2017), no. 1, 245–274, DOI 10.1007/s00029-016-0238-x. MR3595893

Department of Mathematics, University at Buffalo, Buffalo, New York 14260.
Email address: `dsage@buffalo.edu`

Contemporary Mathematics
Volume **837**, 2026
https://doi.org/10.1090/conm/837/16794

Nakajima's quiver varieties and triangular bases of bipartite cluster algebras

Li Li

ABSTRACT. Berenstein and Zelevinsky introduced quantum cluster algebras [Adv. Math. 195 (2005), no. 2, 405–455] and the triangular bases [Int. Math. Res. Not. 6 (2014), 1651–1688]. The support conjecture proposed in [Proc. Natl. Acad. Sci. USA 111 (2014), no. 27, 9712–9716], which asserts that the support of each triangular basis element for a rank-2 cluster algebra is bounded by an explicitly described region, was established in [J. Algebra 634 (2023), 97–164] for skew-symmetric rank-2 cluster algebras. In this paper we extend this result by proving a bound on the support of each triangular basis element for bipartite cluster algebras.

CONTENTS

1. Introduction

Cluster algebras and quantum cluster algebras were introduced by Fomin-Zelevinsky [**4**] and Berenstein-Zelevinsky [**2**], respectively. A primary objective behind the inception of quantum cluster algebras is to understand good bases arising from the representation theory of certain non-associative algebras. Mimicking the construction of the dual canonical basis from a PBW basis [**10**], Berenstein and Zelevinsky [**3**] constructed the triangular bases for acyclic quantum cluster algebras. These triangular bases exhibit many desirable properties. In particular, it is worth mentioning the work by Qin [**13–15**], where he has constructed triangular bases for injective-reachable cluster algebras using a different approach, and showed that his triangular bases coincide with the bases by Berenstein and Zelevinsky for the seeds associated with acyclic quivers.

2020 *Mathematics Subject Classification.* Primary 13F60; Secondary 14F06, 16G20, 32S60.

In [**8**], Lee, Rupel, Zelevinsky and the author proposed a conjecture of a precisely described region to constrain the support of each triangular basis element of any rank-2 quantum cluster algebra. The proof for the skew-symmetric rank-2 case was established in [**9**] using Nakajima's quiver varieties. In this work, we prove a similar result for any bipartite cluster algebra.

The definition of quantum cluster algebra and (Berenstein-Zelevinsky's) triangular basis will be reviewed in §2. Our focus here is on bipartite quantum cluster algebras, that is, those come from bipartite quiver. Additionally, for the sake of clarity and simplicity, we confine our study to quantum cluster algebras with principal quantization.

Consider a rank-n bipartite cluster algebra with principal quantization determined by an $n \times n$ skew-symmetric matrix $B = [b_{ij}]$. Let the initial seed be $(\Lambda, \tilde{B}, \tilde{X})$, where

$$\tilde{B} = \begin{bmatrix} B \\ \mathbf{I}_n \end{bmatrix}, \quad \Lambda = \begin{bmatrix} 0 & -\mathbf{I}_n \\ \mathbf{I}_n & -B \end{bmatrix}, \quad \tilde{X} = (X_1, \dots, X_{2n})$$

where $\mathbf{I}_n$ the $n \times n$ identity matrix. In this setup, the first n variables of $\tilde{X}$ are called cluster variables, while the last n variables are called frozen variables.

For $c \in \mathbb{R}$, we denote $[c]_+ = \max(c, 0)$; the vector b_k is the k-th column of $\tilde{B}$; for a vector b, we use $[b]_+$ to denote the vector obtained from b by applying $[\,]_+$ to each coordinate.

The triangular basis $\{C_a | a \in \mathbb{Z}^{2n}\}$ is introduced in [**3**]. Define a map $\varphi : \mathbb{Z}^{2n} \to \mathbb{Z}^{2n}$,

$$\varphi(a) := a + \sum_{k=1}^{n} [-a_k]_+ [-b_k]_+. \tag{1.1}$$

It is easy to see that φ is bijective and keeps the last n coordinates unchanged. Then C_a are of the following form:

$$C_a = \sum_{v \in \mathbb{Z}^n_{\geq 0}} e_v X^{\varphi(a) + \tilde{B}v}, \quad e_v \in \mathbb{Z}[\mathbf{v}^{\pm}]. \tag{1.2}$$

We define $\deg(e_v)$ to be the largest exponent of $\mathbf{v}$ in e_v. Define a quiver $\mathcal{Q}^{\mathrm{op}} = (I, \Omega)$ with vertex set $I = \{1, \dots, n\} = I_0 \sqcup I_1$ (where vertices in I_0 are sinks, vertices in I_1 are sources) and arrow set Ω determined by B such that there are b_{ij} arrows $h : j \to i$ whenever $b_{ij} > 0$. We denote the source and target of h by $s(h)$ and $t(h)$, respectively. See §2 for details.

THEOREM 1.1. *Let $a = (a_1, \dots, a_{2n}) \in \mathbb{Z}^{2n}$, $(g_1, \dots, g_{2n}) = \varphi(a)$. Each e_v in C_a as in* (1.2) *is symmetric and unimodal, does not depend on the last n coordinates of a, and satisfies*

$$\deg(e_v) \leq f(v) := -\sum_{i=1}^{n} (g_i + v_i) v_i + \sum_{h \in \Omega} v_{s(h)} v_{t(h)}.$$

In particular, $e_v \neq 0$ only if $f(v) \geq 0$.

Together with (3.3), the support

$$\mathrm{Supp}(C_a) := \{v \in \mathbb{Z}^n_{\geq 0} \mid e_v \neq 0\}$$

satisfies the following condition:

$$f(v) \geq 0, \quad 0 \leq v_\alpha \leq [-g_\alpha]_+ \ (\forall \alpha \in I_0),$$

and

$$0 \leq v_\beta \leq [-g_\beta]_+ + \sum_{h:\ s(h)=\beta} v_{t(h)} \ (\forall \beta \in I_1).$$

Note that not every integer point satisfying the above condition are in $\mathrm{Supp}(C_a)$. In an ongoing project, the author is looking for a precise bound of $\mathrm{Supp}(C_a)$ to full generalize the results in [**9**].

The proof of the theorem follows the idea presented in [**9**]. While a significant portion of the methodology used in [**9**] can be extended to the bipartite case, many adjustments are necessary, and the proofs and computations become notably more intricate. Certain results are already established in the work of Nakajima [**11**] and Qin [**13**]; however, our approach endeavors to provide a self-contained treatment, with minimal reliance on representation theory and predominantly using elementary algebraic geometry, in alignment with the approach adopted in [**9**].

The paper is organized as follows. In §2 we revisit the definition and properties of quantum cluster algebra and triangular basis. In §3 we review some facts on Nakajima's quiver varieties. In §4 we prove an algebraic version of the transversal slice theorem, and show that the local systems appeared in the BBDG decomposition theorem are trivial. The BBDG Decomposition theorem [**1**] serves as the primary advanced tool used in this paper. It asserts that, for a proper map of algebraic varieties $f : X \to Y$, the pushforward of the intersection cohomology complex IC_X can be decomposed into a direct sum of $IC_V(L)[d]$ for a collection of irreducible subvarieties $V \subseteq Y$, local systems L on (an open dense subset of) V, and integers d. While the local systems are often trivial, as is the case in our scenario, the proof is not always straightforward. Here, we present a proof using the Algebraic Transversal Slice Theorem. In §5 we identify the dual canonical basis and triangular basis in our specific case. In §6 we prove the main theorem and provide some examples.

2. Quantum cluster algebras and triangular bases

In this section we recall the definition and facts of quantum cluster algebras with principal quantization. See [**2, 3**] for details.

2.1. Quantum cluster algebras. Let n be a positive integer and $m = 2n$. Let B be a skew-symmetrizable integer $n \times n$ matrix, that is, there exists a diagonal matrix D such that DB is a skew-symmetric matrix. Denote the $n \times n$ identity matrix by $\mathbf{I}_n$, or simply by $\mathbf{I}$ if n is clear from the context. Let $\tilde{B} := \begin{bmatrix} B \\ \mathbf{I}_n \end{bmatrix}$ where $\mathbf{I}_n$ the $n \times n$ identity matrix. Let $\Lambda := \begin{bmatrix} 0 & -D \\ D & -DB \end{bmatrix}$, viewed as a bilinear form by defining $\Lambda(\mathbf{u}, \mathbf{u}') = \mathbf{u}^T \Lambda \mathbf{u}'$ for column vectors $\mathbf{u}, \mathbf{u}' \in \mathbb{Z}^m$. Let the quantum torus $\mathcal{T}$ be the $\mathbb{Z}[\mathbf{v}^{\pm 1}]$-algebra with basis $\{X^{\mathbf{e}} : \mathbf{e} \in \mathbb{Z}^m\}$, and the multiplication given by $X^{\mathbf{e}} X^{\mathbf{e}'} = \mathbf{v}^{\Lambda(\mathbf{e},\mathbf{e}')} X^{\mathbf{e}+\mathbf{e}'}$ for $\mathbf{e}, \mathbf{e}' \in \mathbb{Z}^m$. Let $\mathcal{F}$ be the skew-field of fractions of $\mathcal{T}$. Let $e_1, \ldots, e_m \in \mathbb{Z}^m$ be the standard basis, that is, e_i has 1 in the i-th coordinate and 0 elsewhere. Let $\tilde{X} = \{X_1, \ldots, X_m\}$ where $X_i = X^{e_i}$ are called (initial) cluster variables if $1 \leq i \leq n$, or frozen variables if $n + 1 \leq i \leq m$. The *bar-involution* is the $\mathbb{Z}$-linear anti-automorphism of $\mathcal{T}$ determined by $\overline{\mathbf{v}^i X^{\mathbf{e}}} = \mathbf{v}^{-i} X^{\mathbf{e}}$ for $\mathbf{e} \in \mathbb{Z}^m$.

Note that $\overline{XY} = \overline{Y}\,\overline{X}$ for $X, Y \in \mathcal{T}$. An element in $\mathcal{T}$ which is invariant under the bar-involution is said to be *bar-invariant.*

Fix an n-regular tree $\mathbb{T}_n$ with root t_0 and label each edge by a number in $\{1, \ldots, n\}$ such that any two edges sharing an endpoint are labelled differently. Attach to t_0 the tuple $(\Lambda, \tilde{B}, \tilde{X})$ as above, called the initial seed. Given an edge joining t and t' with label k, we denote $t' = \mu_k(t)$ (and $t = \mu_k(t')$); if the seed $(\Lambda, \tilde{B}, \tilde{X})$ at t is constructed, we construct a seed $(\Lambda', \tilde{B}', \tilde{X}')$ attached to t' by mutation, where $\tilde{B}' = [b'_{ij}]$ is obtained from $\tilde{B} = [b_{ij}]$ by

$$b'_{ij} = \begin{cases} -b_{ij}, \text{ if } i = k \text{ or } j = k; \\ b_{ij} + [b_{ik}]_+[b_{kj}]_+ - [-b_{ik}]_+[-b_{kj}]_+, \text{ otherwise.} \end{cases}$$

and $\tilde{X}'$ is obtained from $\tilde{X}$ by replacing X_k with $X'_k = X^{-e_k+[b_k]_+} + X^{-e_k+[-b_k]_+}$. The bilinear form Λ' is determined by

$$\begin{cases} \Lambda'(e_i, e_j) = \Lambda(e_i, e_j), \text{ for } i, j \neq k; \\ \Lambda'(e_k, e_j) = \Lambda(-e_k + [b_k]_+, e_j), \text{ for } j \neq k. \end{cases}$$

Denote $\mathbb{Z}P[\mathbf{v}^{\pm}] := \mathbb{Z}[\mathbf{v}^{\pm}][X^{\pm}_{n+1}, \ldots, X^{\pm}_m]$. The *quantum cluster algebra* $\mathcal{A}$ is the $\mathbb{Z}P[\mathbf{v}^{\pm 1}]$-subalgebra of $\mathcal{F}$ generated by all cluster variables.

Note: In the rest of the paper we assume B is skew-symmetric and take $D = I_n$.

2.2. Triangular bases. Assume that the seed $(\Lambda, \tilde{B}, \tilde{X})$ is acyclic, that is, there exists a linear order $\lhd$ on $\{1, \ldots, n\}$ such that $b_{ij} \leq 0$ if $i \lhd j$. Define $r(a) = \sum_{k=1}^{n}[-a_k]_+$ for $a \in \mathbb{Z}^m$ and define a partial order $\prec$ on $\mathbb{Z}^m$ by $a' \prec a \Leftrightarrow r(a') < r(a)$.

The construction of the triangular basis starts with the standard monomial basis $\{E_a : a \in \mathbb{Z}^m\}$. Let $X'_k = \mu_k(X_k)$. For every $a = (a_1, \ldots, a_m)$, the *standard monomial* E_a is defined as

$$E_a = \mathbf{v}^{v(a)} X^{\sum_{i=n+1}^{m} a_i e_i + \sum_{j=1}^{n} [a_j]_+ e_j} \prod_{1 \leq k \leq n}^{\lhd} (X'_k)^{[-a_k]_+}. \tag{2.1}$$

where $\prod^{\lhd}_{1 \leq k \leq n}$ means taking the product in increasing order with respect to $\lhd$, and $v(a) \in \mathbb{Z}$ is determined by the condition that the leading term of E_a, which is the term obtained by replacing X'_k by $X^{-e_k-[b_k]_+}$ in (2.1), is bar-invariant. (In [**3**], $X^{-e_k+[b_k]_+}$ is used, but as they remarked, it can be replaced by $X^{-e_k-[b_k]_+}$ and the value of $v(a)$ remains the same. We use the latter, because then a gives the g-vector of E_a. This is also the choice in [**13**].)

It is known that $\{E_a\}$ is a $\mathbb{Z}[\mathbf{v}^{\pm 1}]$-basis of the cluster algebra $\mathcal{A}$. However, these E_a are not bar-invariant in general and do not contain all the cluster monomials, and they are inherently dependent on the choice of an initial cluster. These drawbacks provide a motivation to consider the triangular basis (which is introduced in [**3**] and recalled below) constructed from the standard monomial basis with a built-in bar-invariance property.

Definition 2.1 ([**3**]). The triangular basis $\{C_a : a \in \mathbb{Z}^{2n}\}$ is the unique collection of elements in $\mathcal{A}(\tilde{X}, \tilde{B})$ satisfying:

(P1) Each C_a is bar-invariant;

(P2) For each a, $C_a - E_a \in \bigoplus_{a'} \mathbf{v}\mathbb{Z}[\mathbf{v}]E_{a'}$.

Moreover, those a' satisfy $a' \prec a$.

It is shown in [**3**] that the triangular basis exists uniquely, does not depend on the choice of an initial (acyclic) seed, and contains all cluster monomials associated to acyclic seeds.

3. Nakajima's graded quiver varieties

In this section we recall Nakajima's graded quiver varieties. The references for this section are [**11, 13**]. There is some inconsistency in these references. In this paper, we will start with a quiver $\mathcal{Q}^{\mathrm{op}}$ (pardon the notation), since it is used essentially in the paper, for Nakajima's graded quiver varieties. Its opposite quiver $\mathcal{Q}$ is to match the exchange matrix B for quantum cluster algebras in the usual convention.

Let $\mathcal{Q}^{\mathrm{op}}$ be a bipartite quiver with vertex set $I = \{1, \dots, n\} = I_0 \sqcup I_1$ such that vertices in I_0 (resp. I_1) are sinks (resp. sources), and let $\Omega = \{h_1, \dots, h_r\}$ be the set of arrows in $\mathcal{Q}^{\mathrm{op}}$, and denote $\alpha_j := t(h_j)$, $\beta_j := s(h_j)$, that is, h_j is an arrow $\beta_j \to \alpha_j$. Let $\widetilde{\mathcal{Q}^{\mathrm{op}}}$ be the decorated quiver obtained from $\mathcal{Q}^{\mathrm{op}}$ by adding a new vertex i' and an arrow $i' \to i$ if $i \in I_0$ (resp. an arrow $i \to i'$ if $i \in I_1$).

Following the notation in [**11**, §4], denote dimension vectors $w = (w_i, w_i')_{i \in I}$, $v = (v_i)_{i \in I}$, where all the components are nonnegative integers.

Fix $\mathbb{C}$-vector spaces W_i, W_i', V_i with dimensions w_i, w_i', v_i for $i \in I$, respectively. Define

$$W := \bigoplus_{i \in I}(W_i \oplus W_i'), \quad V := \bigoplus_{i \in I} V_i.$$

A quiver representation of $\widetilde{\mathcal{Q}^{\mathrm{op}}}$ consists linear maps $x_i : W_i' \to W_i$ $(i \in I_0)$, $x_i : W_i \to W_i'$ $(i \in I_1)$, and $y_h : W_{s(h)} \to W_{t(h)}$ $(h \in \Omega)$. Below is an example with $I_0 = \{3\}$ and $I_1 = \{1, 2\}$:

(3.1)

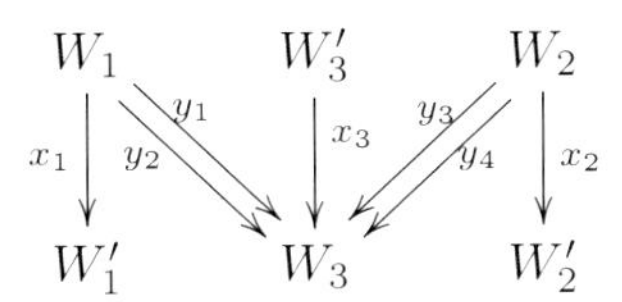

DEFINITION 3.1. For each sink $\alpha \in I_0$, let $r_1 < \dots < r_s$ be the indices such that $h_{r_1}, \dots, h_{r_s}$ are all the arrows with target α, and define a linear map

$$A_\alpha := x_\alpha + \sum_{h:\ t(h)=\alpha} y_h : W_\alpha' \oplus \bigoplus_{h:\ t(h)=\alpha} W_{s(h)} \to W_\alpha,$$
$$(a, b_1, \dots, b_s) \mapsto x_\alpha(a) + y_{r_1}(b_1) + \dots + y_{r_s}(b_r)$$

For each source $\beta \in I_1$, let $r_1' < \dots < r_t'$ be the indices such that $h_{r_1'}, \dots, h_{r_t'}$ are all the arrows with source β, and define a linear map

$$A_\beta := x_\beta \oplus \bigoplus_{h:\ s(h)=\beta} y_h : W_\beta \to W_\beta' \oplus \bigoplus_{h:\ s(h)=\beta} W_{t(h)},$$
$$b \mapsto (x_\beta(b), y_{r_1'}(b), \dots, y_{r_t'}(b))$$

Note that $s, r_1, \dots, r_s$ depend on α so, strictly speaking, we should write them as $s(\alpha)$, $r_1(\alpha)$, etc., but we omit α for simplicity of notation; similarly, $t, r'_1, \dots, r'_t$ depend on β.

Alternatively, we can express these maps by matrices. Fix bases for W_i and W'_i, and by abuse of notation we use x_i (resp. y_h) to denote the matrix representing the corresponding linear map. Then the above A_α and A_β are represented by the following block matrices:

$$A_\alpha = [x_\alpha, y_{r_1}, \dots, y_{r_s}]_{\mathrm{hor}} := [x_\alpha | y_{r_1} | y_{r_2} | \cdots | y_{r_s}],$$

$$A_\beta = [x_\beta, y_{r'_1}, y_{r'_2} \dots, y_{(r'_t)}]_{\mathrm{vert}} := \begin{bmatrix} x_\beta \\ y_{r'_1} \\ y_{r'_2} \\ \vdots \\ y_{r'_t} \end{bmatrix}. \tag{3.2}$$

Below is the list of sizes of various matrices, where $\alpha \in I_0$, $\beta = I_1$:

x_α	x_β	y_h	A_α	A_β
$w_\alpha \times w'_\alpha$	$w'_\beta \times w_\beta$	$w_{t(h)} \times w_{s(h)}$	$w_\alpha \times (w'_\alpha + \sum_{h:\, t(h)=\alpha} w_{s(h)})$	$(w'_\beta + \sum_{h:\, s(h)=\beta} w_{t(h)}) \times w_\beta$

For a vector space E of dimension e, denote by $Gr(d, E)$ or $Gr(d, e)$ the Grassmannian space parametrizing all d-dimensional linear subspaces of E. Let $S_{Gr(d,e)}$ be the tautological subbundle (i.e., the universal subbundle) on the Grassmannian variety $Gr(d, e)$.

Definition 3.2 ([**11**, §4]). Let dimension vectors $w = (w_i, w'_i)_{i\in I}$, $v = (v_i)_{i\in I}$, where all the components are nonnegative integers.

(a) The variety $\mathbf{E}_w$ is defined as the space of quiver representations of the quiver $\widetilde{Q^{\mathrm{op}}}$:

$$\begin{aligned}\mathbf{E}_w &:= \bigoplus_{\alpha\in I_0} \mathrm{Hom}(W'_\alpha, W_\alpha) \oplus \bigoplus_{\beta\in I_1} \mathrm{Hom}(W_\beta, W'_\beta) \oplus \bigoplus_{h\in\Omega} \mathrm{Hom}(W_{s(h)}, W_{t(h)}) \\ &\cong \mathbb{C}^{\sum_{i\in I} w_i w'_i + \sum_{h\in\Omega} w_{s(h)} w_{t(h)}}.\end{aligned}$$

We denote elements of $\mathbf{E}_w$ as $(x_i, y_h)_{i\in I, h\in\Omega}$.

(b) The projective variety $\mathcal{F}_{v,w}$ is a closed subvariety of

$$\prod_{\alpha\in I_0} Gr(v_\alpha, W_\alpha) \times \prod_{\beta\in I_1} Gr\Big(v_\beta, W'_\beta \oplus \bigoplus_{h:\, s(h)=\beta} W_{t(h)}\Big)$$

defined as

$$\begin{aligned}\mathcal{F}_{v,w} := \big\{(X_i)_{i\in I} \mid \dim X_i = v_i \ (\forall i \in I),\ X_\alpha \subseteq W_\alpha (\forall \alpha \in I_0), \\ X_\beta \subseteq W'_\beta \oplus \bigoplus_{h:\, s(h)=\beta} X_{t(h)} (\forall \beta \in I_1)\big\}.\end{aligned}$$

(c) *Nakajima's nonsingular graded quiver variety* $\tilde{\mathcal{F}}_{v,w}$ is given by

$$\begin{aligned}\tilde{\mathcal{F}}_{v,w} := \big\{(x_i, y_h, X_i)_{i\in I, h\in\Omega} \mid (x_i, y_h)_{i\in I, h\in\Omega} \in \mathbf{E}_w, (X_i)_{i\in I} \in \mathcal{F}_{v,w}, \\ \mathrm{im} A_\alpha \subseteq X_\alpha \ (\forall \alpha \in I_0),\ \mathrm{im} A_\beta \subseteq X_\beta \ (\forall \beta \in I_1)\big\}.\end{aligned}$$

(Note that A_α, A_β are determined by $(x_i, y_h)_{i\in I, h\in\Omega}$.)

(d) *Nakajima's affine graded quiver variety* is

$$\mathbf{E}_{v,w} := \{(x_i, y_h)_{i\in I, h\in\Omega} \in \mathbf{E}_w \,|\, \mathrm{rank} A_i \leq v_i \ (\forall i \in I)\}.$$

Note that although the definition of $\tilde{\mathcal{F}}_{v,w}$ looks different from [**11**, Definition 4.2 (2)] where "$\mathrm{im} A_\alpha \subseteq X_\alpha$" is replaced by "$\mathrm{im} x_\alpha \subseteq X_\alpha$", these two definitions are indeed equivalent because for any arrow $h : \beta \to \alpha$, the two conditions $\mathrm{im} A_\beta \subseteq X_\beta$ and $X_\beta \subseteq W'_\beta \oplus \bigoplus_{h:\ s(h)=\beta} X_{t(h)}$ imply $\mathrm{im} y_h \subseteq X_\alpha$.

Nakajima showed in [**11**] that the above definition of Nakajima's nonsingular (resp. affine) graded quiver variety is equivalent to the original definition given by the GIT quotient (resp. algebro-geometric quotient). The following statement is also proved by Nakajima.

LEMMA 3.3. *Given dimension vectors* $w = (w_i, w'_i)_{i\in I} \in \mathbb{Z}^{2n}_{\geq 0}$, $v = (v_i)_{i\in I} \in \mathbb{Z}^n_{\geq 0}$.

(a) *The variety* $\mathbf{E}_{v,w}$ *contains the origin, so is nonempty. The variety* $\mathcal{F}_{v,w}$ *(thus* $\tilde{\mathcal{F}}_{v,w}$*) is nonempty if and only if*

$$0 \leq v_\alpha \leq w_\alpha \ (\forall \alpha \in I_0), \ \ \text{and} \ \ 0 \leq v_\beta \leq w'_\beta + \sum_{h:\ s(h)=\beta} v_{t(h)} \ (\forall \beta \in I_1). \tag{3.3}$$

(b) *If* (3.3) *holds, then the natural projection*

$$\pi' : \tilde{\mathcal{F}}_{v,w} \to \mathcal{F}_{v,w}, \ (x_i, y_h, X_i)_{i\in I, h\in\Omega} \mapsto (X_i)_{i\in I}$$

gives a vector bundle of rank $\sum_{\alpha\in I_0} v_\alpha w'_\alpha + \sum_{\beta\in I_1} v_\beta w_\beta$. *Meanwhile, the natural projection*

$$\pi : \tilde{\mathcal{F}}_{v,w} \to \mathbf{E}_{v,w}, \quad (x_i, y_h, X_i)_{i\in I, h\in\Omega} \mapsto (x_i, y_h)_{i\in I, h\in\Omega} \tag{3.4}$$

has the zero fiber $\pi^{-1}(0) \cong \mathcal{F}_{v,w}$.

PROOF. Part (a) and the second statement of (b) are obvious. For the first statement of (b), following a similar proof of [**9**, Lemma 4.3], we have

$$\tilde{\mathcal{F}}_{v,w} \cong \iota^*\Big(\bigoplus_{\alpha\in I_0} S^{\oplus w'_\alpha}_{Gr(v_\alpha, W_\alpha)} \oplus \bigoplus_{\beta\in I_1} S^{\oplus w_\beta}_{Gr(v_\beta, W'_\beta\oplus\bigoplus_{h:\ s(h)=\beta} W_{t(h)})}\Big) \tag{3.5}$$

where ι is the natural embedding

$$\iota : \mathcal{F}_{v,w} \to \prod_{\alpha\in I_0} Gr(v_\alpha, W_\alpha) \times \prod_{\beta\in I_1} Gr(v_\beta, W'_\beta \oplus \bigoplus_{h:\ s(h)=\beta} W_{t(h)}).$$

It follows that $\tilde{\mathcal{F}}_{v,w}$ is a vector bundle over $\mathcal{F}_{v,w}$ of rank $\sum_{\alpha\in I_0} v_\alpha w'_\alpha + \sum_{\beta\in I_1} v_\beta w_\beta$. □

Consider the following stratification $\mathbf{E}_w = \bigcup_v \mathbf{E}^\circ_{v,w}$ where

$$\mathbf{E}^\circ_{v,w} := \{(x_i, y_h)_{i\in I, h\in\Omega} \in \mathbf{E}_w \,|\, \mathrm{rank} A_i = v_i \ (\forall i \in I)\}. \tag{3.6}$$

For any two vectors $u = (u_1, \ldots, u_n)$ and $v = (v_1, \ldots, v_n)$, we denote $u \leq v$ if $u_i \leq v_i$ for all i. We say that j is adjacent to i if there is an arrow $h \in \Omega$ of the form $i \to j$ or $i \leftarrow j$; we denote this arrow as $h : i - j$ whenever the direction of the arrow is irrelevant. Recall the q-analog C_q of the Cartan matrix in [**11**, (3.2)]:

$$C_q v = (u_i, u'_i)_{i\in I}, \text{where } \ u_i = v_i, \ u'_i = v_i - \sum_{h:i-j} v_j \tag{3.7}$$

where h runs over all arrows with an endpoint i.

DEFINITION 3.4. Given dimension vectors $w = (w_i, w'_i)_{i\in I} \in \mathbb{Z}^{2n}_{\geq 0}$, $v = (v_i)_{i\in I} \in \mathbb{Z}^n_{\geq 0}$. We say that (v, w) is *l-dominant* if $w - C_q v \geq 0$, or equivalently, if

$$\text{For all } \forall i \in I, \quad w_i \geq v_i, \quad w'_i \geq v_i - \sum_{h:i-j} v_j. \tag{3.8}$$

(Note that the condition (3.8) implies (3.3).) Define

$$\mathrm{Dom}(w) := \{v \in \mathbb{Z}^n_{\geq 0} \mid (v, w) \text{ is } l\text{-dominant}\}.$$

$$\mathrm{Dom} := \{(v, w) \in \mathbb{Z}^n_{\geq 0} \times \mathbb{Z}^{2n}_{\geq 0} \mid (v, w) \text{ is } l\text{-dominant}\}.$$

In the following, we define $\bar{v}$ with the property that $(\bar{v}, w)$ is l-dominant and $\mathbf{E}_{v,w} = \mathbf{E}_{\bar{v},w}$ (which will be proved in Proposition 3.7).

LEMMA 3.5. *Let* $w = (w_i, w'_i)_{i\in I} \in \mathbb{Z}^{2n}_{\geq 0}$, $v = (v_i)_{i\in I} \in \mathbb{Z}^n_{\geq 0}$, $V = \{v' \in \mathbb{Z}^n_{\geq 0} \mid v' \leq v\}$. *Then* $V \cap \mathrm{Dom}(w)$ *has a unique maximal element* $\bar{v} = (\bar{v}_i)_{i\in I}$ *in the sense that every* $v' \in V \cap \mathrm{Dom}(w)$ *satisfies* $v' \leq \bar{v}$. *More explicitly,*

$$\bar{v}_i = \min\big(v_i, w_i, w'_i + \sum_{h:i-j} \min(v_j, w_j)\big). \tag{3.9}$$

In particular:

- *if* (v, w) *satisfies* (3.3), *then for* $\beta \in I_1$ *we have* $\bar{v}_\beta = \min(v_\beta, w_\beta)$;
- *if* $v \in \mathrm{Dom}(w)$, *then* $\bar{v} = v$.

PROOF. It suffices to prove (3.9). It is obvious true if "=" is replaced by "$\leq$". We are left to prove that $\bar{v}$ defined in (3.9) satisfies (3.8). By definition, $w_i \geq \bar{v}_i$ for all $i \in I$. So we only need to show $w'_i \geq \bar{v}_i - \sum_{h:i-j} \bar{v}_j$ for all $i \in I$.

We prove by contradiction. Assume for some $i \in I$, $w'_i < \bar{v}_i - \sum_{h:i-j} \bar{v}_j$. If i is an isolated vertex (that is, not incident to any arrow in Ω), then $w'_i \geq \min(v_i, w_i, w'_i) = \bar{v}_i = \bar{v}_i - \sum_{h:i-j} \bar{v}_j$, a contradiction. So we can assume that i is adjacent to at least one vertex. Let $m_i = \min(v_i, w_i)$ for $i \in I$. Then $\bar{v}_i = \min(m_i, w'_i + \sum_{h:j-i} m_j)$, and

$$\min(m_i - w'_i, \sum_{h:i-j} m_j) = \bar{v}_i - w'_i > \sum_{h:i-j} \bar{v}_j = \sum_{h:i-j} \min(m_j, w'_j + \sum_{h':j-k} m_k) \tag{3.10}$$
$$\geq \sum_{h:i-j} \min(m_j, w'_j + m_i),$$

where the last inequality is because $\sum_{h':j-k} m_k \geq m_i$ (since h is one of the possible h').

If there exists an arrow $h : i - j_0$ such that $m_{j_0} \geq w'_{j_0} + m_i$, then

$$m_i - w'_i \geq \min_i(m_i - w'_i, \sum_{h:i-j} m_j) \overset{(3.10)}{>} \sum_{h:i-j} \min(m_j, w'_j + m_i)$$
$$\geq \min(m_{j_0}, w'_{j_0} + m_i) = w'_{j_0} + m_i$$

which gives a contradiction. So we can assume $\min(m_j, w'_j + m_i) = m_j$ for all j in the last $\sum$ of (3.10). Thus (3.10) becomes

$$\min(m_i - w'_i, \sum_{h:i-j} m_j) > \sum_{h:i-j} m_j$$

which again gives a contradiction. $\square$

LEMMA 3.6. *Fix* $w = (w_i, w'_i)_{i\in I} \in \mathbb{Z}^{2n}_{\geq 0}$, $v = (v_i)_{i\in I} \in \mathbb{Z}^n_{\geq 0}$. *Let* $\pi : \tilde{\mathcal{F}}_{v,w} \to \mathbf{E}_{v,w}$ *be the map defined in* (3.4).

(a) *For each* $v' \leq v$, *the variety* $\mathbf{E}^\circ_{v',w}$ *defined in* (3.6) *is nonempty if and only if* (v', w) *is* l*-dominant. In this case, it is nonsingular and locally closed in* $\mathbf{E}_w$ *(so is also locally closed in* $\mathbf{E}_{v,w}$*), and is irreducible and rational. Thus the variety* $\mathbf{E}_{v,w}$ *has a stratification*

$$\mathbf{E}_{v,w} = \bigcup_{v'} \mathbf{E}^\circ_{v',w}, \quad \text{where } v' \leq v \text{ and } (v', w) \text{ are } l\text{-dominant.}$$

(b) *For each* $v' \leq v$ *with* (v', w) *being* l*-dominant, the restricted projection* $\pi^{-1}(\mathbf{E}^\circ_{v',w}) \to \mathbf{E}^\circ_{v',w}$ *is a Zariski locally trivial* $\mathcal{M}$*-bundle, where* $\mathcal{M}$ *itself is a*

$$\Big(\prod_{\beta\in I_1} Gr(v_\beta - v'_\beta, w'_\beta - v'_\beta + \sum_{h:s(h)=\beta} v_{t(h)}\Big)\text{-bundle over } \prod_{\alpha\in I_0} Gr(v_\alpha - v'_\alpha, w_\alpha - v'_\alpha)$$

defined as

$$\begin{aligned}\mathcal{M} := \Big\{(X'_i)_{i\in I}\Big|\, & X'_\alpha \in Gr(v_\alpha - v'_\alpha, w_\alpha - v'_\alpha)\ (\forall \alpha \in I_0),\\ & X'_\beta \in Gr\big(v_\beta - v'_\beta, \mathbb{A}^{w'_\beta - v'_\beta + \sum_{h:s(h)=\beta} v'_{t(h)}} \oplus \bigoplus_{h:s(h)=\beta} (X'_{t(h)})\ (\forall \beta \in I_1)\Big\}.\end{aligned}$$

In particular, $\mathcal{F}_{v,w}$ *is a* $\prod_{\beta\in I_1} Gr(v_\beta, w'_\beta + \sum_{h:s(h)=\beta} v_{t(h)})$*-bundle over* $\prod_{\alpha\in I_0} Gr(v_\alpha, w_\alpha)$.

PROOF. The proof is similar to the proof of [**9**, Lemma 4.6].

(a) $\mathbf{E}^\circ_{v',w} \subseteq \mathbf{E}_w$ is locally closed because

$$\begin{aligned}\mathbf{E}^\circ_{v',w} =& \{(x_i, y_h)_{i\in I, h\in\Omega} \in \mathbf{E}_w \mid \operatorname{rank} A_i \leq v'_i (\forall i)\}\\ &- \bigcup_{j\in I} \{(x_i, y_h)_{i\in I, h\in\Omega} \in \mathbf{E}_w \mid \operatorname{rank} A_j \leq v'_j - 1,\ \operatorname{rank} A_i \leq v'_i (\forall i)\}\end{aligned}$$

To show $\mathbf{E}^\circ_{v',w}$ is nonsingular, we consider the following open covering

$$\mathbf{E}^\circ_{v',w} = \bigcup U_{\mathbf{J}} \tag{3.11}$$

where $\mathbf{J} = (J_i)_{i\in I}$, $J_i \subseteq \{1, \ldots, w_i\}$ and $|J_i| = v'_i$, $U_{\mathbf{J}} \subseteq \mathbf{E}^\circ_{v',w}$ consists of those (x_i, y_h) such that for all $i \in I_0$ (resp. $i \in I_1$), the rows (resp. column) of A_i indexed by J_i are linearly independent. We will show that $U_{\mathbf{J}}$ is isomorphic to an open subset $\tilde{\mathbb{A}}^\circ$ of an affine space $\tilde{\mathbb{A}}$, which implies that $\mathbf{E}^\circ_{v',w}$ is nonsingular.

Without loss of generality, assume $J_i = \{1, \ldots, v'_i\}$. Denote

$$\begin{aligned}\tilde{\mathbb{A}} :=& \prod_{\alpha\in I_0} \mathbb{A}^{(w_\alpha - v'_\alpha)\times v'_\alpha} \times \prod_{\beta\in I_1} \mathbb{A}^{v'_\beta \times (w_\beta - v'_\beta)} \times \prod_{\alpha\in I_0} \mathbb{A}^{v'_\alpha \times w'_\alpha}\\ &\times \prod_{\beta\in I_1} \mathbb{A}^{w'_\beta \times v'_\beta} \times \prod_{h\in\Omega} \mathbb{A}^{v'_{t(h)} \times v'_{s(h)}}\\ \cong& \mathbb{A}^{\sum_{i\in I}(w_i + w'_i - v'_i)v'_i + \sum_{h\in\Omega} v'_{t(h)} v'_{s(h)}}\end{aligned}$$

and write its elements as $(P_i, Z_i, C_h)_{i\in I, h\in\Omega}$ where the size of each matrix is given in the following table: (where $\alpha \in I_0$, $\beta \in I_1$, $h \in \Omega$)

P_α	P_β	Z_α	Z_β	C_h
$(w_\alpha - v'_\alpha)\times v'_\alpha$	$v'_\beta \times (w_\beta - v'_\beta)$	$v'_\alpha \times w'_\alpha$	$w'_\beta \times v'_\beta$	$v'_{t(h)} \times v'_{s(h)}$

Then define a morphism

$$\varphi : \tilde{\mathbb{A}} \to \mathbf{E}_w$$
$$(P_i, Z_i, C_h)_{i\in I, h\in\Omega} \mapsto (x_i, y_h)_{i\in I, h\in\Omega}$$

where, $x_\alpha = \begin{bmatrix} Z_\alpha \\ P_\alpha Z_\alpha \end{bmatrix}$ for $\alpha \in I_0$; $x_\beta = \begin{bmatrix} Z_\beta & Z_\beta P_\beta \end{bmatrix}$ for $\beta \in I_1$; $y_h = \begin{bmatrix} C_h & C_h P_\beta \\ P_\alpha C_h & P_\alpha C_h Q_\beta \end{bmatrix}$ for $h : \beta \to \alpha \in \Omega$. It is easy to see that $\operatorname{im}\varphi \subseteq \mathbf{E}_{v',w}$.

For $\alpha \in I_0$, $\beta \in I_1$, let $r_1, \dots, r_s$ and $r'_1, \dots, r'_t$ be given as in Definition 3.1. Define

$$A'_\alpha := [Z_\alpha\ C_{r_1},\ \cdots,\ C_{r_s}]_{\text{hor}}, \quad A'_\beta := [Z_\beta\ C_{r'_1},\ \cdots,\ C_{r'_t}]_{\text{vert}} \tag{3.12}$$

Note that the size of A'_α is $v'_\alpha \times (w'_\alpha + \sum_{h:\ t(h)=\alpha} v'_{s(h)})$, so its number of rows $\le$ its number of columns when (v', w) is l-dominant; similarly, the size of A'_β is $(w'_\beta + \sum_{h:\ s(h)=\beta} v'_{t(h)}) \times v'_\beta$, so its number of columns $\le$ its number of rows when (v', w) is l-dominant.

Define $\tilde{\mathbb{A}}^\circ := \Big\{ (P_i, Z_i, C_h) \in \tilde{\mathbb{A}} \ \Big|\ \operatorname{rank} A'_i = v'_i \ (\forall i \in I) \Big\}$ to be an open subset of $\tilde{\mathbb{A}}$.

Similar to the proof of [**9**, Lemma 4.6], it can be shown that:

(i) φ restricts to an isomorphism $\varphi|_{\tilde{\mathbb{A}}^\circ} :\ \tilde{\mathbb{A}}^\circ \xrightarrow{\cong} U_{\mathbf{J}}$;
(ii) $\mathbf{E}^\circ_{v',w}$ is nonempty if and only if (v', w) is l-dominant;
(iii) if $\mathbf{E}^\circ_{v',w}$ is nonempty, it is irreducible.

It follows that $\mathbf{E}^\circ_{v',w}$ is rational.

As pointed out in the proof of [**9**, Lemma 4.6], the proof of the "if" part of (ii) and the proof of (iii) therein make use of some generic choice. Alternatively, We give the new proofs below that do not use the generic choice.

For the "if" part of (ii): it suffices to show that $\tilde{\mathbb{A}}^\circ$ is nonempty by constructing a point on it. Note that P_i are irrelevant so we let them be zero matrices. We construct Z_i, C_h as follows: start with all Z_i, C_h being zero matrices, then for each A'_α and A'_β defined in (3.12), change the diagonal entries (that is, the (j, j)- entry for all j) to 1 and make the corresponding change in Z_i and C_h. We showed that the resulting (P_i, Z_i, C_h) satisfies the equality $\operatorname{rank} A'_i = v'_i$ $(\forall i \in I)$. Without loss of generality, we only need to show $\operatorname{rank} A'_\alpha = v'_\alpha$ for $\alpha \in I_0$. We claim the A'_α is an "upper triangular matrix" in the sense that its (i, j)-entries are 0 whenever $i > j$; then together with the fact that its diagonal entries are 1, we can conclude that $\operatorname{rank} A'_\alpha = v'_\alpha$ (that is, A'_α has full row rank). Assume the contrary that some (i, j)-entry of A'_α is 1 with $i > j$. Then it corresponds to the (i, j')-entry of C_h for some arrow $h : \beta \to \alpha$ and $j' \le j$; moreover it must correspond to a diagonal entry of A'_β. So it is simultaneously the (j', j')-entry of A'_β and the (i, j')-entry of A'_α. In particular, $i = j'$. However, $i > j \ge j'$, a contradiction.

For (iii), assuming (v', w) is l-dominant (so $\mathbf{E}^\circ_{v',w}$ is nonempty), we shall prove that $\mathbf{E}^\circ_{v',w}$ is irreducible. It suffices to show that $U_{\mathbf{J}} \cap U_{\mathbf{J}'} \neq \emptyset$ where $\mathbf{J} = (J_i)_{i\in I}$ and $\mathbf{J}' = (J'_i)_{i\in I}$ satisfy the following condition: there exists $i_0 \in I$ such that J_{i_0} and J'_{i_0} only differ by one element, and $J_i = J'_i$ for $i \neq i_0$. Without loss of generality, we can assume $i_0 = \alpha \in I_0$, $J_\alpha = \{1, \ldots, v'_\alpha\}$, $J'_\alpha = \{1, \ldots, v'_{\alpha-1}, v'_\alpha\}$. Define $Z_i (i \in I), C_h (h \in \Omega)$ to be the same as the last paragraph, $P_i (i \in I)$ be any matrices, under the condition that the first row of P_α is $[0, \ldots, 0, 1]$. Then in the v'_α-th row and the $(v'_\alpha + 1)$-th row of A'_α are identical. It is then obvious that $\varphi((P_i, Z_i, C_h))$ is a point in $U_{\mathbf{J}} \cap U_{\mathbf{J}'}$, thus the latter is nonempty.

(b) Similar to the proof of [**9**, Lemma 4.6]: let

$$\mathcal{X} := \big\{(X_i, Y_i)_{i\in I} \mid (Y_i)_{i\in I} \in \mathcal{F}_{v',w}, (X_i)_{i\in I} \in \mathcal{F}_{v,w}, Y_i \subseteq X_i (\forall i \in I)\big\},$$
$$f : (x_i, y_h)_{i\in I, h\in\Omega} \mapsto (X_i)_{i\in I}, \quad \text{where } X_i = \operatorname{im} A_i,$$
$$q : (X_i, Y_i)_{i\in I} \mapsto (Y_i)_{i\in I}.$$

Then $\pi^{-1}(\mathbf{E}^\circ_{v',w}) \to \mathbf{E}^\circ_{v',w}$ is the pullback of the locally trivial bundle $q : \mathcal{X} \to \mathcal{F}_{v',w}$:

$$\begin{array}{ccc} \pi^{-1}(\mathbf{E}^\circ_{v',w}) & \longrightarrow & \mathcal{X} \\ \downarrow{\scriptstyle \pi} & & \downarrow{\scriptstyle q} \\ \mathbf{E}^\circ_{v',w} & \xrightarrow{\ f\ } & \mathcal{F}_{v',w} \end{array}$$

and the fibers of q are isomorphic to $\mathcal{M}$. □

PROPOSITION 3.7. *Let v, w, π be the same as Lemma 3.6 and $\bar{v}$ be as in Lemma 3.5.*

(a) *$\mathbf{E}_{v,w} = \mathbf{E}_{\bar{v},w}$ is irreducible. Moreover, $\mathbf{E}^\circ_{\bar{v},w}$ is its largest stratum; so $\mathbf{E}_{v,w} = \overline{\mathbf{E}^\circ_{\bar{v},w}}$, the Zariski closure of $\mathbf{E}^\circ_{\bar{v},w}$.*

(b) *Assume (3.3). Then $\tilde{\mathcal{F}}_{v,w}$ and $\mathcal{F}_{v,w}$ are irreducible and nonsingular and π is surjective. Moreover,*

$$\begin{aligned}
\dim \mathcal{F}_{v,w} = d_{v,w} &:= \sum_{\beta\in I_1} (w'_\beta + \sum_{h:s(h)=\beta} v_{t(h)} - v_\beta) v_\beta + \sum_{\alpha\in I_0} (w_\alpha - v_\alpha) v_\alpha \\
&= \sum_{\alpha\in I_0} w_\alpha v_\alpha + \sum_{\beta\in I_1} w'_\beta v_\beta - \sum_{i\in I} v_i^2 + \sum_{h\in\Omega} v_{s(h)} v_{t(h)} \\
\dim \tilde{\mathcal{F}}_{v,w} = \tilde{d}_{v,w} &:= \dim \mathcal{F}_{v,w} + \sum_{\alpha\in I_0} v_\alpha w'_\alpha + \sum_{\beta\in I_1} v_\beta w_\beta \\
&= \sum_{\alpha\in I_0} (w_\alpha + w'_\alpha) v_\alpha + \sum_{\beta\in I_1} (w_\beta + w'_\beta) v_\beta - \sum_{i\in I} v_i^2 + \sum_{h\in\Omega} v_{s(h)} v_{t(h)} \\
\dim \mathbf{E}_{v,w} &= \dim \mathbf{E}_{\bar{v},w} = \tilde{d}_{\bar{v},w}
\end{aligned}$$

(c) *Further assume that (v, w) is l-dominant. Then π is birational. In this case, it restricts to an isomorphism $\pi^{-1}(\mathbf{E}^\circ_{v,w}) \xrightarrow{\cong} \mathbf{E}^\circ_{v,w}$.*

PROOF. Similar to the proof of [**9**, Proposition 4.7]. □

4. Decomposition theorem

The following fact is proved by Nakajima in [**12**, Theorem 14.3.2] using the representation theory of quantum affine algebras. We will give a proof similar to [**9**, Theorem 5.1]; instead of representation theory, we use the algebraic version of the Transversal Slice Theorem.

THEOREM 4.1. *Assume $\tilde{\mathcal{F}}_{v,w} \neq \emptyset$. The local system appeared in the BBDG decomposition for $\pi_* IC_{\tilde{\mathcal{F}}_{v,w}}$ are all trivial. Thus,*

$$\pi_*(IC_{\tilde{\mathcal{F}}_{v,w}}) = \bigoplus_{v'} \bigoplus_{d} a^d_{v,v';w} IC_{\mathbf{E}_{v',w}}[d] \tag{4.1}$$

where $v' \leq v$ satisfies the condition that (v', w) is l-dominant, and $a^d_{v,v';w} \in \mathbb{Z}_{\geq 0}$.

The proof is delayed to §4.2. First, recall:

THEOREM 4.2 (BBDG Decomposition Theorem). [**1**] *Let $f : X \to Y$ be a proper algebraic morphism between complex algebraic varieties. Then there is a finite list of triples (Y_a, L_a, n_a), where for each a, Y_a is locally closed smooth irreducible algebraic subvariety of Y, L_a is a semisimple local system on Y_a, n_a is an integer, such that:*

$$f_* IC_X \cong \bigoplus_a IC_{\overline{Y_a}}(L_a)[n_a] \tag{4.2}$$

Moreover, even though the isomorphism is not necessarily canonical, the direct summands appeared on the right hand side are canonical.

4.1. Algebraic Transversal Slice Theorem. Nakajima proves an analytic transversal slice theorem [**12**, §3] and we strengthen the result to an algebraic version for the varieties studied in this paper using an explicit construction similar to the one in [**9**].

LEMMA 4.3 (Algebraic Transversal Slice Theorem). *Let $p \in \mathbf{E}_{v,w}$ be a point in the stratum $\mathbf{E}^\circ_{v^0,w}$. So $v^0 \leq v$. Define*

$$w^\perp = w - C_q v^0 = (w_i^\perp, w_i^{\perp\prime})_{i \in I}, \quad v^\perp = v - v^0 = (v_i^\perp)_{i \in I},$$

that is, $w_i^\perp = w_i - v_i^0$, $w_i^{\perp\prime} = w'_i - v_i^0 + \sum_{h:j-i} v_j^0$, $v_i^\perp = v_i - v_i^0$. Take $U^\perp = \mathbf{E}_{v^\perp, w^\perp}$. Then there exist Zariski open neighborhoods $U \subseteq \mathbf{E}_{v,w}$ of p, $U^0 \subseteq \mathbf{E}^\circ_{v^0,w}$ of p, and isomorphisms φ, ψ making the following diagram commute:

$$\begin{array}{ccccccc}
\tilde{\mathcal{F}}_{v,w} & \supseteq & \pi^{-1}(U) & \xrightarrow[\cong]{\varphi} & U^0 \times \pi^{-1}(U^\perp) & \subseteq & \mathbf{E}^\circ_{v^0,w} \times \tilde{\mathcal{F}}_{v^\perp,w^\perp} \\
 & & \Big\downarrow \pi & & \Big\downarrow 1\times\pi & & \\
\mathbf{E}_{v,w} & \supseteq & U & \xrightarrow[\cong]{\psi} & U^0 \times U^\perp & \subseteq & \mathbf{E}^\circ_{v^0,w} \times \mathbf{E}_{v^\perp,w^\perp}
\end{array}$$

We can further assume U and U^0 are both invariant under the natural $\mathbb{C}^*$-action (by multiplying all entries with the same scalar).

Moreover, $\psi(p) = (p, 0)$, and the diagram is compatible with the stratifications in the sense that $\psi(\mathbf{E}^\circ_{u,w} \cap U) = U_0 \times \mathbf{E}^\circ_{u^\perp,w^\perp}$ for each u satisfying $v^0 \leq u \leq v$, where $u^\perp = u - v^0$.

PROOF. (1) We will define ψ. Recall that the point p determines n matrices as in (3.2), denoted $A_i(p)$ $(i \in I)$. Since $p \in \mathbf{E}^\circ_{v^0,w}$, without loss of generality, we can assume that, for $i \in I_0$, the first v_i^0 row vectors of $A_i(p)$ are linearly independent; and for $i \in I_1$, the first v_i^0 column vectors of $A_i(p)$ are linearly independent. Then for $\alpha \in I_0$, $\beta \in I_1$, there exist matrices with the indicated sizes:

$P_\alpha(p)$	$P_\beta(p)$	$Z_\alpha(p)$	$Z_\beta(p)$	$C_h(p)$
$(w_\alpha - v_\alpha^0) \times v_\alpha^0$	$v_\beta^0 \times (w_\beta - v_\beta^0)$	$v_\alpha^0 \times w'_\alpha$	$w'_\beta \times v_\beta^0$	$v_{t(h)}^0 \times v_{s(h)}^0$

such that (where for simplicity we drop the parameter p for Z_α, C_{r_1}, etc.)

$$A_\alpha(p) = \begin{bmatrix} Z_\alpha & C_{r_1} & C_{r_1}P_{\beta_{r_1}} & \cdots & C_{r_s} & C_{r_s}P_{\beta_{r_s}} \\ P_\alpha Z_\alpha & P_\alpha C_{r_1} & P_\alpha C_{r_1}P_{\beta_{r_1}} & \cdots & P_\alpha C_{r_s} & P_\alpha C_{r_s}P_{\beta_{r_s}} \end{bmatrix},$$

$$A_\beta(p) = \begin{bmatrix} Z_\beta & Z_\beta P_\beta \\ C_{r'_1} & C_{r'_1}P_\beta \\ P_{\alpha_{r'_1}}C_{r'_1} & P_{\alpha_{r'_1}}C_{r'_1}P_\beta \\ \vdots & \vdots \\ C_{r'_t} & C_{r'_t}P_\beta \\ P_{\alpha_{r'_t}}C_{r'_t} & P_{\alpha_{r'_t}}C_{r'_t}P_\beta \end{bmatrix},$$

and that the corresponding $A'_\alpha(p)$ of size $v^0_\alpha \times (w'_\alpha + \sum_{h:\ t(h)=\alpha} v^0_{s(h)})$ and $A'_\beta(p)$ of size $(w'_\beta + \sum_{h:\ s(h)=\beta} v^0_{t(h)}) \times v^0_\beta$, as defined in (3.12), are full rank.

STEP 1 (define U and U^0). For $\alpha \in I_0$, define M_α to be the index set of the columns of $A_\alpha(p)$ that correspond to the columns of $A'_\alpha(p)$. The submatrix $A'_\alpha(p) := A_\alpha(p)_{[1,\dots,v^0_\alpha; M_\alpha]}$ has rank v^0_α, thus there exists a subset $J_\alpha \subseteq M_\alpha$ of cardinality v^0_α such that $A_\alpha(p)_{[1,\dots,v^0_\alpha; J_\alpha]}$ is an invertible square matrix. Similarly, for $\beta \in I_1$, define M_β to be the index set of the rows of $A_\beta(p)$ that correspond to the rows of $A'_\beta(p)$, and there exists $J_\beta \subseteq M_\beta$ of cardinality v^0_β such that $A_\beta(p)_{[J_\beta;1,\dots,v^0_\beta]}$ is an invertible square matrix. Define an open subset U (which depends on J_α) of $\mathbf{E}_{v,w}$ as

$$U := \big\{q \in \mathbf{E}_{v,w} \ \big|\ A_\alpha(q)_{[1,\dots,v^0_\alpha; J_\alpha]}\ (\forall \alpha \in I_0) \\ \text{and } A_\beta(q)_{[J_\beta;1,\dots,v^0_\beta]}\ (\forall \beta \in I_1) \text{ are invertible}\big\}.$$

Define $U^0 := U \cap \mathbf{E}_{v^0,w}$ which is an open subset of $\mathbf{E}^\circ_{v^0,w}$. So

$$U^0 = \big\{q \in \mathbf{E}_{v^0,w} \ \big|\ A_\alpha(q)_{[1,\dots,v^0_\alpha; J_\alpha]}\ (\forall \alpha \in I_0) \\ \text{and } A_\beta(q)_{[J_\beta;1,\dots,v^0_\beta]}\ (\forall \beta \in I_1) \text{ are invertible}\big\}.$$

Obviously, both U and U^0 are invariant under the natural $\mathbb{C}^*$-action.

STEP 2. In the rest we assume $q \in \mathbf{E}^\circ_{u,w} \cap U$. We construct q^0 as follows. Denote the following (where we drop the parameter q of Z_α, Z'_α, etc.)

$$A_\alpha(q) = \begin{bmatrix} Z_\alpha & C_{r_1} & C'_{r_1} & \cdots & C_{r_s} & C'_{r_s} \\ Z'_\alpha & C''_{r_1} & C'''_{r_1} & \cdots & C''_{r_s} & C'''_{r_s} \end{bmatrix}, \quad A_\beta(q) = \begin{bmatrix} Z_\beta & Z'_\beta \\ C_{r'_1} & C'_{r'_1} \\ C''_{r'_1} & C'''_{r'_1} \\ \vdots & \vdots \\ C_{r'_t} & C'_{r'_t} \\ C''_{r'_t} & C'''_{r'_t} \end{bmatrix}$$

where the sizes of the blocks are as follows:

block	$Z_\alpha(q)$	$Z'_\alpha(q)$	$Z_\beta(q)$	$Z'_\beta(q)$
size	$v^0_\alpha \times w'_\alpha$	$(w_\alpha - v^0_\alpha) \times w'_\alpha$	$w'_\beta \times v^0_\beta$	$w'_\beta \times (w_\beta - v^0_\beta)$

block	$C_j(q)$	$C'_j(q)$	$C''_j(q)$	$C'''_j(q)$
size	$v^0_{\alpha_j} \times v^0_{\beta_j}$	$v^0_{\alpha_j} \times (w_{\beta_j} - v^0_{\beta_j})$	$(w'_{\alpha_j} - v^0_{\alpha_j}) \times v^0_{\beta_j}$	$(w'_{\alpha_j} - v^0_{\alpha_j}) \times (w_{\beta_j} - v^0_{\beta_j})$

Since $A_\alpha(q)_{[1,\ldots,v^0_\alpha;J_\alpha]}$ $(\forall\alpha\in I_0)$ and $A_\beta(q)_{[J_\beta;1,\ldots,v^0_\beta]}$ $(\forall\beta\in I_1)$ are invertible, there are unique $(w_\alpha-v^0_\alpha)\times v^0_\alpha$ -matrix $P_\alpha(q)$ $(\forall\alpha\in I_0)$ and $v^0_\beta\times(w_\beta-v^0_\beta)$ -matrix $P_\beta(q)$ $(\forall\beta\in I_1)$ such that (where we omit the parameter q of P_α, P_β)

$$(4.3)\quad A_\alpha(q)_{[-;J_\alpha]}=\begin{bmatrix}A_\alpha(q)_{[1,\ldots,v^0_\alpha;J_\alpha]}\\ P_\alpha A_\alpha(q)_{[1,\ldots,v^0_\alpha;J_\alpha]}\end{bmatrix},$$

$$A_\beta(q)_{[J_\beta;-]}=\begin{bmatrix}A_\beta(q)_{[J_\beta;1,\ldots,v^0_\beta]} & A_\beta(q)_{[J_\beta;1,\ldots,v^0_\beta]}P_\beta\end{bmatrix}.$$

Define $q^0=(x^0_i,y^0_h)_{i\in I,h\in\Omega}$ where

$$x^0_\alpha=\begin{bmatrix}Z_\alpha\\ P_\alpha Z_\alpha\end{bmatrix},x^0_\beta=\begin{bmatrix}Z_\beta & Z_\beta P_\beta\end{bmatrix},y^0_j=\begin{bmatrix}C_j & C_jP_{\beta_j}\\ P_{\alpha_j}C_j & P_{\alpha_j}C_jP_{\beta_j}\end{bmatrix}$$

(where we omitted the parameter q). Then q^0 is in U^0.

Note that the projection $U\to U^0$, $q\mapsto q^0$ is compatible with the $\mathbb{C}^*$-action.

STEP 3. We let $U^\perp=\mathbf{E}^\circ_{v^\perp,w^\perp}$ and construct $q^\perp=(x^\perp_i,y^\perp_h)_{i\in I,h\in\Omega}$ as follows. Define

$$\widetilde{P}_\alpha:=\begin{bmatrix}\mathbf{I}_{v^0_\alpha} & 0\\ -P_\alpha & \mathbf{I}_{w_\alpha-v^0_\alpha}\end{bmatrix}(\forall\alpha\in I_0),\quad \widetilde{P}_\beta:=\begin{bmatrix}\mathbf{I}_{v^0_\beta} & -P_\beta\\ 0 & \mathbf{I}_{w_\beta-v^0_\beta}\end{bmatrix}(\forall\beta\in I_1)$$

and for $\alpha\in I_0$, define (where we omit the parameter q)

$$\widetilde{A}_\alpha(q):=[\widetilde{P}_\alpha x_\alpha,\widetilde{P}_\alpha y_{r_1}\widetilde{P}_{\beta_{r_1}},\cdots,\widetilde{P}_\alpha y_{r_s}\widetilde{P}_{\beta_{r_s}}]_{\mathrm{hor}}.$$

It is easy to see that $\operatorname{rank}\widetilde{A}_\alpha(q)=\operatorname{rank}A_\alpha(q)=u_\alpha$, and $\widetilde{A}_\alpha(q)_{[1,\ldots,v^0_\alpha;J_\alpha]}=\begin{bmatrix}A_\alpha(q)_{[1,\ldots,v^0_\alpha;J_\alpha]}\\ 0\end{bmatrix}$.

Note that the space spanned by the top v^0_α rows of $\widetilde{A}_\alpha(q)$ intersects only at the origin with the space spanned by the rest rows. Since

$$\operatorname{rank}\widetilde{A}_\alpha(q)=\operatorname{rank}A_\alpha(q)=u_\alpha\le v_\alpha,$$

and that $\operatorname{rank}\widetilde{A}_\alpha(q)_{[1,\ldots,v^0_\alpha;-]}=v^0_\alpha$ (because $\widetilde{A}_\alpha(q)_{[1,\ldots,v^0_\alpha;J_\alpha]}=A_\alpha(q)_{[1,\ldots,v^0_\alpha;J_\alpha]}$ is invertible), we have

$$\operatorname{rank}\widetilde{A}_\alpha(q)_{[v^0_\alpha+1,\ldots,w_\alpha;-]}=\operatorname{rank}\widetilde{A}_\alpha(q)-\operatorname{rank}\widetilde{A}_\alpha(q)_{[1,\ldots,v^0_\alpha;-]}=u_\alpha-v^0_\alpha=u^\perp_\alpha\le v^\perp_\alpha.$$

We want $A_\alpha(q)^\perp$ to be $\widetilde{A}_\alpha(q)_{[v^0_\alpha+1,\ldots,w_\alpha;\{1,\ldots,w'_\alpha+\sum_{h:\,t(h)=\alpha}w_{s(h)}\}\setminus J_\alpha]}$ with an appropriate rearrangement of the columns; to be more precise (recall that M_α is defined in Step 1):

$$A_\alpha(q)^\perp=\left[\widetilde{A}_\alpha(q)_{[v^0_\alpha+1,\ldots,w_\alpha;M_\alpha\setminus J_\alpha]}\,\middle|\,P_\alpha C_{r_1}P_{\beta_{r_1}}-P_\alpha C'_{r_1}-C''_{r_1}P_{\beta_{r_1}}+C'''_{r_1}\,\middle|\,\cdots\right].$$

For this, we denote $J'_\alpha=\iota^{-1}(J_\alpha)$ where $\iota:\{1,\ldots,|M_\alpha|\}\to M_\alpha$ is the order preserving bijection, and define

$x^\perp_\alpha:=$the matrix obtained from $[-P_\alpha Z_\alpha+Z'_\alpha,-P_\alpha C_{r_1}+C''_{r_1},\cdots,-P_\alpha C_{r_s}+C''_{r_s}]_{\mathrm{hor}}$ by deleting columns of indices in J'_α (which are zero columns),

$y^\perp_h:=P_\alpha C_hP_\beta-P_\alpha C'_h-C''_hP_\beta+C'''_h$ for arrow $h:\beta\to\alpha$.

Note that the column spaces of $A_\alpha(q)^\perp$ and $\widetilde{A}_\alpha(q)_{[v_\alpha^0+1,\dots,w_\alpha;-]}$ are the same, which implies

$$\text{rank}\, A_\alpha(q)^\perp = u_\alpha^\perp. \tag{4.4}$$

Similarly, we define $x_\beta^\perp$ to be obtained from

$$[-Z_\beta P_\beta + Z'_\beta, -C_{r'_1}P_\beta + C'_{r'_1}, \cdots, -C_{r'_t}P_\beta + C'_{r'_t}]_{\text{vert}}$$

by deleting rows of indices in J_β where $J'_\beta = \iota^{-1}(J_\beta)$ (here $\iota : \{1,\dots,|M_\beta|\} \to M_\beta$ is the order preserving bijection). We have

$$\text{rank}\, A_\beta(q)^\perp = u_\beta^\perp. \tag{4.5}$$

This complete the construction of $q^\perp$. Note that the projection $U \to U^\perp$, $q \mapsto q^\perp$ is also compatible with the $\mathbb{C}^*$-action.

Note that ψ is a morphism because it can be expressed in terms of matrix additions, multiplications, and inverses, so can be expressed as rational functions.

(2) We show that ψ is an isomorphism by constructing its inverse ψ^{-1}. Assume $q^0, q^\perp$ are given. From q^0 we can recover Z_i, P_i, C_h by rational functions. From $q^\perp$ we can recover $Z'_i, C'_h, C''_h, C'''_h$ by rational functions. This uniquely determines q. It is routine to check that q is well-defined, that ψ^{-1} is a morphism and that ψ^{-1} is indeed the two-sided inverse of ψ.

(3) We define φ. Given $(q = (x_i, y_h), X_i) \in \pi^{-1}(U)$, define

$$\varphi(q, Xf_i) := (q^0, (q^\perp, X_i^\perp)),$$

where q^0 and $q^\perp$ are defined in (1), and $X_i^\perp$ are defined as follows:

We define $X_\alpha^\perp$ ($\alpha \in I_0$) as follows. For $\mathbf{u} \in X_\alpha$, define

$$\rho^0(\mathbf{u}) := \begin{bmatrix} \mathbf{I} & 0 \\ P_\alpha & 0 \end{bmatrix} \mathbf{u}, \quad \rho^\perp(\mathbf{u}) := \mathbf{u} - \rho^0(\mathbf{u}) = \begin{bmatrix} 0 & 0 \\ -P_\alpha & \mathbf{I} \end{bmatrix} \mathbf{u}.$$

Then the top v_α^0 entries of $\rho^\perp(\mathbf{u})$ are 0. Define

$$X_\alpha^0 := \{\rho^0(\mathbf{u}) \mid \mathbf{u} \in X_\alpha\},$$
$$X_\alpha^\perp := \{\rho^\perp(\mathbf{u}) \mid \mathbf{u} \in X_\alpha\} = \left\{ \begin{bmatrix} 0 \\ -P_\alpha \mathbf{u}_1 + \mathbf{u}_2 \end{bmatrix} \;\middle|\; \begin{bmatrix} \mathbf{u}_1 \\ \mathbf{u}_2 \end{bmatrix} \in X_\alpha \right\}.$$

Then $\dim X_\alpha^0 = v_\alpha^0$, $\dim X_\alpha^\perp = v_\alpha^\perp$. Note that $X_\alpha^\perp = X_\alpha \cap (0 \oplus \mathbb{A}^{w_\alpha - v_\alpha^0})$ is the set of vectors in X_α with the first v_α^0 entries being 0.

We define $X_\beta^\perp$ ($\beta \in I_1$) as follows. Let $r'_1, \dots, r'_t$ be defined as in Definition 3.1. Define

$$P_\beta^* := \begin{bmatrix} \mathbf{I} & 0 & \cdots & 0 \\ 0 & \widetilde{P}_{\alpha_{r'_1}} & \cdots & 0 \\ \vdots & \vdots & \ddots & \vdots \\ 0 & 0 & \cdots & \widetilde{P}_{\alpha_{r'_t}} \end{bmatrix}, \quad \widetilde{X}_\beta := P_\beta^* X_\beta.$$

Denote $\gamma|_{J_\beta} \in \mathbb{A}^{v_\beta^0}$ to be the vector obtained from γ by only keeping its entries of indices in J_β, where J_β is defined in (1) Step 1. Define

$$X_\beta^\perp := \{\gamma \in \widetilde{X}_\beta \text{ such that } \gamma|_{J_\beta} = \mathbf{0}\}.$$

We now show that $\dim X_\beta^\perp = v_\beta^\perp$. Note that $\dim \widetilde{X}_\beta = \dim X_\beta = v_\beta$. Note that

$$\{\gamma|_{J_\beta} \text{ for } \gamma \in \widetilde{X}_\beta\} = \{\gamma|_{J_\beta} \text{ for } \gamma \in X_\beta\} = \mathbb{A}^{v_\beta^0}.$$

So there is a subspace V of $\widetilde{X}_\beta$ of dimension v_β^0 such that $\{\gamma|_{J_\beta}$ for $\gamma \in V\}$ has full dimension v_β^0. Then $V \cap X_\beta^\perp = 0$ and $V + X_\beta^\perp = \widetilde{X}_\beta$, so $\dim X_\beta^\perp = v_\beta^\perp$.

Next we shall show that $X_\beta^\perp \subseteq W_\beta'^\perp \oplus \bigoplus_{j=1}^t X_{\alpha_{r_j'}}^\perp$, where

$$W_\beta'^\perp \cong \mathbb{A}^{w_\beta' + \sum_{h:s(h)=\beta} v_{t(h)}^0 - v_\beta^0}$$

consists of all vectors in $\mathbb{A}^{w_\beta' + \sum_{h:\, s(h)=\beta} w_{t(h)}}$ whose j-th entry are 0 for $j \in J_\beta$ or $j \notin M_\beta$. Take an arbitrary element $P_\beta^* \mathbf{u} \in X_\beta^\perp$ for $u \in X_\beta$. We can write

$$\mathbf{u} = \begin{bmatrix} \mathbf{u}_0 \\ \mathbf{u}_1 \\ \vdots \\ \mathbf{u}_t \end{bmatrix} \quad \text{where } \mathbf{u}_0 \in W_\beta',\ \mathbf{u}_i \in X_{\alpha_{r_i'}} \text{ for } i = 1, \dots, t.$$

Then

$$P_\beta^* \mathbf{u} = \begin{bmatrix} \mathbf{u}_0 \\ \widetilde{P}_{\alpha_{r_1'}} \mathbf{u}_1 \\ \vdots \\ \widetilde{P}_{\alpha_{r_t'}} \mathbf{u}_t \end{bmatrix} = \mathbf{v}_1 + \mathbf{v}_2, \text{ where } \mathbf{v_1} = \begin{bmatrix} \mathbf{u}_0 \\ \begin{bmatrix} \mathbf{I} & 0 \\ 0 & 0 \end{bmatrix} \mathbf{u}_1 \\ \vdots \\ \begin{bmatrix} \mathbf{I} & 0 \\ 0 & 0 \end{bmatrix} \mathbf{u}_t \end{bmatrix}, \ \mathbf{v}_2 = \begin{bmatrix} 0 \\ \begin{bmatrix} 0 & 0 \\ -P_{\alpha_{r_1'}} & \mathbf{I} \end{bmatrix} \mathbf{u}_1 \\ \vdots \\ \begin{bmatrix} 0 & 0 \\ -P_{\alpha_{r_t'}} & \mathbf{I} \end{bmatrix} \mathbf{u}_t \end{bmatrix}$$

Note that $\mathbf{v}_1 \in W_\beta'^\perp$ and $\mathbf{v}_2 \in \mathbf{0} \oplus \bigoplus_{j=1}^t X_{\alpha_{r_j'}}^\perp$. (Indeed, the second one is immediate from the definition of $\mathbf{v}_2$. To see the first one: if $j \notin M_\beta$, the j-th entry of $\mathbf{v}_1$ is 0 by the definition of $\mathbf{v}_1$; if $j \in J_\beta$, the j-th entry of $\mathbf{v}_2$ is 0 and the j-th entry of $\mathbf{v}_1 + \mathbf{v}_2$ is 0 since $P^* u|_{J_\beta} = \mathbf{0}$, which implies that the j-th entry of $\mathbf{v}_1$ is 0). This proves $X_\beta^\perp \subseteq W_\beta'^\perp \oplus \bigoplus_{j=1}^t X_{\alpha_{r_j'}}^\perp$.

Similar as in (1), we see that φ is a morphism.

(4) To show φ is an isomorphism, we construct $\varphi^{-1}(q^0, (q^\perp, X_i^\perp)) = (q, X_i)$ as follows: we let $q = \psi^{-1}(q^0, q^\perp)$, and for $\alpha \in I_0$, $\beta \in I_1$, let

$$X_\alpha := (\text{ column space of } A_\alpha(q^0)) + X_\alpha^\perp = (\text{ column space of } \begin{bmatrix} \mathbf{I} \\ P_\alpha \end{bmatrix}) + X_\alpha^\perp,$$

$$X_\beta := (\text{ column space of } A_\beta(q^0) + (P_\beta^*)^{-1} X_\beta^\perp.$$

It is routine to check that φ and φ^{-1} are indeed inverse to each other.

(5) To check that the diagram commutes:

$$\psi \circ \pi(q, X_i) = \psi(q) = (q^0, q^\perp) = (1 \times \pi)(q^0, (q^\perp, X_i^\perp)) = (1 \times \pi) \circ \varphi(q, X_i).$$

(6) To check $\psi(p) = (p, 0)$: let $q = p$, and use the same assumption as above, we need to show $q^0 = q$ and $q^\perp = 0$. For $\alpha \in I_0$, since $A_\alpha(q)$ has rank v_α^0, all other rows are linear combinations of its first v_α^0 rows, that is, $A_\alpha(q) = \begin{bmatrix} A_\alpha(q)_{[1,\dots,v_\alpha^0;-]} \\ P' A_\alpha(q)_{[1,\dots,v_\alpha^0;-]} \end{bmatrix}$ for some matrix P'. Comparing with (4.3) we see that $P' = P_\alpha$. Thus $Z_\alpha' = P_\alpha Z_\alpha$,

$C''_{r_j} = P_\alpha C_{r_j}$, $C'''_{r_j} = P_\alpha C'_{r_j}$. Argue similarly for $A_\beta(q)$. We then conclude that $q = q^0$. Next, note that $\widetilde{A}_\alpha(q) = \begin{bmatrix} Z_\alpha & C_{r_1} & 0 & \cdots & C_{r_s} & 0 \\ 0 & 0 & 0 & \cdots & 0 & 0 \end{bmatrix}$, so $A_\alpha(q)^\perp = 0$, and similarly $A_\beta(q)^\perp = 0$. Thus $q^\perp = 0$.

The fact that $\psi(\mathbf{E}^\circ_{u,w} \cap U) = U_0 \times (\mathbf{E}^\circ_{u^\perp, w^\perp} \cap U^\perp)$, follows from (4.4) and (4.5). □

4.2. Proof of trivial local systems. It is the same as the proof of [**9**, Theorem 5.1]. We reproduce it here for the readers' convenience. As pointed out in [**9**, Remark 5.5], for the proof to work, it is essential to have the algebraic version, instead of the analytic version, of the Transversal Slice Theorem (Lemma 4.3).

LEMMA 4.4. *Given $f : X \to Y$, Y_a, L_a, n_a and the decomposition in (4.2). Let V be a nonsingular variety. Consider the following Cartesian diagram*

$$\begin{array}{ccc} V \times X & \xrightarrow{p'} & X \\ \Big\downarrow{\scriptstyle 1\times f} & & \Big\downarrow{\scriptstyle f} \\ V \times Y & \xrightarrow{p} & Y \end{array}$$

where p and p' are the natural projections. Define the pullback $\tilde{L}_a := p^* L_a$. Then

$$(1 \times f)_* IC_{V \times X} \cong \bigoplus_a IC_{V \times \overline{Y_a}}(\tilde{L}_a)[n_a].$$

PROOF. Same as [**9**, Lemma 5.4]. □

PROOF OF THEOREM 4.1. Let $IC_Z(L)[n]$ be a direct summand that appears in the decomposition of $\pi_*(IC_{\tilde{\mathcal{F}}_{v,w}})$. Assume that a nonempty open subset of Z is contained in a stratum $\mathbf{E}^\circ_{v^0,w}$ (thus $Z \subseteq \mathbf{E}_{v^0,w}$) and apply Lemma 4.3. Then U^0 is a Zariski open subset of $\mathbf{E}^\circ_{v^0,w}$. By uniqueness of the BBDG Decomposition, $IC_Z(L)[n]|_U$ is a direct summand of the decomposition of $\pi_*(IC_{\pi^{-1}(U)})$. By Lemma 4.4, we have $IC_Z(L)[n]|_U \cong IC_{U^0 \times \overline{Y_a}}(\tilde{L}_a)[n_a]$ for some a. Thus $Z \cap U = U^0 \times \overline{Y_a}$, $L \cong \tilde{L}_a$ on $Z \cap U$, and $n = n_a$. But $Z \subseteq \mathbf{E}_{v^0,w}$ implies $Z \cap U \subseteq U^0 \times \{0\}$. So we must have $Y_a = \{0\}$ and $Z \cap U = U^0 \times \{0\}$, thus $Z = \mathbf{E}_{v^0,w}$. Moreover, L_a is the trivial local system $\mathbb{Q}$. So the local system $\tilde{L}_a$ on U^0 is also trivial since it is the pullback of L_a under the map $U^0 \times \{0\} \to \{0\}$. Then L is trivial on $Z \cap U$, and we see that $IC_Z(L)[n] \cong IC_{\mathbf{E}_{v^0,w}}[n]$. Let $v' = v^0$. Then $v' \leq v$ as seen in Lemma 4.3, and (v', w) is l-dominant because $\mathbf{E}^\circ_{v^0,w}$ is nonempty. □

5. Dual canonical basis

The identification between dual canonical basis and triangular basis is explained to us by Fan Qin using results in [**7**, **13**].

5.1. Definitions of $L(w)$, $M(w)$, $\mathbf{R}^{\text{finite}}_t$, χ. Let $\mathcal{D}(\mathbf{E}_w)$ be the bounded derived category of constructible sheaves of $\mathbb{Q}$-vector spaces on $\mathbf{E}_w$. For simplicity of notation, denote $IC_w(v) := IC_{\mathbf{E}_{v,w}}$. Define $\mathcal{P}_w := \{IC_w(v) \mid v \in \mathrm{Dom}(w)\}$. Define a full subcategory $\mathcal{Q}_w$ of $\mathcal{D}(\mathbf{E}_w)$ whose objects are finite direct sums of $IC_w(v)[k]$ for various $v \in \mathrm{Dom}(w)$, $k \in \mathbb{Z}$.

Define an abelian group $\mathcal{K}_w$ to be generated by isomorphism classes (L) of objects of $\mathcal{Q}_w$ and quotient by the relations $(L) + (L') = (L'')$ whenever $L \oplus L' \cong L''$.

By abuse of notation we denote (L) as L. We can view $\mathcal{K}_w$ as a free $\mathbb{Z}[t^{\pm}]$-module with a basis $\mathcal{P}_w$, by defining $t^i L = L[i]$ for $i \in \mathbb{Z}$. The duality on $\mathcal{D}(\mathbf{E}_w)$ induces the bar involution on $\mathcal{K}_w$ satisfying $\overline{tL} = t^{-1}\overline{L}$, $\overline{IC_w(v)} = IC_w(v)$. By Theorem 4.1, for arbitrary v, w,

(5.1)
$$\pi_w(v) := \pi_*(IC_{\widetilde{\mathcal{F}}_{v,w}}) = \sum_{v' \in \mathrm{Dom}(w)} \sum_d a^d_{v,v';w} IC_w(v')[d] = \sum_{v' \in \mathrm{Dom}(w)} a_{v,v';w} IC_w(v')$$

where $a_{v,v';w} := \sum_d a^d_{v,v';w} t^d$. By Lemma 4.3,

$$a_{v,v';w} = a_{v^\perp,v'^\perp;w^\perp} \text{ for any } v^0 \in \mathrm{Dom}(w). \tag{5.2}$$

Then $\{\pi_w(v) \mid v \in \mathrm{Dom}(w)\}$ is also a $\mathbb{Z}[t^{\pm}]$-basis for $\mathcal{K}_w$, and $a_{v,v;w} = 1$ for $(v, w) \in \mathrm{Dom}$. Define the dual $\mathcal{K}_w^* := \mathrm{Hom}_{\mathbb{Z}[t^{\pm}]}(\mathcal{K}_w, \mathbb{Z}[t^{\pm}])$, which is a free $\mathbb{Z}[t^{\pm}]$-module with a basis $\{L_w(v) \mid v \in \mathrm{Dom}(w)\}$, the dual basis to $\mathcal{P}_w$, that is, $\langle L_w(v), IC_w(v')\rangle = \delta_{v,v'}$. The pairing $\langle -, -\rangle : \mathcal{K}_w^* \times \mathcal{K}_w \to \mathbb{Z}[t^{\pm}]$ satisfies the condition $\langle L, C[1]\rangle = \langle L[-1], C\rangle = t\langle L, C\rangle$, for all $L \in \mathcal{K}_w^*, C \in \mathcal{K}_w$. Then $\langle L_w(v), IC_w(v')\rangle = \delta_{v,v'} = \delta_{v^\perp,v'^\perp} = \langle L_{w^\perp}(v^\perp), IC_{w^\perp}(v'^\perp)\rangle$ for any $v^0 \leq v$, where v^0 is used to define $w^\perp = w - C_q v^0, v^\perp = v - v^0, v'^\perp = v' - v^0$. Define

$$\mathbf{R}_t := \{(f_w) \in \prod_w \mathcal{K}_w^* \mid \langle f_w, IC_w(v)\rangle = \langle f_{w^\perp}, IC_{w^\perp}(v^\perp)\rangle$$
$$\text{whenever } v \in \mathrm{Dom}(w), v^0 \leq v\}.$$

For $(f_w) \in \prod_w \mathcal{K}_w^*$, define $c_{wv} \in \mathbb{Z}[t^{\pm}]$ to satisfy $f_w = \sum_{v\in\mathrm{Dom}(w)} c_{wv} L_w(v)$. Then $c_{wv} = \langle f_w, IC_w(v)\rangle$, and $(f_w) \in \mathbf{R}_t$ if and only if $c_{wv} = c_{w-C_q v,0}$ for every l-dominant pair (w, v). So an element in $\mathbf{R}_t$ is uniquely determined by $\{c_{w0}\}_w$.

Define $L(w) \in \mathbf{R}_t$ to be induced by $L_w(0)$, that is, for any $(v', w') \in \mathrm{Dom}$, $\langle L(w), IC_{w'}(v')\rangle = \delta_{w,w'-C_q v'}$. Define $\mathbf{R}_t^{\text{finite}}$ to be the $\mathbb{Z}[t^{\pm}]$-submodule of $\mathbf{R}_t$ with the basis $\{L(w)\}_w$.

Define $\{M_w(v) \mid v \in \mathrm{Dom}(w)\} \in \mathcal{K}_w^*$ to be the functional

$$(L) \mapsto \sum_k t^{\dim \mathbf{E}^\circ_{v,w} - k} \dim H^k(i^!_{x_{v,w}} L)$$

where $x_{v,w}$ is a point in $\mathbf{E}^\circ_{v,w}$, and $i_{x_{v,w}} : \{x_{v,w}\} \to \mathbf{E}_w$ is the natural embedding.

Define $M(w) \in \mathbf{R}_t$ to be induced by $M_w(0)$, that is, denoting $i_0 : \{0\} \to \mathbf{E}_w$ to be the natural embedding, then for any $(v', w') \in \mathrm{Dom}$,

(5.3)
$$\langle M(w), IC_{w'}(v')\rangle$$
$$= \begin{cases} \langle M_w(0), IC_w(v)\rangle = \sum_k t^{-k} \dim H^k(i^!_0 IC_w(v)), & \text{if } \exists v \in \mathbb{Z}^n_{\geq 0} : w - w' = C_q(v - v'); \\ 0, \text{ otherwise.} \end{cases}$$

Similar to [**9**, §7], denote a partial order $w' \leq_{\mathrm{w}} w \Leftrightarrow w' - w = C_q u$ for some $u \geq 0$ and denote $w' <_{\mathrm{w}} w$ if $w' \leq_{\mathrm{w}} w$ and $w' \neq w$. Then $\{M(w)\}_w$ is a basis of $\mathbf{R}_t^{\text{finite}}$, and

$$L(w) = \sum_{w'' \leq_{\mathrm{w}} w} b'_{ww''} M(w'') \in M(w) + \sum_{w'' <_{\mathrm{w}} w} t^{-1}\mathbb{Z}[t^{-1}] M(w''). \tag{5.4}$$

Define a map

$$\Phi : \mathbb{Z}^{2n} \mapsto \mathbb{Z}^{2n}, \quad \Phi(w) = \sum_{\beta \in I_1} (w_\beta - w'_\beta) e_\beta + \sum_{\alpha \in I_0} (w'_\alpha - w_\alpha) e_\alpha. \tag{5.5}$$

Define a map $\chi : \mathbf{R}_t^{\text{finite}} \to \mathcal{T}$ by defining it on the basis $\{M(w)\}$ (note that we do not require $v \in \mathrm{Dom}(w)$):

$$\begin{aligned} \chi(M(w)) &= \sum_{v \in \mathbb{Z}^n_{\geq 0}} \left(\langle M(w), \pi_w(v) \rangle_{t \to \mathbf{v}} \right) X^{\Phi(w) + \tilde{B}v} \\ &= \sum_{v \in \mathbb{Z}^n_{\geq 0}} \sum_k \mathbf{v}^{-k} \dim H^k(i_0^! \pi_w(v)) X^{\Phi(w) + \tilde{B}v} \end{aligned} \tag{5.6}$$

and extend it by the rules $\chi(L_1 + L_2) = \chi(L_1) + \chi(L_2)$ and $\chi(tL) = \mathbf{v}\chi(L)$. The following is shown in [**9**, §7.3]:

$$\chi(L(w)) = \sum_{v \in \mathbb{Z}^n_{\geq 0}} a_{v,0;w}(\mathbf{v}) X^{\Phi(w) + \tilde{B}v}. \tag{5.7}$$

5.2. $\{\overline{\chi(M(w))}\}$ and the standard monomial basis. All cohomology groups in this paper are taken with coefficient $\mathbb{Q}$, thus we write $\dim H^i(X)$ for $\dim H^i(X, \mathbb{Q})$.

Denote $P_{\mathbf{v}}(X) := \sum_i \dim H^i(X) \mathbf{v}^i$. For $n, k \in \mathbb{Z}_{\geq 0}$, define the q-binomial coefficient as follows (where q is replaced by $\mathbf{v}$).

$$[n] := \frac{\mathbf{v}^n - \mathbf{v}^{-n}}{\mathbf{v} - \mathbf{v}^{-1}}, \qquad \begin{bmatrix} n \\ k \end{bmatrix} := \frac{[n][n-1]\cdots[n-k+1]}{[k][k-1]\cdots[1]}.$$

The cohomology groups of a (complex) Grassmannian is well-known; from which we have

$$P_{\mathbf{v}}(Gr(k,n)) = \mathbf{v}^{\dim_{\mathbb{C}} Gr(k,n)} \begin{bmatrix} n \\ k \end{bmatrix} = \mathbf{v}^{k(n-k)} \begin{bmatrix} n \\ k \end{bmatrix}.$$

The following is also well-known.

LEMMA 5.1. *Let V be a complex vector bundle of rank n over a complex variety X, $1 \leq d \leq n-1$, and $E = Gr_X(d, V)$ be the associated Grassmannian bundle over X, $F \cong Gr(d,n)$ be a fiber. Then*

$$P_{\mathbf{v}}(E) = P_{\mathbf{v}}(F) \cdot P_{\mathbf{v}}(X).$$

PROOF. It follows the Leray–Hirsch theorem [**6**, §4.D] and the surjectivity of $H^*(E) \to H^*(F)$. For the latter, recall that $A^*(X)$ is the Chow ring of X and $c_k(E)$ (for $k = 1, \ldots, n$) are the Chern classes of E. By [**5**, Example 14.6.6],

$$A^*(E) = A^*(X)[a_1, \ldots, a_d, b_1, \ldots, b_{n-d}] / (\sum_{i=0}^{k} a_i b_{k-i} - c_k(E))_{k=1,\ldots,n},$$

$$A^*(F) = \mathbb{Q}[a_1, \ldots, a_d, b_1, \ldots, b_{n-d}] / (\sum_{i=0}^{k} a_i b_{k-i} - \delta_{k,0})_{k=1,\ldots,n},$$

thus the restriction $A^k(E) \to A^k(F)$ is surjective for any k. So the composition $A^k(E) \to H^{2k}(E) \to H^{2k}(F) \cong A^k(F)$ is surjective. Thus $H^{2k}(E) \to H^{2k}(F)$ is surjective. (Alternatively, one can prove an analog of [**5**, Example 14.6.6] for H^* and not use Chow rings.) □

LEMMA 5.2.

(a) *We have*

$$P_{\mathbf{v}}(\mathcal{F}_{v,w}) = \prod_{\beta\in I_1} P_{\mathbf{v}}(Gr(v_\beta, w'_\beta + \sum_{h:s(h)=\beta} v_{t(h)})) \cdot \prod_{\alpha\in I_0} P_{\mathbf{v}}(Gr(v_\alpha, w_\alpha))$$
$$= \mathbf{v}^{d_{v,w}} \prod_{\beta\in I_1} \begin{bmatrix} w'_\beta + \sum_{h:s(h)=\beta} v_{t(h)} \\ v_\beta \end{bmatrix} \cdot \prod_{\alpha\in I_0} \begin{bmatrix} w_\alpha \\ v_\alpha \end{bmatrix}.$$

(b) *Recall that* $d_{v,w} = \dim \mathcal{F}_{v,w}$, $\tilde{d}_{v,w} = \dim \widetilde{\mathcal{F}}_{v,w}$. *Then*

(5.8)
$$\chi(M(w)) = \sum_{v\in\mathbb{Z}^n_{\geq 0}} \mathbf{v}^{d_{v,w}-\tilde{d}_{v,w}} \prod_{\beta\in I_1} \begin{bmatrix} w'_\beta + \sum_{h:s(h)=\beta} v_{t(h)} \\ v_\beta \end{bmatrix} \cdot \prod_{\alpha\in I_0} \begin{bmatrix} w_\alpha \\ v_\alpha \end{bmatrix} X^{\Phi(w)+\tilde{B}v}$$
$$= \sum_{v\in\mathbb{Z}^n_{\geq 0}} \mathbf{v}^{-\tilde{d}_{v,w}} P_{\mathbf{v}}(\mathcal{F}_{v,w}) X^{\Phi(w)+\tilde{B}v}.$$

PROOF.

(a) follows from iteratively applying Lemma 5.1.
(b) Denote the embedding $i_0 : \{0\} \to \mathbf{E}_{v,w}$. Then

$$\langle M(w), \pi_w(v)\rangle_{t\to\mathbf{v}} = \mathbf{v}^{2d_{v,w}-\tilde{d}_{v,w}} \sum_k \mathbf{v}^{-k} \dim H^k(\mathcal{F}_{v,w})$$
$$= \mathbf{v}^{2d_{v,w}-\tilde{d}_{v,w}} \dim P_{\mathbf{v}^{-1}}(\mathcal{F}_{v,w})$$

(5.9)
$$\overset{(a)}{=} \mathbf{v}^{2d_{v,w}-\tilde{d}_{v,w}} \mathbf{v}^{-\dim_{\mathbb{C}} \mathcal{F}_{v,w}} \prod_{\beta\in I_1} \begin{bmatrix} w'_\beta + \sum_{h:s(h)=\beta} v_{t(h)} \\ v_\beta \end{bmatrix} \cdot \prod_{\alpha\in I_0} \begin{bmatrix} w_\alpha \\ v_\alpha \end{bmatrix}$$
$$= \mathbf{v}^{d_{v,w}-\tilde{d}_{v,w}} \prod_{\beta\in I_1} \begin{bmatrix} w'_\beta + \sum_{h:s(h)=\beta} v_{t(h)} \\ v_\beta \end{bmatrix} \cdot \prod_{\alpha\in I_0} \begin{bmatrix} w_\alpha \\ v_\alpha \end{bmatrix} = \mathbf{v}^{-\tilde{d}_{v,w}} P_{\mathbf{v}}(\mathcal{F}_{v,w}).$$

This together with (5.6) imply (5.8). □

Without loss of generality, assume $I_1 = \{1, \ldots, k\}$, $I_0 = \{k+1, \ldots, n\}$. We denote $t' = \mu_{I_1}(t_0) := \mu_1 \cdots \mu_k(t_0)$ and the attached seed is $(\Lambda', \tilde{B}', \tilde{X}')$. Thus $\tilde{B}' = \begin{bmatrix} -B \\ J \end{bmatrix}$ where J is the diagonal matrix with the first k diagonal entries being -1 and the rest being 1, $\tilde{\Lambda}' = \begin{bmatrix} 0 & -J \\ J & -B \end{bmatrix}$. The standard monomial basis elements with respect to this initial seed is

$$E_a(t') = \mathbf{v}^{v'(a)} X^{\sum_{i=n+1}^{2n} a_i e_i} \prod_{i=n,\ldots,1} \left(X_i(t')^{[a_i]_+} X'_i(t')^{[-a_i]_+} \right), \text{ for } a \in \mathbb{Z}^{2n}.$$

where the index i runs decreasingly from n to 1, $X'_i(t') = \mu_i(X_i(t'))$, and $v'(a) \in \mathbb{Z}$ is determined by the condition that the leading term of $E_a(t')$ in seed t' is bar-invariant. Then $X'_i(t') = X_i$ for $i \in I_1$, $X'_i(t') = X(t')^{-e_i+[b'_i]_+} + X(t')^{-e_i+[-b'_i]_+}$ for $i \in I_0$. Note that in the above expression of $E_a(t')$, the two factors in $X_i(t')^{[a_i]_+} X'_i(t')^{[-a_i]_+}$ can be swapped because one of them is 1.

We modify the above expression of $E_a(t')$ into E^*_a as follows:

$$E^*_{w,a_{n+1},\dots,a_{2n}} := \mathbf{v}^{v'} X^{\sum_{i=n+1}^{2n} a_i e_i} \prod_{i=n,\dots,k+1} \left(X'_i(t')^{w_i} X_i(t')^{w'_i}\right) \cdot \prod_{i=k,\dots,1} \left(X_i(t')^{w'_i} X'_i(t')^{w_i}\right)$$

where v' is similar as above. Note a technical subtlety that in the first product (for $i = n, \dots, k+1$) we have to write the two factors of $X'_i(t')^{w_i} X_i(t')^{w'_i}$ in that specific order; if we swap the two factors, then the power of $\mathbf{v}$ in Lemma 5.4 will be incorrect.

Each $w \in \mathbb{Z}^{2n}_{\geq 0}$ can be uniquely written as $w = {}^f w + {}^\phi w$ where both ${}^f w$ and ${}^\phi w$ are in $\mathbb{Z}^{2n}_{\geq 0}$ and ${}^\phi w$ satisfies ${}^\phi w_i {}^\phi w'_i = 0$ for all $1 \leq i \leq n$. We have

$$\text{if } w = {}^\phi w, \text{ then } E^*_{w,a_{n+1},\dots,a_{2n}} = E_a(t'). \tag{5.10}$$

The following is easy to show so omit skip the proof.

LEMMA 5.3. *Assume $u_i \geq 0$ for $i \in I_1$. Then*

$$X(t')^u = \sum_v \left(\prod_{i \in I_1} \begin{bmatrix} u_i \\ v_i \end{bmatrix}\right) X^{\sum_{i \in I_1}(-u_i e_i + v_i b_i) + \sum_{i \notin I_1} u_i e_i}.$$

LEMMA 5.4. *For any $w \in \mathbb{Z}^{2n}_{\geq 0}$, we have*

$$\begin{aligned} E^*_{w,0,\dots,0} &= \overline{\chi(M(w))} \\ &= \sum_{v \in \mathbb{Z}^n_{\geq 0}} \mathbf{v}^{\tilde{d}_{v,w} - d_{v,w}} \left(\prod_{\alpha \in I_0} \begin{bmatrix} w_\alpha \\ v_\alpha \end{bmatrix} \prod_{\beta \in I_1} \begin{bmatrix} w'_\beta + \sum_{\alpha \in I_0} b_{\alpha\beta} v_\alpha \\ v_\beta \end{bmatrix}\right) X^{\Phi(w) + \sum_{1 \leq i \leq n} v_i b_i}. \end{aligned}$$

where Φ is defined in (5.5). In particular, if $w = {}^\phi w$, then $\overline{\chi(M(w))}$ is the standard basis element $E_{(w'_i - w_i)_{1 \leq i \leq n}}(t')$ with the initial seed at $t' = \mu_{I_1}(t_0)$.

PROOF. For $i \in I_0$, we have $b'_i \in \mathbb{Z}^m_{\geq 0}$, thus

$$\begin{aligned} X'_i(t')^{w_i} X_i(t')^{w'_i} &= \left(X(t')^{-e_i + b'_i} + X(t')^{-e_i}\right)^{w_i} X(t')^{w'_i e_i} \\ &= \sum_{v_i} \begin{bmatrix} w_i \\ v_i \end{bmatrix} X(t')^{-w_i e_i + v_i b'_i} X(t')^{w'_i e_i} \\ &= \sum_{v_i} \mathbf{v}^{\Lambda'(-w_i e_i + v_i b'_i, w'_i e_i)} \begin{bmatrix} w_i \\ v_i \end{bmatrix} X(t')^{(w'_i - w_i)e_i + v_i b'_i} \\ &= \sum_{v_i} \mathbf{v}^{w'_i v_i} \begin{bmatrix} w_i \\ v_i \end{bmatrix} X(t')^{(w'_i - w_i)e_i + v_i b'_i}. \end{aligned}$$

For $i \in I_1$, $X_i(t')^{w'_i} X'_i(t')^{w_i} = X(t')^{w'_i e_i} X_i^{w_i}$.

By comparing the leading terms, it is easy to see that the power of $\mathbf{v}$ to normalize E^*_w in t' is the same as the one to normalize E^*_w in t. Thus we ignore factor of powers of $\mathbf{v}$ in the following computation, and denote $f \sim \mathbf{v}^i f$ for $i \in \mathbb{Z}$.

$$E^*_{w,0,\dots,0} \sim \Big(\prod_{i=n,\dots,k+1} \sum_{v_i} \mathbf{v}^{w'_i v_i} \begin{bmatrix} w_i \\ v_i \end{bmatrix} X(t')^{(w'_i - w_i)e_i + v_i b'_i} \Big) \cdot \Big(\prod_{i=k,\dots,1} X(t')^{w'_i e_i} \, X_i^{w_i} \Big)$$

$$\sim \sum_{v_n,\dots,v_{k+1}} \mathbf{v}^{p_1} \Big(\prod_{i=n,\dots,k+1} \begin{bmatrix} w_i \\ v_i \end{bmatrix} \Big) X(t')^{\sum_{i=k+1}^{n} \left((w'_i - w_i)e_i + v_i b'_i\right) + \sum_{i=1}^{k} w'_i e_i} \prod_{i=k,\dots,1} X_i^{w_i}$$

$$\sim \sum_{v_n,\dots,v_{k+1}} \mathbf{v}^{p_1} \Big(\prod_{i=n,\dots,k+1} \begin{bmatrix} w_i \\ v_i \end{bmatrix} \Big)$$

$$\cdot X(t')^{\sum_{i \in I_1} (w'_i + \sum_{j \in I_0} b_{ji} v_j) e_i + \sum_{i \in I_0} (w'_i - w_i) e_i + \sum_{i \in I_0} v_i e_{n+i}} \prod_{i=k,\dots,1} X_i^{w_i}$$

$$\sim \sum_{v} \mathbf{v}^{p_1} \Big(\prod_{\alpha \in I_0} \begin{bmatrix} w_\alpha \\ v_\alpha \end{bmatrix} \prod_{\beta \in I_1} \begin{bmatrix} w'_\beta + \sum_{\alpha \in I_0} b_{\alpha\beta} v_\alpha \\ v_\beta \end{bmatrix} \Big)$$

$$\cdot X^{\sum_{\beta \in I_1} ((-w'_\beta - \sum_{\alpha \in I_0} v_\alpha b_{\beta\alpha}) e_\beta + v_\beta b_\beta) + \sum_{\alpha \in I_0} (w'_\alpha - w_\alpha) e_\alpha + \sum_{\alpha \in I_0} v_\alpha e_{n+\alpha}}$$

$$\cdot \prod_{\beta=k,\dots,1} X_\beta^{w_\beta} \text{ (Lemma 5.3)}$$

$$\sim \sum_{v} \mathbf{v}^{p_1} \Big(\prod_{\alpha \in I_0} \begin{bmatrix} w_\alpha \\ v_\alpha \end{bmatrix} \prod_{\beta \in I_1} \begin{bmatrix} w'_\beta + \sum_{\alpha \in I_0} b_{\alpha\beta} v_\alpha \\ v_\beta \end{bmatrix} \Big)$$

$$\cdot X^{\sum_{\beta \in I_1} (-w'_\beta e_\beta) + \sum_{\alpha \in I_0} (w'_\alpha - w_\alpha) e_\alpha + \sum_{1 \le i \le n} v_i b_i}$$

$$\cdot \prod_{\beta=k,\dots,1} X_\beta^{w_\beta} \text{ (used } v_\alpha e_{n+\alpha} - \sum_{\beta \in I_1} v_\alpha b_{\beta\alpha} e_\beta = v_\alpha b_\alpha)$$

$$\sim \sum_{v} \mathbf{v}^{p_2} \Big(\prod_{\alpha \in I_0} \begin{bmatrix} w_\alpha \\ v_\alpha \end{bmatrix} \prod_{\beta \in I_1} \begin{bmatrix} w'_\beta + \sum_{\alpha \in I_0} b_{\alpha\beta} v_\alpha \\ v_\beta \end{bmatrix} \Big)$$

$$\cdot X^{\sum_{\beta \in I_1} (w_\beta - w'_\beta) e_\beta + \sum_{\alpha \in I_0} (w'_\alpha - w_\alpha) e_\alpha + \sum_{1 \le i \le n} v_i b_i}$$

where $p_1 = \sum_{\alpha \in I_0} w'_\alpha v_\alpha$, $p_2 = \sum_{\alpha \in I_0} w'_\alpha v_\alpha + \sum_{\beta \in I_1} w_\beta v_\beta = \tilde{d}_{v,w} - d_{v,w}$. Since the last expression is already normalized, it gives a correct formula for $E^*_{w,0,\dots,0}$. This completes the proof. □

Lemma 5.5. *Let $w \in \mathbb{Z}^{2n}_{\ge 0}$, $a = (w'_1 - w_1, \dots, w'_n - w_n, a_{n+1}, \dots, a_{2n}) \in \mathbb{Z}^{2n}$.*

(a) *If $\min(w_\alpha, w'_\alpha) > 0$ for some $\alpha \in I_0$, then*

$$E^*_{w,a_{n+1},\dots,a_{2n}} = E^*_{w^{(1)},a_{n+1},\dots,a_{2n}} + \mathbf{v}^p E^*_{w^{(2)},a'_{n+1},\dots,a_{n+\alpha}+1,\dots,a_{2n}}$$

where p is a positive integer, $w^{(1)}$ is obtained from w by $w_\alpha \mapsto w_\alpha - 1$ and $w'_\alpha \mapsto w'_\alpha - 1$; $w^{(2)}$ is obtained from w by $w_\alpha \mapsto w_\alpha - 1$, $w'_\alpha \mapsto w'_\alpha - 1$, $w'_\beta \mapsto w'_\beta + b'_{\beta\alpha}$.

(b) *If $\min(w_\beta, w'_\beta) > 0$ for some $\beta \in I_0$, then*

$$E^*_{w,a_{n+1},\dots,a_{2n}} = E^*_{w^{(1)},a_{n+1},\dots,a_{2n}} + \mathbf{v}^p E^*_{w^{(2)},a'_{n+1},\dots,a_{n+\beta}+1,\dots,a_{2n}}$$

where p is a positive integer, $w^{(1)}$ is obtained from w by $w_\beta \mapsto w_\beta - 1$ and $w'_\beta \mapsto w'_\beta - 1$; $w^{(2)}$ is obtained from w by $w_\beta \mapsto w_\beta - 1$, $w'_\beta \mapsto w'_\beta - 1$, $w'_\alpha \mapsto w'_\alpha - b'_{\alpha\beta}$.

(c) *We have*

$$E^*_{w,a_{n+1},\dots,a_{2n}} \in E_a(t') + \sum_{r(b) \le r(a)} \mathbf{v}\mathbb{Z}[\mathbf{v}] E_b(t').$$

PROOF.

(a) If $\min(w_\alpha, w'_\alpha) > 0$ for $\alpha \in I_0$, then

$$\begin{aligned}
&E^*_{w,a_{n+1},\dots,a_{2n}}\\
&\sim X^{\sum_{i=n+1}^{2n} a_i e_i} \prod_{i=n,\dots,\alpha+1} \left(X'_i(t')^{w_i} X_i(t')^{w'_i}\right) \cdot X'_\alpha(t')^{w_\alpha - 1} \cdot (X'_\alpha(t') X_\alpha(t'))\\
&\quad \cdot X_\alpha(t')^{w'_\alpha - 1} \cdot \prod_{i=\alpha-1,\dots,k+1} \left(X'_i(t')^{w_i} X_i(t')^{w'_i}\right) \prod_{i=k,\dots,1} \left(X_i(t')^{w'_i} X'_i(t')^{w_i}\right)\\
&\sim X^{\sum_{i=n+1}^{2n} a_i e_i} \prod_{i=n,\dots,\alpha+1} \left(X'_i(t')^{w_i} X_i(t')^{w'_i}\right) \cdot X'_\alpha(t')^{w_\alpha - 1} \cdot (1 + \mathbf{v} X(t')^{b'_\alpha})\\
&\quad \cdot X_\alpha(t')^{w'_\alpha - 1} \cdot \prod_{i=\alpha-1,\dots,k+1} \left(X'_i(t')^{w_i} X_i(t')^{w'_i}\right) \prod_{i=k,\dots,1} \left(X_i(t')^{w'_i} X'_i(t')^{w_i}\right)\\
&\sim \mathbf{v}^{p_1} E^*_{w^{(1)},a_{n+1},\dots,a_{2n}} + \mathbf{v}^{p_2} E^*_{w^{(2)},a_{n+1},\dots,a_{n+\alpha}+1,\dots,a_{2n}}
\end{aligned}$$

where $w^{(1)}$ is obtained from w by $w_\alpha \mapsto w_\alpha - 1$ and $w'_\alpha \mapsto w'_\alpha - 1$; $w^{(2)}$ is obtained from w by $w_\alpha \mapsto w_\alpha - 1$, $w'_\alpha \mapsto w'_\alpha - 1$, $w'_\beta \mapsto w'_\beta + b'_{\beta\alpha}$. Note that $(w^{(2)}, a_{n+1}, \dots, a_{n+\alpha} + 1, \dots, a_{2n})$ corresponds to $b = (a_1 + b'_{1\alpha}, \dots, a_k + b'_{k\alpha}, a_{k+1}, \dots, a_n, a_{n+1}, \dots, a_{n+\alpha} + 1, \dots, a_{2n})$, which satisfies $b \prec a$ or, unpleasantly, $r(b) = r(a)$. If the latter case happens, at least we see that $\sum_i w_i$ strictly decreases.

Next, we show that $p > 0$. Indeed, note that $e'_\alpha = -e_\alpha$ for $\alpha \in I_0$, $e'_\beta = -e_\beta + (-b'_\beta)$ for $\beta \in I_0$, and $p_1 = \Lambda'\big(\sum_{n+1}^{2n} a_i e_i, w_n e'_n, w'_n e_n, \dots, (w_\alpha - 1)e'_\alpha, (w'_\alpha - 1)e_\alpha, \dots, w_{k+1}e'_{k+1}, w'_{k+1}e_{k+1}, w'_k e_k, w_k e'_k, \dots, w'_1 e_1, w_1 e'_1\big)$, while $p_2 = 1 + \Lambda'\big(\sum_{n+1}^{2n} a_i e_i, w_n e'_n, w'_n e_n, \dots, (w_\alpha - 1)e'_\alpha, b'_\alpha, (w'_\alpha - 1)e_\alpha, \dots, w_{k+1}e'_{k+1}, w'_{k+1}e_{k+1}, w'_k e_k, w_k e'_k, \dots, w'_1 e_1, w_1 e'_1\big)$. It is routine to verify that

$$p = p_2 - p_1 = w_\alpha + w'_\alpha - 1 + \sum_{\beta \in I_1} w_\beta(-b'_{\alpha\beta}) \geq 1$$

where the last inequality is because $b'_{\alpha\beta} \leq 0$ for $\alpha \in I_0, \beta \in I_1$.

(b) is proved similarly to (a).

(c) is obtained by applying (a) and (b) recursively.

□

5.3. $\{\chi(L(w))\}$ and the triangular basis.

LEMMA 5.6. *For any $w \in \mathbb{Z}^{2n}_{\geq 0}$, $\chi(L(w)) = C_{\Phi(w)}$ is a triangular basis element. In particular, $\chi(L({}^\phi w)) = \chi(L(w))$ (therefore $a_{v,0;w} = a_{v,0;{}^\phi w}$ for any $v \in \mathbb{Z}^n_{\geq 0}$) and $\chi(L({}^f w)) = 1$.*

PROOF. It follows from (5.7) that $\chi(L(w))$ is bar-invariant. Meanwhile,

$$\chi(L(w)) - E_{\Phi(w)}(t') \in \chi(L(w)) - E^*_{w,0,\dots,0} + \sum \mathbf{v}\mathbb{Z}[\mathbf{v}]E_b(t') \quad \text{(by Lemma 5.5)}$$
$$= \chi(L(w)) - \overline{\chi(M(w))} + \sum \mathbf{v}\mathbb{Z}[\mathbf{v}]E_b(t') \quad \text{(by Lemma 5.4)}$$
$$\subseteq \overline{\chi(M(w))} + \sum \mathbf{v}\mathbb{Z}[\mathbf{v}]\overline{\chi(M(w''))} - \overline{\chi(M(w))} + \sum \mathbf{v}\mathbb{Z}[\mathbf{v}]E_b(t') \quad \text{(by (5.4))}$$
$$= \sum \mathbf{v}\mathbb{Z}[\mathbf{v}]E^*_{w'',0,\dots,0} + \sum \mathbf{v}\mathbb{Z}[\mathbf{v}]E_b(t') \quad \text{(by Lemma 5.4)}$$
$$\subseteq \sum \mathbf{v}\mathbb{Z}[\mathbf{v}](\sum \mathbb{Z}[\mathbf{v}]E_{b'}(t') + \sum \mathbf{v}\mathbb{Z}[\mathbf{v}]E_b(t')$$
$$= \sum \mathbf{v}\mathbb{Z}[\mathbf{v}]E_b(t').$$

By [**3**, Theorem 1.1], $\chi(L(w))$ equals to a triangular basis element for the seed at t'. Since the triangular basis does not depend on the chosen acyclic seed (also proved in [**3**]), $\chi(L(w))$ is also a triangular basis element for initial seed at t_0. □

6. The proof of the main result

We cite the following lemma in [**9**].

LEMMA 6.1. *Let $f : Y \to X$ be a proper morphism between complex algebraic varieties, Y be nonsingular, let 0 be a point in X. Let $d = \dim Y$, $Y_0 = f^{-1}(0)$ and $d_0 = \dim Y_0$. Write the BBDG decomposition in the form*

$$f_* IC_Y = \bigoplus_b IC_0^{\oplus s_{0,b}}[b] \oplus \bigoplus_{V,L,b} IC_V^{\oplus s_{V,L,b}}(L)[b] \tag{6.1}$$

where $V \neq 0$ are subvarieties of X, each L is a local system on an open dense subset of V, $s_{0,b}, s_{V,L,b} \in \mathbb{Z}_{\geq 0}$ are multiplicities of the corresponding IC-sheaves. Then $\{s_{0,b}\}$ satisfy the following conditions:

(i) *$s_{0,b} = s_{0,-b}$ for every $b \in \mathbb{Z}$.*
(ii) *$s_{0,b} \geq s_{0,b+2}$ for every $b \in \mathbb{Z}_{\geq 0}$.*
(iii) *$s_{0,b} = 0$ if $|b| > 2d_0 - d$. In particular, if $2d_0 < d$, then $s_{0,b} = 0$ for all b.*

We also need the following lemma to prove the main theorem.

LEMMA 6.2. *Assume (1.2) and that $a_\circ \in \mathbb{Z}^{2n}$ has its first n coordinates equal to 0, then $C_{a+a_\circ} = \sum_v e_v X^{\varphi(a)+a_\circ+\tilde{B}v}$. As a consequence, each e_v does not depend on the last n coordinates of a.*

PROOF. Write $C_a = \sum_{a'} f_{aa'}E_{a'}$ where $f_{aa} = 1$ and other $f_{aa'} \in \mathbf{v}\mathbb{Z}[\mathbf{v}]$. Those a, a' must satisfy the condition that $\varphi(a') - \varphi(a) = \sum_i \lambda_i b_i = \tilde{B}\lambda$ where $\lambda = (\lambda_1, \dots, \lambda_n)^T \in \mathbb{Z}^n$, because $E_{a'}$ has a term $X^{\varphi(a')}$, and the exponents of any two terms in $E_{a'}$ satisfy a similar condition. For any a' such that $\varphi(a') = \varphi(a) + \tilde{B}\lambda$, we have $\Lambda(a_\circ, \varphi(a')) = \Lambda(a_\circ, \varphi(a)) =: p$, because

$$\Lambda(a_\circ, \varphi(a')) - \Lambda(a_\circ, \varphi(a)) = \Lambda(a_\circ, \tilde{B}\lambda) = a_\circ^T \begin{bmatrix} 0 & -\mathbf{I}_n \\ \mathbf{I}_n & -B \end{bmatrix} \begin{bmatrix} B \\ \mathbf{I}_n \end{bmatrix} \lambda$$
$$= a_\circ^T \begin{bmatrix} -\mathbf{I}_n \\ 0 \end{bmatrix} \lambda = 0\lambda = 0.$$

Moreover,

$$E_{a'+a_\circ} = \mathbf{v}^{-p} X^{a_\circ} E_{a'} = \mathbf{v}^p E_{a'} X^{a_\circ} \tag{6.2}$$

(to show the second equality, note that $X^{a_\circ}$ and $E_{a'}$ quasi-commute, thus $E_{a'}X^{a_\circ} = \mathbf{v}^i X^{a_\circ} E_{a'}$ for some integer i; then compare the coefficient of $X^{\varphi(a')+a_\circ}$ on both sides, we see $i = \mathbf{v}^{-2p}$). Taking bar on two sides, we have

$$\overline{E}_{a'+a_\circ} = \mathbf{v}^{-p} X^{a_\circ} \overline{E}_{a'} \tag{6.3}$$

We claim that $\sum_{a'} f_{aa'} E_{a'+a_\circ}$ is bar-invariant, which immediately implies the lemma. Indeed, since C_a is bar-invariant, we have $\sum_{a'} \bar{f}_{aa'} \overline{E}_{a'} = \sum_{a'} f_{aa'} E_{a'}$. Left multiplying both sides by $\mathbf{v}^{-p} X^{a_\circ}$ and using (6.2) and (6.3), we get $\sum_{a'} \bar{f}_{aa'} \overline{E}_{a'+a_\circ} = \sum_{a'} f_{aa'} E_{a'+a_\circ}$, which proves the claim. □

PROOF OF THEOREM 1.1. Lemma 6.2 shows that e_v is independent of the last n coordinates of a. Define $w = (w_i, w_i')_{1\le i\le n} \in \mathbb{Z}^{2n}_{\ge 0}$ as follows: for $\alpha \in I_0$, define $w_\alpha = [-g_\alpha]_+$, $w'_\alpha = [g_\alpha]_+$; for $\beta \in I_1$, define $w_\beta = [g_\beta]_+$, $w'_\beta = [-g_\beta]_+$. Then $\Phi(w) = (g_1, \dots, g_n, 0, \dots, 0)$. By (5.7), $e_v = a_{v,0;w}$. Apply Lemma 6.1 to the projection $\pi : \tilde{\mathcal{F}}_{v,w} \to \mathbf{E}_{v,w}$ defined in (3.4), we conclude that e_v is symmetric, unimodal, and

$$\deg(e_v) = \deg a_{v,0;w} \le 2d_{v,w} - \tilde{d}_{v,w} = f(v)$$

by Proposition 3.7 (b). □

We end the paper with two examples, computed using SageMath [**16**].

EXAMPLE 6.3.

(a) Consider the initial seed determined by $B = \begin{bmatrix} 0 & -3 \\ 3 & 0 \end{bmatrix}$.

The corresponding quiver $\mathcal{Q}^{\mathrm{op}}$ is $1 \Rrightarrow 2$.

We compute the triangle basis element C_a for $a = (-3, -4, 0, 0)$. The corresponding $g = \varphi(a) = (9, -4, 0, 0)$, $w = (w_1, w_1', w_2, w_2') = (9, 0, 4, 0)$. Then

$$\begin{aligned} C_{(-3,-4,0,0)} = {} & X^{(-3,-4,0,4)} + (\mathbf{v}^2 + 1 + \mathbf{v}^{-2})X^{(-3,-1,1,4)} \\ & + (\mathbf{v}^2 + 1 + \mathbf{v}^{-2})X^{(-3,2,2,4)} \\ & + X^{(-3,5,3,4)} + (\mathbf{v}^3 + \mathbf{v} + \mathbf{v}^{-1} + \mathbf{v}^{-3})X^{(0,-4,0,3)} \\ & + (\mathbf{v}^2 + 1 + \mathbf{v}^{-2})X^{(0,-1,1,3)} \\ & + (\mathbf{v}^4 + \mathbf{v}^2 + 2 + \mathbf{v}^{-2} + \mathbf{v}^{-4})X^{(3,-4,0,2)} \\ & + X^{(3,-1,1,2)} + (\mathbf{v}^3 + \mathbf{v} + \mathbf{v}^{-1} + \mathbf{v}^{-3})X^{(6,-4,0,1)} + X^{(9,-4,0,0)}, \end{aligned}$$

$$f(v) = -(9 + v_1)v_1 - (-4 + v_2)v_2 + 3v_1v_2 = -v_1^2 + 3v_1v_2 - v_2^2 - 9v_1 + 4v_2.$$

$$\mathrm{Supp}(C_a) = \{(0,0), ((0,1), (0,2), (0,3), (0,4), (1,2), (1,3), (1,4), (2,4), (3,4)\}.$$

(See Figure 1 Left. The inequality $f(v) \ge 0$ gives the upper left side of the blue curve. This triangular basis element corresponds to $C[3,4]$ given in [**9**, §10.1(a)].)

(b) Consider the initial seed determined by $= \begin{bmatrix} 0 & 0 & -2 \\ 0 & 0 & -2 \\ 2 & 2 & 0 \end{bmatrix}$.

The corresponding quiver $\mathcal{Q}^{\mathrm{op}}$ is $1 \rightrightarrows 3 \leftleftarrows 2$.

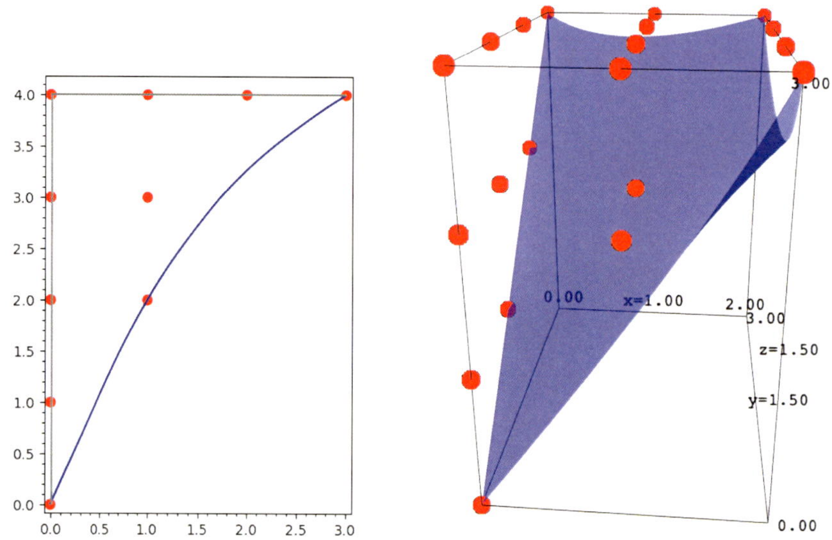

FIGURE 1. Left: the support of $C_{(-3,-4,0,0)}$ in (a). Right: the support of $C_{(4,3,-3,0,0,0)}$ in (b). The points in $\mathrm{Supp}(C_a)$ are in red color.

We compute the triangle basis element at $a=(-2,-3,-3,0,0,0)$. The corresponding $g=\varphi(a)=(4,3,-3,0,0,0)$, $w=(w_1,w_1',w_2,w_2',w_3,w_3')=(4,0,3,0,3,0)$. Then

$$\begin{aligned}
C_{(-2,-3,-3,0,0,0)} = {} & X^{(-2,-3,-3,0,0,3)} + (\mathbf{v}^2+1+\mathbf{v}^{-2})X^{(-2,-3,-1,0,1,3)} \\
& + (\mathbf{v}+\mathbf{v}^{-1}) * X^{(-2,-3,-1,1,0,3)} + (\mathbf{v}^2+1+\mathbf{v}^{-2})X^{(-2,-3,1,0,2,3)} \\
& + (\mathbf{v}^3+2\mathbf{v}+2\mathbf{v}^{-1}+\mathbf{v}^{-3})X^{(-2,-3,1,1,1,3)} + X^{(-2,-3,1,2,0,3)} \\
& + X^{(-2,-3,3,0,3,3)} + (\mathbf{v}^3+2\mathbf{v}+2\mathbf{v}^{-1}+\mathbf{v}^{-3})X^{(-2,-3,3,1,2,3)} \\
& + (\mathbf{v}^2+1+\mathbf{v}^{-2})X^{(-2,-3,3,2,1,3)} + (\mathbf{v}+\mathbf{v}^{-1})X^{(-2,-3,5,1,3,3)} \\
& + (\mathbf{v}^2+1+\mathbf{v}^{-2})X^{(-2,-3,5,2,2,3)} + X^{(-2,-3,7,2,3,3)} \\
& + (\mathbf{v}^2+1+\mathbf{v}^{-2})X^{(0,-1,-3,0,0,2)} + (\mathbf{v}^2+2+\mathbf{v}^{-2})X^{(0,-1,-1,0,1,2)} \\
& + (\mathbf{v}+\mathbf{v}^{-1})X^{(0,-1,-1,1,0,2)} + X^{(0,-1,1,0,2,2)} \\
& + (\mathbf{v}+\mathbf{v}^{-1})X^{(0,-1,1,1,1,2)} + (\mathbf{v}^2+1+\mathbf{v}^{-2})X^{(2,1,-3,0,0,1)} \\
& + X^{(2,1,-1,0,1,1)} + X^{(4,3,-3,0,0,0)},
\end{aligned}$$

$$
\begin{aligned}
f(v) &= -(4+v_1)v_1 - (3+v_2)v_2 - (-3+v_3)v_3 + 2v_1v_3 + 2v_2v_3 \\
&= -v_1^2 - v_2^2 - v_3^2 + 2v_1v_3 + 2v_2v_3 - 4v_1 - 3v_2 + 3v_3.
\end{aligned}
$$

$$
\begin{aligned}
\mathrm{Supp}(C_a) = \{&(0,0,0),(0,0,1),(0,0,2),(0,0,3),(0,1,1),(0,1,2),(0,1,3),(0,2,2),\\
&(0,2,3),(0,3,3),(1,0,2),(1,0,3),(1,1,2),(1,1,3),(1,2,3),(1,3,3),\\
&(2,0,3),(2,1,3),(2,2,3),(2,3,3)\}.
\end{aligned}
$$

(See Figure 1 Right. The inequality $f(v) \geq 0$ gives the upper region of the blue surface.)

Acknowledgments

The author thanks the organizers, Yiqiang Li and Changlong Zhong, of the AMS Special Session held at the University of Buffalo in September 2023, during which numerous inspiring conversations took place. He is deeply grateful to the anonymous referee for many detailed and insightful comments, which significantly improved the clarity and readability of this paper. Computer calculations were performed using SageMath [**16**].

References

[1] A. A. Beĭlinson, J. Bernstein, and P. Deligne, *Faisceaux pervers* (French), Analysis and topology on singular spaces, I (Luminy, 1981), Astérisque, vol. 100, Soc. Math. France, Paris, 1982, pp. 5–171. MR751966

[2] Arkady Berenstein and Andrei Zelevinsky, *Quantum cluster algebras*, Adv. Math. **195** (2005), no. 2, 405–455, DOI 10.1016/j.aim.2004.08.003. MR2146350

[3] Arkady Berenstein and Andrei Zelevinsky, *Triangular bases in quantum cluster algebras*, Int. Math. Res. Not. IMRN **6** (2014), 1651–1688, DOI 10.1093/imrn/rns268. MR3180605

[4] Sergey Fomin and Andrei Zelevinsky, *Cluster algebras. I. Foundations*, J. Amer. Math. Soc. **15** (2002), no. 2, 497–529, DOI 10.1090/S0894-0347-01-00385-X. MR1887642

[5] William Fulton, *Intersection theory*, 2nd ed., Ergebnisse der Mathematik und ihrer Grenzgebiete. 3. Folge. A Series of Modern Surveys in Mathematics [Results in Mathematics and Related Areas. 3rd Series. A Series of Modern Surveys in Mathematics], vol. 2, Springer-Verlag, Berlin, 1998, DOI 10.1007/978-1-4612-1700-8. MR1644323

[6] Allen Hatcher, *Algebraic topology*, Cambridge University Press, Cambridge, 2002. MR1867354

[7] Yoshiyuki Kimura and Fan Qin, *Graded quiver varieties, quantum cluster algebras and dual canonical basis*, Adv. Math. **262** (2014), 261–312, DOI 10.1016/j.aim.2014.05.014. MR3228430

[8] Kyungyong Lee, Li Li, Dylan Rupel, and Andrei Zelevinsky, *Greedy bases in rank 2 quantum cluster algebras*, Proc. Natl. Acad. Sci. USA **111** (2014), no. 27, 9712–9716, DOI 10.1073/pnas.1313078111. MR3263303

[9] Li Li, *Nakajima's quiver varieties and triangular bases of rank-2 cluster algebras*, J. Algebra **634** (2023), 97–164, DOI 10.1016/j.jalgebra.2023.06.028. MR4620671

[10] Bernard Leclerc, *Dual canonical bases, quantum shuffles and q-characters*, Math. Z. **246** (2004), no. 4, 691–732, DOI 10.1007/s00209-003-0609-9. MR2045836

[11] Hiraku Nakajima, *Quiver varieties and cluster algebras*, Kyoto J. Math. **51** (2011), no. 1, 71–126, DOI 10.1215/0023608X-2010-021. MR2784748

[12] Hiraku Nakajima, *Quiver varieties and finite-dimensional representations of quantum affine algebras*, J. Amer. Math. Soc. **14** (2001), no. 1, 145–238, DOI 10.1090/S0894-0347-00-00353-2. MR1808477

[13] Fan Qin, *t-analog of q-characters, bases of quantum cluster algebras, and a correction technique*, Int. Math. Res. Not. IMRN **22** (2014), 6175–6232, DOI 10.1093/imrn/rnt115. MR3283002

[14] Fan Qin, *Triangular bases in quantum cluster algebras and monoidal categorification conjectures*, Duke Math. J. **166** (2017), no. 12, 2337–2442, DOI 10.1215/00127094-2017-0006. MR3694569

[15] Fan Qin, *Compare triangular bases of acyclic quantum cluster algebras*, Trans. Amer. Math. Soc. **372** (2019), no. 1, 485–501, DOI 10.1090/tran/7610. MR3968777

[16] Sage Developers, Sagemath, the Sage Mathematics Software System (Version 9.2), `https://www.sagemath.org`.

Department of Mathematics and Statistics, Oakland University, Rochester, Michigan 48309-4479

Email address: `li2345@oakland.edu`

Contemporary Mathematics
Volume **837**, 2026
https://doi.org/10.1090/conm/837/16795

Elliptic Schubert classes and the Poincaré duality

Changlong Zhong

ABSTRACT. In this expository note, by using the Kostant-Kumar method, we prove the Poincaré duality of the elliptic classes associated to Schubert varieties.

0. Introduction

Equivariant elliptic cohomology has become a significant area of study in topology, with connections to mathematical physics, integrable systems, representation theory, and Schubert calculus. It complements rich theories like equivariant cohomology and equivariant K-theory. Recent progress has been made in the equivariant elliptic cohomology of symplectic resolutions, starting with [**1**]. For example, the elliptic stable envelope has been studied and shown to have a close connection to 3d mirror symmetry in physics. In algebraic geometry, elliptic Schubert classes associated with Schubert varieties are defined in [**5**, **7**]. Indeed, it is known that the elliptic fundamental classes of Schubert varieties are not well defined due to the singularties. However, following Borisov-Libgober's work, Rimányi-Weber were able to define these classes after introducing a new parameter known as the dynamical or Kähler parameter. This parameter emerges from the action of the Weyl group of the Langlands dual root system. For example, it has been shown that the classical elliptic Demazure-Lusztig operators (without the dynamical parameter) do not satisfy the braid relations, whereas the modified operators, which include the dynamical parameter, do. This result is established through geometric methods in [**7**].

In [**6**, **7**], elliptic Schubert classes are defined and shown to be permuted by the elliptic Demazure-Lusztig operators. A natural next question is to find a Billey-AJS type formula for the restriction coefficients, along with the combinatorial properties of the elliptic Schubert classes. This will be the subject of joint work by the author, Lenart, and Xiong.

The torus fixed point method has been a key tool in studying equivariant cohomology theories. One of the most important approaches using this method is the one developed in Kostant and Kumar's groundbreaking work from 1986 and 1990 [**3**, **4**]. In this paper, we develop the elliptic Schubert classes using the Kostant-Kumar method, along with the axiomatic approach introduced in [**2**]. In simple terms, this method begins by studying the elliptic Demazure-Lusztig operators.

The author was supported by the Institut des Hautes Études Scientifiques during his visit in July 2024.

The elliptic Schubert classes are then defined as the classes obtained by applying these operators.

Our main results include Theorems 1.4 and 3.10. The first theorem examines the transition matrix between the (dynamical) Weyl group and the elliptic DL operators, while the second proves the Poincaré duality between the elliptic Schubert classes and their opposite counterparts. Although a similar result was obtained in [**6**] using a more conceptual approach, the method in this paper is more algebraic and concrete. This makes it a more practical framework for applications in combinatorics.

The note is organized as follows: In Section 1, we define the elliptic twisted group algebra and the elliptic DL operators, and examine their transition matrix under the Weyl group action. In Section 2, we define the dual of the elliptic twisted group algebra, an auxiliary object that still captures the dynamical Weyl group. Finally, in Section 3, we define the elliptic Schubert classes and show that they are Poincaré dual to their opposite counterparts.

1. The elliptic Demazure-Lusztig operators

In this section, we define the elliptic twisted group algebra together with the elliptic DL operators.

1.1. Elliptic curve. Let E be an elliptic curve over $\mathbb{C}$. We have isomorphisms

$$E \cong \mathbb{C}/(\mathbb{Z}+\mathbb{Z}\tau) \cong \mathbb{C}^*/q^{\mathbb{Z}},$$

where $\tau \in \mathbb{C}$ such that $\operatorname{im}\tau > 0$, and $q = e^{2\pi i\tau}$. Recall the Jacobi-Theta function

$$\vartheta(u) = \frac{1}{2\pi i}(u^{1/2} - u^{-1/2})\prod_{s>0}(1-q^s u)(1-q^s u^{-1})\cdot\prod_{s>0}(1-q^s)^{-2}, \quad u \in \mathbb{C}^*.$$

Since $|q| < 1$, this series converges and defines a holomorphic function on a double cover of $\mathbb{C}^*$ and vanishes of order 1 at $q^{\mathbb{Z}}$. Write $\theta(x) = \vartheta(e^{2\pi i x})$ with $x \in \mathbb{C}$.

Let G be a reductive group with a Borel subgroup B and a maximal torus T. Let $\mathbb{X}^*(T)$ (resp. $\mathbb{X}_*(T)$) be the group of characters (resp. cocharacters) of the torus T. Let $\Sigma \subset \mathbb{X}^*(T)$ be the set of roots, $\Pi = \{\alpha_1, \dots, \alpha_n\}$ be the set of simple roots, and W be the Weyl group. Define $\mathfrak{A} = \mathbb{X}_*(T)\otimes E$, and $\mathfrak{A}^\vee = \mathbb{X}^*(T)\otimes E \cong \operatorname{Pic}^0(\mathfrak{A})$, which is the dual abelian variety of $\mathfrak{A}$. Each $\mu \in \mathbb{X}^*(T)$ defines a map $\chi_\mu : \mathfrak{A} \to E$. Denote $z_\mu = \chi_\mu(z), z \in \mathfrak{A}$. Dually, each $\mu^\vee \in \mathbb{X}_*(T)$ defines a map $\chi_{\mu^\vee} : \mathfrak{A}^\vee \to E$, and denote $\lambda_{\mu^\vee} = \chi_{\mu^\vee}(\lambda), \lambda \in \mathfrak{A}^\vee$. Note that

$$\mathbb{X}_*(T)\otimes\mathbb{C} \cong \mathbb{C}^n, \ \mathbb{X}^*(T)\otimes\mathbb{C} \cong \mathbb{C}^n.$$

and $\theta(z_\mu), \theta(\lambda_{\mu^\vee})$ are considered as meromorphic functions on complex spaces.

1.2. The twisted group algebra. We fix $\hbar \in \mathbb{C}$. Let $\mathcal{Q}$ be the field of meromorphic functions on

$$\mathbb{X}_*(T)\otimes\mathbb{C}\times\mathbb{X}^*(T)\otimes\mathbb{C} \cong \mathbb{C}^n\times\mathbb{C}^n.$$

The actions of W on $\mathbb{X}_*(T)$ and on $\mathbb{X}^*(T)$ induce two commuting actions on $\mathcal{Q}$, denoted by W and W^{d}, respectively. That is, W acts on $z \in \mathfrak{A}$ and W^{d} acts on $\lambda \in \mathfrak{A}^\vee$. W^{d} is usually referred as the dynamical action, which is also called the Kähler action. Let $\mathcal{Q}_{W^{\mathrm{d}}}$ be the twisted product of $\mathcal{Q}$ with W^{d}, and $\mathcal{Q}_{W\times W^{\mathrm{d}}}$ be the twisted product of $\mathcal{Q}_{W^{\mathrm{d}}}$ and W. Then $\mathcal{Q}_{W^{\mathrm{d}}}$ is a free left $\mathcal{Q}$-module with basis

$\{\delta^{\mathrm{d}}_w | w \in W\}$, and $\mathcal{Q}_{W\times W^{\mathrm{d}}}$ is a free left $\mathcal{Q}_{W^{\mathrm{d}}}$-module with basis $\{\delta_w, w\in W\}$. The product in $\mathcal{Q}_{W^{\mathrm{d}}\times W}$ can be written as

(1.1)

$$\delta_w\delta^{\mathrm{d}}_v = \delta^{\mathrm{d}}_v\delta_w, \quad (a\delta^{\mathrm{d}}_w\delta_v)(a'\delta^{\mathrm{d}}_{w'}\delta_{v'}) = a\cdot {}^{vw^{\mathrm{d}}}a'\delta^{\mathrm{d}}_{ww'}\delta_{vv'}, \quad w,v,w',v'\in W, a,a'\in\mathcal{Q}.$$

Denote $\delta_\alpha = \delta_{s_\alpha}$.

We fix some notations. Let $\Sigma_w = \{\alpha > 0 | w^{-1}\alpha < 0\}$ be the left inversion set of w. Denote

$$\theta_\Pi(\hbar\pm z) = \prod_{\alpha>0}\theta(\hbar\pm z_\alpha), \quad \theta_\Pi(\hbar\pm\lambda) = \prod_{\alpha>0}\theta(\hbar\pm\lambda_{\alpha^\vee}),$$

$$\theta_\Pi(z) = \prod_{\alpha>0}\theta(z_\alpha), \quad \theta_\Pi(\lambda) = \prod_{\alpha>0}\theta(\lambda_{\alpha^\vee}).$$

Denote $\epsilon_w = (-1)^{\ell(w)}$. Note that

$${}^w\theta_\Pi(z) = \epsilon_w\theta_\Pi(z), \quad {}^{w^{\mathrm{d}}}\theta_\Pi(\lambda) = \epsilon_w\theta_\Pi(\lambda).$$

Let $w_0\in W$ be the longest element. Denote

$$\mathbf{g} = \prod_{\alpha>0}\frac{\theta(\hbar - z_\alpha)}{\theta(z_\alpha)} = \frac{\theta_\Pi(\hbar - z)}{\theta_\Pi(z)}, \qquad \mathbf{h} = \prod_{\alpha>0}\frac{\theta(\hbar-\lambda_{\alpha^\vee})}{\theta(\lambda_{\alpha^\vee})} = \frac{\theta_\Pi(\hbar-\lambda)}{\theta_\Pi(\lambda)}.$$

It is easy to see that

$${}^{w_0}\mathbf{g} = \epsilon_{w_0}\frac{\theta_\Pi(\hbar+z)}{\theta_\Pi(z)}, \qquad {}^{w_0^{\mathrm{d}}}\mathbf{h} = \epsilon_{w_0}\frac{\theta_\Pi(\hbar+\lambda)}{\theta_\Pi(\lambda)}.$$

1.3. The elliptic DL operators. Let $z\in\mathfrak{A}, \lambda\in\mathfrak{A}^\vee$. For any simple root α, define the elliptic Demazure-Lusztig (DL) operator

$$T^\lambda_\alpha = \frac{\theta(\hbar - z_\alpha)\theta(\lambda_{\alpha^\vee})}{\theta(z_\alpha)\theta(\hbar-\lambda_{\alpha^\vee})}\delta_\alpha\delta^{\mathrm{d}}_\alpha + \frac{\theta(\hbar)\theta(z_\alpha-\lambda_{\alpha^\vee})}{\theta(z_\alpha)\theta(\hbar-\lambda_{\alpha^\vee})}\delta^{\mathrm{d}}_\alpha \in \mathcal{Q}_{W^{\mathrm{d}}\times W}.$$

Since $(-\lambda)_{\alpha^\vee} = -\lambda_{\alpha^\vee}$, so

$$T^{-\lambda}_\alpha = -\frac{\theta(\hbar - z_\alpha)\theta(\lambda_{\alpha^\vee})}{\theta(z_\alpha)\theta(\hbar+\lambda_{\alpha^\vee})}\delta_\alpha\delta^{\mathrm{d}}_\alpha + \frac{\theta(\hbar)\theta(z_\alpha+\lambda_{\alpha^\vee})}{\theta(z_\alpha)\theta(\hbar+\lambda_{\alpha^\vee})}\delta^{\mathrm{d}}_\alpha, \tag{1.2}$$

and it also satisfies that $T^\lambda_\alpha = \delta^{\mathrm{d}}_\alpha T^{-\lambda}_\alpha\delta^{\mathrm{d}}_\alpha$. Furthermore, there is an anti-involution that switches the two elements as well. See (1.9) below. It is easy to get that

$$\delta_\alpha = -\frac{\theta(z_\alpha)\theta(\hbar+\lambda_{\alpha^\vee})}{\theta(\hbar-z_\alpha)\theta(\lambda_{\alpha^\vee})}\delta^{\mathrm{d}}_\alpha T^\lambda_\alpha + \frac{\theta(\hbar)\theta(z_\alpha+\lambda_{\alpha^\vee})}{\theta(\hbar-z_\alpha)\theta(\lambda_{\alpha^\vee})}.$$

For simplicity, we will denote

$$\delta_i = \delta_{\alpha_i}, \ \delta^{\mathrm{d}}_i = \delta^{\mathrm{d}}_{\alpha_i}, \ T^{\pm\lambda}_i = T^{\pm\lambda}_{\alpha_i}.$$

For the remaining calculation, we sometimes write

$$\delta^{\mathrm{d}}_i T^\lambda_i = p_i\delta_i + q_i, \ p_i, q_i\in\mathcal{Q}. \tag{1.3}$$

REMARK 1.1. Let $\mathcal{L}_\alpha = T\times_B\mathbb{C}_{-\alpha}$ be the line bundle over G/B. If one identifies $c_1^{coh}(\mathcal{L}_\alpha) = z_\alpha, -\ln h = \hbar, \ln h^{\alpha^\vee} = \lambda_{\alpha^\vee}$, then $-T^\lambda_\alpha$ corresponds to the operator in [**7**, Theorem 1.3]. The notion $T^{\pm\lambda}_\alpha$ was denoted by $T^{z,\pm\lambda}_\alpha$ in [**6**].

One can show that

(i) $(T^\lambda_\alpha)^2 = 1$. This can be proved by direct computations by using Fay's Trisecant Identity.

(ii) The braid relations are satisfied. This is proved in [**7**] by using geometric arguments. Define $T^\lambda_w = T^\lambda_{i_1}\cdots T^\lambda_{i_k}$ if $w = s_{i_1}\cdots s_{i_k}$ is a reduced decomposition. Because the braid relations are satisfied, the definition of T^λ_w does not depend on the choice of reduced decompositions.

We develop some basic properties of the elliptic DL operators.

LEMMA 1.2. *For any $a \in \mathcal{Q}$, we have the Bernstein relation*

$$T^\lambda_\alpha a = {}^{s_\alpha s^{\mathrm{d}}_\alpha}a T^\lambda_\alpha + \frac{\theta(\hbar)\theta(z_\alpha - \lambda_{\alpha^\vee})}{\theta(z_\alpha)\theta(\hbar - \lambda_{\alpha^\vee})}\cdot {}^{s^{\mathrm{d}}_\alpha}\left(a - {}^{s_\alpha}a\right)\delta^{\mathrm{d}}_\alpha.$$

PROOF. Following the notation in (1.3), we have

$$\begin{aligned}
T^\lambda_\alpha a - {}^{s_\alpha s^{\mathrm{d}}_\alpha}a T^\lambda_\alpha &= ({}^{s^{\mathrm{d}}_\alpha}p_\alpha\delta_\alpha + {}^{s^{\mathrm{d}}_\alpha}q_\alpha)\delta^{\mathrm{d}}_\alpha a - {}^{s_\alpha s^{\mathrm{d}}_\alpha}a({}^{s^{\mathrm{d}}_\alpha}p_\alpha\delta_\alpha + {}^{s^{\mathrm{d}}_\alpha}q_\alpha)\delta^{\mathrm{d}}_\alpha\\
&= {}^{s^{\mathrm{d}}_\alpha}p_\alpha\cdot {}^{s_\alpha s^{\mathrm{d}}_\alpha}a\delta_\alpha\delta^{\mathrm{d}}_\alpha + {}^{s^{\mathrm{d}}_\alpha}q_\alpha\cdot {}^{s^{\mathrm{d}}_\alpha}a\delta^{\mathrm{d}}_\alpha - {}^{s_\alpha s^{\mathrm{d}}_\alpha}a\cdot {}^{s^{\mathrm{d}}_\alpha}p_\alpha\delta_\alpha\delta^{\mathrm{d}}_\alpha\\
&\quad - {}^{s_\alpha s^{\mathrm{d}}_\alpha}a\cdot {}^{s^{\mathrm{d}}_\alpha}q_\alpha\delta^{\mathrm{d}}_\alpha\\
&= {}^{s^{\mathrm{d}}_\alpha}q_\alpha\cdot {}^{s^{\mathrm{d}}_\alpha}(a - {}^{s_\alpha}a)\delta^{\mathrm{d}}_\alpha\\
&= \frac{\theta(\hbar)\theta(z_\alpha - \lambda_{\alpha^\vee})}{\theta(z_\alpha)\theta(\hbar - \lambda_{\alpha^\vee})}\cdot {}^{s^{\mathrm{d}}_\alpha}(a - {}^{s_\alpha}a)\delta^{\mathrm{d}}_\alpha.
\end{aligned}$$

□

LEMMA 1.3. *For any two simple roots α_i, α_j, we have*

$$\delta^{\mathrm{d}}_i T^\lambda_i \delta^{\mathrm{d}}_j T^\lambda_j = \frac{p_i}{{}^{s^{\mathrm{d}}_j}p_i}\delta^{\mathrm{d}}_{ji}T^\lambda_{ij} + (q_i - \frac{p_i\cdot {}^{s^{\mathrm{d}}_j}q_i}{{}^{s^{\mathrm{d}}_j}p_i})\delta^{\mathrm{d}}_j T^\lambda_j.$$

Here $\delta^{\mathrm{d}}_{ij} = \delta^{\mathrm{d}}_{s_i s_j}$ and $T^\lambda_{ij} = T^\lambda_{s_i s_j}$.

PROOF. Following the notation in (1.3), we have

$$\begin{aligned}
\frac{p_i}{{}^{s^{\mathrm{d}}_j}p_i}\delta^{\mathrm{d}}_{ji}T^\lambda_{ij} &= \frac{p_i}{{}^{s^{\mathrm{d}}_j}p_i}\delta^{\mathrm{d}}_{ji}\delta^{\mathrm{d}}_i(p_i\delta_i + q_i)\delta^{\mathrm{d}}_j(p_j\delta_j + q_j)\\
&= \frac{p_i}{{}^{s^{\mathrm{d}}_j}p_i}({}^{s^{\mathrm{d}}_j}p_i\delta_i + {}^{s^{\mathrm{d}}_j}q_i)(p_j\delta_j + q_j)\\
&= \frac{p_i}{{}^{s^{\mathrm{d}}_j}p_i}({}^{s^{\mathrm{d}}_j}p_i\cdot {}^{s_i}p_j\delta_{ij} + {}^{s^{\mathrm{d}}_j}q_i p_j\delta_j + {}^{s^{\mathrm{d}}_j}p_i\cdot {}^{s_i}q_j\delta_i + {}^{s^{\mathrm{d}}_j}q_i q_j)\\
&= p_i\cdot {}^{s_i}p_j\delta_{ij} + \frac{p_i\cdot {}^{s^{\mathrm{d}}_j}q_i p_j}{{}^{s^{\mathrm{d}}_j}p_i}\delta_j + p_i\cdot {}^{s_i}q_j\delta_i + \frac{p_i\cdot {}^{s^{\mathrm{d}}_j}q_i q_j}{{}^{s^{\mathrm{d}}_j}p_i}\\
\delta^{\mathrm{d}}_i T^\lambda_i \delta^{\mathrm{d}}_j T^\lambda_j &= (p_i\delta_i + q_i)(p_j\delta_j + q_j)\\
&= p_i\cdot {}^{s_i}p_j\delta_{ij} + q_i p_j\delta_j + p_i\cdot {}^{s_i}q_j\delta_i + q_i q_j\\
&= \frac{p_i}{{}^{s^{\mathrm{d}}_j}p_i}\delta^{\mathrm{d}}_{ji}T^\lambda_{ij} + (q_i - \frac{p_i\cdot {}^{s^{\mathrm{d}}_j}q_i}{{}^{s^{\mathrm{d}}_j}p_i})(p_j\delta_j + q_j)\\
&= \frac{p_i}{{}^{s^{\mathrm{d}}_j}p_i}\delta^{\mathrm{d}}_{ji}T^\lambda_{ij} + (q_i - \frac{p_i\cdot {}^{s^{\mathrm{d}}_j}q_i}{{}^{s^{\mathrm{d}}_j}p_i})\delta^{\mathrm{d}}_j T^\lambda_j.
\end{aligned}$$

The lemma is proved. □

THEOREM 1.4.

(a) *We have the following transition matrices:*

$$T_w^\lambda=\sum_{v\le w}a^\lambda_{w,v}\delta^{\mathrm{d}}_w\delta_v,\quad \delta_w=\sum_{v\le w}b^\lambda_{w,v}\delta^{\mathrm{d}}_{v^{-1}}T^\lambda_v,\quad a^\lambda_{w,v},b^\lambda_{w,v}\in\mathcal{Q},\tag{1.4}$$

where the coefficients $a^\lambda_{w,v},b^\lambda_{w,v}$ satisfy

$$\sum_v a^\lambda_{w,v}\cdot {}^{w^{\mathrm{d}}}b^\lambda_{v,u}=\delta^{\mathrm{Kr}}_{w,u},\quad \sum_v b^\lambda_{w,v}\cdot {}^{(v^{-1})^{\mathrm{d}}}a^\lambda_{v,u}=\delta^{\mathrm{Kr}}_{w,u}.\tag{1.5}$$

Here $\delta^{\mathrm{Kr}}_{w,u}$ is the Kronecker symbol. Moreover,

$$a^\lambda_{w,w}=\prod_{\alpha\in\Sigma_w}\frac{\theta(\hbar-z_\alpha)\theta(\lambda_{\alpha^\vee})}{\theta(z_\alpha)\theta(\hbar-\lambda_{\alpha^\vee})},\quad b^\lambda_{w,w}=\prod_{\alpha\in\Sigma_w}\frac{\theta(z_\alpha)}{\theta(\hbar-z_\alpha)}{}^{(w^{-1})^{\mathrm{d}}}\Big(\frac{\theta(\hbar-\lambda_{\alpha^\vee})}{\theta(\lambda_{\alpha^\vee})}\Big).\tag{1.6}$$

(b) *The set $\{T^\lambda_w\mid w\in W\}$ is a basis of the left $\mathcal{Q}_{W^{\mathrm{d}}}$-module $\mathcal{Q}_{W^{\mathrm{d}}\times W}$.*

PROOF. From

$$T^\lambda_\alpha={}^{s^{\mathrm{d}}_\alpha}p_\alpha\delta_\alpha\delta^{\mathrm{d}}_\alpha+{}^{s^{\mathrm{d}}_\alpha}q_\alpha\delta^{\mathrm{d}}_\alpha,$$

if $w=s_{i_1}s_{i_1}\cdots s_{i_k}$, we have

$$T^\lambda_w=T^\lambda_{i_1}\cdots T^\lambda_{i_k}=({}^{s^{\mathrm{d}}_{i_1}}p_{i_1}\delta_{i_1}\delta^{\mathrm{d}}_{i_1}+{}^{s^{\mathrm{d}}_{i_1}}q_{i_1}\delta^{\mathrm{d}}_{i_1})\cdots({}^{s^{\mathrm{d}}_{i_k}}p_{i_k}\delta_{i_k}\delta^{\mathrm{d}}_{i_k}+{}^{s^{\mathrm{d}}_{i_k}}q_{i_k}\delta^{\mathrm{d}}_{i_k}).$$

Opening the brackets, it is standard to obtain the formula in the first part of (1.4). Moreover, the leading coefficient $a^\lambda_{w,w}$ is given by

$$\prod_{\alpha\in\Sigma_w}\frac{\theta(\hbar-z_\alpha)\theta(\lambda_{\alpha^\vee})}{\theta(z_\alpha)\theta(\hbar-\lambda_{\alpha^\vee})}.$$

This gives the first identity of (1.6).

We consider the second part of (1.4), that is, we express δ_v as a linear combination of T^λ_w. Write

$$\delta_i=a_i+b_i\delta^{\mathrm{d}}_iT^\lambda_i,\ a_i,b_i\in\mathcal{Q}.$$

From Lemmas 1.2 and 1.3, we can write

$$T^\lambda_ia={}^{s_is^{\mathrm{d}}_i}aT^\lambda_i+f_i(a)\delta^{\mathrm{d}}_i,\quad \delta^{\mathrm{d}}_iT^\lambda_i\delta^{\mathrm{d}}_jT^\lambda_j=f_{ij}\delta^{\mathrm{d}}_{ji}T^\lambda_{ji}+g_{ij}\delta^{\mathrm{d}}_jT^\lambda_j,\ f_i(a),f_{ij},g_{ij}\in\mathcal{Q}.$$

We have

$$\begin{aligned}\delta_{ij}&=\delta_i\delta_j\\&=(a_i+b_i\delta^{\mathrm{d}}_iT^\lambda_i)(a_j+b_j\delta^{\mathrm{d}}_jT^\lambda_j)\\&=a_ia_j+a_ib_j\delta^{\mathrm{d}}_jT^\lambda_j+b_i\delta^{\mathrm{d}}_iT^\lambda_ia_j+b_i\delta^{\mathrm{d}}_iT^\lambda_ib_j\delta^{\mathrm{d}}_jT^\lambda_j\\&=a_ia_j+a_ib_j\delta^{\mathrm{d}}_jT^\lambda_j+b_i({}^{s_i}a_j\delta^{\mathrm{d}}_iT^\lambda_i+{}^{s^{\mathrm{d}}_i}f_i(a_j))+b_i({}^{s_i}b_j\delta^{\mathrm{d}}_iT^\lambda_i+{}^{s^{\mathrm{d}}_i}f_i(b_j))\delta^{\mathrm{d}}_jT^\lambda_j\\&=a_ia_j+a_ib_j\delta^{\mathrm{d}}_jT^\lambda_j+b_i({}^{s_i}a_j\delta^{\mathrm{d}}_iT^\lambda_i+{}^{s^{\mathrm{d}}_i}f_i(a_j))+b_i\cdot{}^{s_i}b_j\delta^{\mathrm{d}}_iT^\lambda_i\delta^{\mathrm{d}}_jT^\lambda_j\\&\quad+b_i\cdot{}^{s^{\mathrm{d}}_i}f_i(b_j)\delta^{\mathrm{d}}_jT^\lambda_j.\end{aligned}$$

By induction, the second expression in (1.4) follows.

Since the matrix $(a^\lambda_{w,v}\delta^{\mathrm{d}}_w)_{(w,v)\in W^2}$ is upper triangular with diagonal entries all invertible in $\mathcal{Q}_{W^{\mathrm{d}}}$, so it is invertible in $\mathcal{Q}_{W^{\mathrm{d}}}$, and part (b) then follows.

We now consider (1.5). We have

$$T_w^\lambda = \sum_v a_{w,v}^\lambda \delta_w^{\mathrm{d}} \delta_v = \sum_v a_{w,v}^\lambda \delta_w^{\mathrm{d}} \sum_u b_{v,u}^\lambda \delta_{u^{-1}}^{\mathrm{d}} T_u^\lambda = \sum_u (\sum_v a_{w,v}^\lambda \cdot {}^{w^{\mathrm{d}}} b_{v,u}^\lambda \delta_{wu^{-1}}^{\mathrm{d}}) T_u^\lambda.$$

so

$$\delta_{w,u}^{\mathrm{Kr}} = \sum_v a_{w,v}^\lambda \cdot {}^{w^{\mathrm{d}}} b_{v,u}^\lambda.$$

The second identity in (1.5) can be proved easily.

Lastly, we consider the second part of (1.6). From (1.5), we see that

$$b_{w,w}^\lambda \cdot {}^{(w^{-1})^{\mathrm{d}}} a_{w,w}^\lambda = 1,$$

so the expression for $b_{w,w}^\lambda$ follows. This concludes the proof of the theorem. □

As a special case, we have

$$(1.7)\qquad a_{w_0,w_0}^\lambda = \frac{\theta_\Pi(\hbar - z)\theta_\Pi(\lambda)}{\theta_\Pi(z)\theta_\Pi(\hbar-\lambda)} = \frac{\mathbf{g}}{\mathbf{h}}, \qquad b_{w_0,w_0}^\lambda = \epsilon_{w_0} \frac{\theta_\Pi(z)\theta_\Pi(\hbar+\lambda)}{\theta_\Pi(\hbar - z)\theta_\Pi(\lambda)} = \frac{{}^{w_0^{\mathrm{d}}}\mathbf{h}}{\mathbf{g}},$$

$$(1.8)\qquad a_{w_0,w_0}^{-\lambda} = \frac{\mathbf{g}}{{}^{w_0^{\mathrm{d}}}\mathbf{h}}, \qquad b_{w_0,w_0}^{-\lambda} = \frac{\mathbf{h}}{\mathbf{g}}.$$

For the root system of type A_2, the 'a'-coefficients are computed in [**6**, Example 8.3].

1.4. The anti-involution. It turns out that one can relate T_α^λ and $T_\alpha^{-\lambda}$ by using an anti-involution as follows. We will also show that these two operators are adjoint to each other. We define an anti-involution $\iota : \mathcal{Q}_{W^{\mathrm{d}} \times W} \to \mathcal{Q}_{W^{\mathrm{d}} \times W}$,

$$a\delta_v^{\mathrm{d}} \delta_w \mapsto \delta_{v^{-1}}^{\mathrm{d}} \delta_{w^{-1}} a \frac{{}^{w}\mathbf{g}}{\mathbf{g}} = {}^{(v^{-1})^{\mathrm{d}} w^{-1}} a \cdot \frac{\mathbf{g}}{{}^{w^{-1}}\mathbf{g}} \delta_{v^{-1}}^{\mathrm{d}} \delta_{w^{-1}}.$$

This anti-involution was defined in [**6**]. For any simple root α, it is easy to see that

$${}^{s_\alpha}\mathbf{g} = \mathbf{g} \cdot (-1) \cdot \frac{\theta(\hbar + z_\alpha)}{\theta(\hbar - z_\alpha)},$$

so one can verify

$$(1.9)\qquad \iota(T_i^\lambda) = T_i^{-\lambda}, \quad \iota(T_w^\lambda) = T_{w^{-1}}^{-\lambda}.$$

Lemma 1.5 deals with the coefficients after applying the anti-involution.

LEMMA 1.5. *For any $w, v \in W$, we have*

$${}^{v}\Big(\frac{a_{w^{-1},v^{-1}}^\lambda}{\mathbf{g}}\Big) = \frac{{}^{(w^{-1})^{\mathrm{d}}} a_{w,v}^{-\lambda}}{\mathbf{g}}.$$

PROOF. We have

$$\begin{aligned}
\sum_v a_{w^{-1},v^{-1}}^\lambda \delta_{w^{-1}}^{\mathrm{d}} \delta_{v^{-1}} &= T_{w^{-1}}^\lambda \\
&= \iota(T_w^{-\lambda}) \\
&= \iota(\sum_v a_{w,v}^{-\lambda} \delta_w^{\mathrm{d}} \delta_v) \\
&= \sum_v {}^{v^{-1}(w^{-1})^{\mathrm{d}}} a_{w,v}^{-\lambda} \cdot \frac{\mathbf{g}}{{}^{v^{-1}}\mathbf{g}} \delta_{w^{-1}}^{\mathrm{d}} \delta_{v^{-1}}.
\end{aligned}$$

The formula then follows from the comparison of the coefficients of $\delta_{w^{-1}}^{\mathrm{d}} \delta_{v^{-1}}$. □

EXAMPLE 1.6. If $w = v = w_0$, then

$$ {}^{w}\Big(\frac{a^{\lambda}_{w^{-1},w^{-1}}}{\mathbf{g}}\Big) = {}^{w}\Big(\frac{\mathbf{g}}{\mathbf{h}}\cdot\frac{1}{\mathbf{g}}\Big) = \frac{1}{\mathbf{h}}, \quad \frac{{}^{(w^{-1})^{\mathrm{d}}}a^{-\lambda}_{w,w}}{\mathbf{g}} = {}^{(w_0^{-1})^{\mathrm{d}}}\left(\frac{\mathbf{g}}{{}^{w_0^{\mathrm{d}}}\mathbf{h}}\right)\frac{1}{\mathbf{g}} = \frac{1}{\mathbf{h}}, $$

so Lemma 1.5 is verified in this case.

2. The dual as a left $\mathcal{Q}_{W^{\mathrm{d}}}$-module

In this section, we define a dual of the twisted group algebra $\mathcal{Q}_{W^{\mathrm{d}}\times W}$. This serves as an intermediate step in the development of the algebraic model for the elliptic Schubert classes.

2.1. We define the following dual:

$$ \mathcal{Q}^{\star}_{W^{\mathrm{d}}\times W} = \mathrm{Hom}_{\mathcal{Q}_{W^{\mathrm{d}}}}(\mathcal{Q}_{W^{\mathrm{d}}\times W}, \mathcal{Q}_{W^{\mathrm{d}}}), \tag{2.1} $$

where $\mathcal{Q}_{W^{\mathrm{d}}\times W}$ is considered as a free left $\mathcal{Q}_{W^{\mathrm{d}}}$-module, with two bases T^{λ}_{w} and δ_w. Then $\mathcal{Q}^{\star}_{W^{\mathrm{d}}\times W}$ becomes a right $\mathcal{Q}_{W^{\mathrm{d}}}$-module coming from the right multiplication of $\mathcal{Q}_{W^{\mathrm{d}}}$ on itself. That is,

$$ (gp)(x) = g(x)p, \quad g \in \mathcal{Q}^{\star}_{W^{\mathrm{d}}\times W}, x \in \mathcal{Q}_{W^{\mathrm{d}}\times W}, p \in \mathcal{Q}_{W^{\mathrm{d}}}. $$

So $\mathcal{Q}^{\star}_{W^{\mathrm{d}}\times W}$ is a free right $\mathcal{Q}_{W^{\mathrm{d}}}$-module with basis g_w dual to the basis $\delta_w \in \mathcal{Q}_{W^{\mathrm{d}}\times W}$. We also denote by $(T^{\lambda}_{w})^{\star}$ the dual basis of T^{λ}_{w}. Following from Theorem 1.4, we have

$$ (T^{\lambda}_{w})^{\star} = \sum_{v} g_v[(T^{\lambda}_{w})^{\star}(\delta_v)] = \sum_{v} g_v[(T^{\lambda}_{w})^{\star}(\sum_{u} b^{\lambda}_{v,u}\delta^{\mathrm{d}}_{u^{-1}}T^{\lambda}_{u})] = \sum_{v} g_v b^{\lambda}_{v,w}\delta^{\mathrm{d}}_{w^{-1}}. \tag{2.2} $$

The right $\mathcal{Q}_{W^{\mathrm{d}}}$-module $\mathcal{Q}^{\star}_{W^{\mathrm{d}}\times W}$ is also a left $\mathcal{Q}_{W^{\mathrm{d}}\times W}$-module induced by the right multiplication of $\mathcal{Q}_{W^{\mathrm{d}}\times W}$ on itself. More precisely, we have

$$ (x \bullet g)(x') = g(x'x), \quad x, x' \in \mathcal{Q}_{W^{\mathrm{d}}\times W},\ g \in \mathcal{Q}^{\star}_{W^{\mathrm{d}}\times W}. $$

LEMMA 2.1. *The $\bullet$-action is right $\mathcal{Q}_{W^{\mathrm{d}}}$-linear.*

PROOF. This follows from the fact the $\bullet$-action comes from the domain in (2.1), and the right $\mathcal{Q}_{W^{\mathrm{d}}}$-action comes from the codomain of (2.1). One can also check directly: for any $x, x' \in \mathcal{Q}_{W^{\mathrm{d}}\times W}, p \in \mathcal{Q}_{W^{\mathrm{d}}}, g \in \mathcal{Q}^{\star}_{W^{\mathrm{d}}\times W}$, we have

$$ (x \bullet (gp))(x') = (gp)(x'x) = g(x'x)p = (x \bullet g)(x')p = ((x \bullet g)p)(x'), $$

so $(x \bullet (gp)) = (x \bullet g)p$. □

LEMMA 2.2. *We have the following explicit formulas of the $\bullet$-action:*

$$ \delta_w \bullet g_v b\delta^{\mathrm{d}}_{u} = g_{vw^{-1}} b\delta^{\mathrm{d}}_{u}, \quad a\delta^{\mathrm{d}}_{w} \bullet g_v b\delta^{\mathrm{d}}_{u} = g_v \cdot {}^{v}a \cdot {}^{w^{\mathrm{d}}}b\delta^{\mathrm{d}}_{wu}, \quad a, b \in \mathcal{Q}. $$

PROOF. Since the $\bullet$-action is right $\mathcal{Q}_{W^{\mathrm{d}}}$-linear, it suffices to evaluate the elements on $\delta_x, x \in W$. We have

$$ (\delta_w \bullet g_v b\delta^{\mathrm{d}}_{u})(\delta_x) = g_v(\delta_x\delta_w)b\delta^{\mathrm{d}}_{u} = \delta^{\mathrm{Kr}}_{xw,v} b\delta^{\mathrm{d}}_{u}, $$

$$ (a\delta^{\mathrm{d}}_{w} \bullet g_v b\delta^{\mathrm{d}}_{u})(\delta_x) = g_v(\delta_x a\delta^{\mathrm{d}}_{w})b\delta^{\mathrm{d}}_{u} = \delta^{\mathrm{Kr}}_{x,v} \cdot {}^{x}a \cdot {}^{w^{\mathrm{d}}}b\delta^{\mathrm{d}}_{wu}, $$

so the identities follow. □

The bimodule $\mathcal{Q}^\star_{W^{\mathrm{d}}\times W}$ has a non-commutative product structure, defined as follows:

$$(g_w p)\cdot(g_v q) = g_v\delta^{\mathrm{Kr}}_{w,v} pq, \quad p,q\in \mathcal{Q}_{W^{\mathrm{d}}}.$$

The multiplicative identity is $\mathbf{1}=\sum_{w\in W} g_w$. Concerning the compatibility of the product with the $\bullet$-action, we have the following special cases:

LEMMA 2.3. *For any $p\in\mathcal{Q}_{W^{\mathrm{d}}}, w\in W, g,g'\in\mathcal{Q}^\star_{W^{\mathrm{d}}\times W}$, we have*

$$p\bullet(g\cdot g') = (p\bullet g)\cdot g', \quad \delta_w\bullet(g\cdot g') = (\delta_w\bullet g)\cdot(\delta_w\bullet g').$$

Note that in general $p\bullet(g\cdot g')\neq g\cdot(p\bullet g')$.

PROOF. Let $p=b\delta^{\mathrm{d}}_{w_1}, g=g_v a\delta^{\mathrm{d}}_u, g'=g_{v'}a'\delta^{\mathrm{d}}_{u'}$, then

$$\begin{aligned}p\bullet(g\cdot g') &= b\delta^{\mathrm{d}}_{w_1}\bullet(g_v a\delta^{\mathrm{d}}_u\cdot g_{v'}a'\delta^{\mathrm{d}}_{u'}) = b\delta^{\mathrm{d}}_{w_1}\bullet(g_v\delta^{\mathrm{Kr}}_{v,v'}a\delta^{\mathrm{d}}_u a'\delta^{\mathrm{d}}_{u'})\\ &= g_v\delta^{\mathrm{Kr}}_{v,v'}\,{}^v b\cdot\delta^{\mathrm{d}}_{w_1}a\delta^{\mathrm{d}}_u a'\delta^{\mathrm{d}}_{u'} = (p\bullet g)\cdot g'.\end{aligned}$$

Similarly,

$$\delta_w\bullet(g\cdot g') = \delta_w\bullet((g_v\delta^{\mathrm{Kr}}_{v,v'}a\delta^{\mathrm{d}}_u a'\delta^{\mathrm{d}}_{u'})) = g_{vw^{-1}}\delta^{\mathrm{Kr}}_{v,v'}a\delta^{\mathrm{d}}_u a'\delta^{\mathrm{d}}_{u'}) = (\delta_w\bullet g)\cdot(\delta_w\bullet g').$$

□

2.2. We will be considering the class $T^{-\lambda}_{w^{-1}w_0}\bullet g_{w_0}$, which is related with the opposite Schubert classes defined below. It coincides with the dual basis $(T^{-\lambda}_w)^\star$ up to a normalization.

LEMMA 2.4. *In $\mathcal{Q}^\star_{W^{\mathrm{d}}\times W}$, we have*

$$T^{\lambda}_{w^{-1}w_0}\bullet g_{w_0} = (T^{\lambda}_w)^\star\frac{\mathbf{g}}{\mathbf{h}}\delta^{\mathrm{d}}_{w_0}, \qquad T^{-\lambda}_{w^{-1}w_0}\bullet g_{w_0} = (T^{-\lambda}_w)^\star\frac{\mathbf{g}}{w_0^{\mathrm{d}}\mathbf{h}}\delta^{\mathrm{d}}_{w_0}.$$

PROOF. By definition of the $\bullet$-action, we have

$$\begin{aligned}(T^{\lambda}_{w^{-1}w_0}\bullet g_{w_0})(T^{\lambda}_u) &= g_{w_0}(T^{\lambda}_u T^{\lambda}_{w^{-1}w_0})\\ &= g_{w_0}(T^{\lambda}_{uw^{-1}w_0})\\ &\overset{\text{Thm. } 1.4}{=} g_{w_0}(\sum_{v\le uw^{-1}w_0} a^{\lambda}_{uw^{-1}w_0,v}\delta^{\mathrm{d}}_{uw^{-1}w_0}\delta_v)\\ &= \delta^{\mathrm{Kr}}_{w_0,uw^{-1}w_0}a^{\lambda}_{w_0,w_0}\delta^{\mathrm{d}}_{w_0}\\ &\overset{(1.7)}{=} \delta^{\mathrm{Kr}}_{u,w}\frac{\mathbf{g}}{\mathbf{h}}\delta^{\mathrm{d}}_{w_0} = ((T^{\lambda}_w)^\star\frac{\mathbf{g}}{\mathbf{h}}\delta^{\mathrm{d}}_{w_0})(\delta_u).\end{aligned}$$

Therefore, the first identity is proved. The second one can be obtained by replacing λ by $-\lambda$. □

Below we will use the $-\lambda$-version of this lemma. Following from (2.2) and Lemma 2.4, we have

$$T^{-\lambda}_{w^{-1}w_0}\bullet g_{w_0} \overset{\text{Lem. } 2.4}{=} (T^{-\lambda}_w)^\star\frac{\mathbf{g}}{w_0^{\mathrm{d}}\mathbf{h}}\delta^{\mathrm{d}}_{w_0} \tag{2.3}$$

$$\overset{(2.2)}{=} \sum_v g_v b^{-\lambda}_{v,w}\delta^{\mathrm{d}}_{w^{-1}}\frac{\mathbf{g}}{w_0^{\mathrm{d}}\mathbf{h}}\delta^{\mathrm{d}}_{w_0} \tag{2.4}$$

$$= \sum_{v\ge w} g_v b^{-\lambda}_{v,w}\cdot\frac{\mathbf{g}}{(w^{-1}w_0)^{\mathrm{d}}\mathbf{h}}\delta^{\mathrm{d}}_{w^{-1}w_0}. \tag{2.5}$$

EXAMPLE 2.5. If $w = w_0$, then $T^{-\lambda}_{w^{-1}w_0} \bullet g_{w_0} = g_{w_0}$, and the right hand side of the identity in (2.5) is

$$g_{w_0} b^{-\lambda}_{w_0,w_0} \frac{\mathbf{g}}{\mathbf{h}} \overset{(1.7)}{=} g_{w_0} \frac{\mathbf{h}}{\mathbf{g}} \frac{\mathbf{g}}{\mathbf{h}} = g_{w_0}.$$

3. The dual as a $\mathcal{Q}$-module

In this section, we define the elliptic Schubert classes and the opposite elliptic Schubert classes, and show that they are dual to each other.

3.1. The $\mathcal{Q}$-dual. Define

$$\mathcal{Q}^*_W = \mathrm{Hom}(W, \mathcal{Q}),$$

which is a free right $\mathcal{Q}$-module with basis f_w such that $f_w(v) = \delta_{w,v}$. Indeed, because $\mathcal{Q}$ is commutative, it does not matter whether we write the coefficients on the left or on the right. Note that we use $()^\star$ for the dual of left $\mathcal{Q}_{W^{\mathrm{d}}}$-module and $()^*$ for the dual of left $\mathcal{Q}$-module. The module $\mathcal{Q}^*_W$ has a commutative product coming from that of $\mathcal{Q}$. That is,

$$(f_w a) \cdot (f_v b) = \delta^{\mathrm{Kr}}_{w,v} f_w ab, \quad a, b \in \mathcal{Q}.$$

The multiplicative identity is also denoted by $\mathbf{1} = \sum_w f_w$.

Similar as above, there is also an action of $\mathcal{Q}_{W^{\mathrm{d}} \times W}$ on $\mathcal{Q}^*_W$:

$$a\delta^{\mathrm{d}}_w \delta_v \bullet (f_u b) = f_{uv^{-1}} \cdot {}^{uv^{-1}}a \cdot {}^{w^{\mathrm{d}}}b, \quad a, b \in \mathcal{Q}. \tag{3.1}$$

The action of $\delta^{\mathrm{d}}_w \in \mathcal{Q}_{W^{\mathrm{d}} \times W}$ is not $\mathcal{Q}$-linear, but the action of $a\delta_v \in \mathcal{Q}_W \subset \mathcal{Q}_{W^{\mathrm{d}} \times W}$ is. Note that it is different from that in [**6**], but it is compatible with the action considered in [**7**]. Other than the dynamical Weyl group action, this $\bullet$-action behaves similarly as the corresponding actions in general oriented cohomology theories (cohomology and K-theory) considered in Kostant-Kumar and Calmès-Zainoulline-Zhong [**2**–**4**].

Indeed, one can view $\mathcal{Q}^*_W$ as the commutative subring inside $\mathcal{Q}^\star_{W^{\mathrm{d}} \times W}$ induced by $\mathcal{Q} \to \mathcal{Q}_{W^{\mathrm{d}}}, a \mapsto a\delta^{\mathrm{d}}_e$. Or equivalently, we have an embedding of rings

$$\phi : \mathcal{Q}^*_W \to \mathcal{Q}^\star_{W^{\mathrm{d}} \times W}, \quad f_w a \mapsto g_w a \delta^{\mathrm{d}}_e, \quad a \in \mathcal{Q}.$$

If we view the right $\mathcal{Q}_{W^{\mathrm{d}}}$-module $\mathcal{Q}^\star_{W^{\mathrm{d}} \times W}$ as a right $\mathcal{Q}$-module via the embedding $\mathcal{Q} \to \mathcal{Q}_{W^{\mathrm{d}}}$, then ϕ is right $\mathcal{Q}$-linear.

We also have a surjection of abelian groups

$$\pi : \mathcal{Q}^\star_{W^{\mathrm{d}} \times W} \to \mathcal{Q}^*_W, \quad g_w a \delta^{\mathrm{d}}_v \mapsto f_w a, \quad a \in \mathcal{Q}.$$

This map does not preserve products, and it is not right $\mathcal{Q}$-linear. However,

LEMMA 3.1. *The map π is $\mathcal{Q}_{W^{\mathrm{d}} \times W}$-equivariant via the '$\bullet$'-action, that is, $\pi(x \bullet g) = x \bullet \pi(g)$ for any $x \in \mathcal{Q}_{W^{\mathrm{d}} \times W}, g \in \mathcal{Q}^\star_{W^{\mathrm{d}} \times W}$.*

PROOF. Let $x = a\delta^{\mathrm{d}}_w \delta_v \in \mathcal{Q}_{W^{\mathrm{d}} \times W}, g = g_u b \delta^{\mathrm{d}}_{u'} \in \mathcal{Q}^\star_{W^{\mathrm{d}} \times W}$, then

$$\begin{aligned}\pi(a\delta^{\mathrm{d}}_w \delta_v \bullet g_u b \delta^{\mathrm{d}}_{u'}) &= \pi(g_{uv^{-1}} \cdot {}^{uv^{-1}}a \cdot {}^{w^{\mathrm{d}}}b \delta^{\mathrm{d}}_{wu'}) = f_{uv^{-1}} \cdot {}^{uv^{-1}}a \cdot {}^{w^{\mathrm{d}}}b \\ &= a\delta^{\mathrm{d}}_w \delta_v \bullet f_u b = a\delta^{\mathrm{d}}_w \delta_v \bullet \pi(g).\end{aligned}$$

□

Starting from now, we only work with $\mathcal{Q}^*_W$.

3.2. Elliptic Schubert classes. We define the main objects of this paper.

DEFINITION 3.2. Let $\lambda \in \mathfrak{A}^\vee$. We define the elliptic Schubert class associated to λ to be $\mathbf{E}_w^\lambda = T_{w^{-1}}^\lambda \bullet f_e \in \mathcal{Q}_W^*$, and the opposite elliptic Schubert class to be $\mathcal{E}_w^{-\lambda} = T_{w^{-1}w_0}^{-\lambda} \bullet f_{w_0} \in \mathcal{Q}_W^*$.

REMARK 3.3. Recall that our operator T_α^λ agrees with the operator $T_\lambda^{z,\lambda}$ in [**6**], but the action $\bullet$ is different from that in [**6**] (see § 3). Moreover, our elliptic class $\mathbf{E}_e^\lambda = f_e$ where the elliptic class $\mathbf{E}_e^{z,\lambda}$ in [**6**, Appendix A] is equal to $\mathfrak{c} f_e$ where $\mathfrak{c}$ is a rational section of the twisted Poincaré line bundle $\mathbb{L}$.

On the other hand, by using the identification spelled out in Remark 1.1, one can identify the restriction $\mathbf{E}_w^\lambda|_\sigma$ with the classes $E_\sigma(X_w, \lambda)$ in [**7**] with $\sigma \in W$.

The elliptic Schubert classes $\mathbf{E}_w^\lambda$ were defined in [**7**]. We will show below that they are dual to the opposite Schubert classes $\mathcal{E}_w^{-\lambda}$.

By definition and the relation that $T_w^\lambda T_v^\lambda = T_{wv}^\lambda$, we have the following recursive formulas:

$$T_i^\lambda \bullet \mathbf{E}_w^\lambda = \mathbf{E}_{ws_i}^\lambda, \quad T_i^{-\lambda} \bullet \mathcal{E}_w^{-\lambda} = \mathcal{E}_{ws_i}^{-\lambda}.$$

The first one was referred as the Bott-Samelson recursion in [**7**].

REMARK 3.4. We can define a right action of T_i^λ on $\mathbf{E}_w^\lambda$ by

$$\mathbf{E}_w^\lambda \odot T_i^\lambda := T_{w^{-1}}^\lambda T_i^\lambda \bullet \mathbf{E}_e^\lambda = \mathbf{E}_{s_i w}^\lambda.$$

This gives the R-matrix recursion of the elliptic Schubert classes in [**7**]. This action permutes the elliptic Schubert classes, but they do not extend to actions of $\mathcal{Q}_W$ or $\mathcal{Q}_{W^{\mathrm{d}} \times W}$ on $\mathcal{Q}_W^*$.

We define the auxillary elliptic Schubert classes

$$(T_w^{-\lambda})^* := \pi((T_w^{-\lambda})^\star) = \sum_{v \geq w} f_v b_{v,w}^{-\lambda},$$

and we have

$$\begin{aligned} \mathcal{E}_w^{-\lambda} = T_{w^{-1}w_0}^{-\lambda} \bullet f_{w_0} &\overset{\text{Lem. 3.1}}{=} \pi(T_{w^{-1}w_0}^{-\lambda} \bullet g_{w_0}) \\ &\overset{(2.5)}{=} \sum_{v \geq w} f_v b_{v,w}^{-\lambda} \frac{\mathbf{g}}{(w^{-1}w_0)^{\mathrm{d}}\mathbf{h}} = (T_w^{-\lambda})^* \frac{\mathbf{g}}{(w^{-1}w_0)^{\mathrm{d}}\mathbf{h}}. \end{aligned} \tag{3.2}$$

From Theorem 1.4, we have the following restriction formula:

$$\mathbf{E}_w^\lambda = \sum_{v \leq w} f_v \cdot {}^v a_{w^{-1},v^{-1}}^\lambda. \tag{3.3}$$

The following lemma shows the relation between the 'a'-coefficients and the 'b'-coefficients.

LEMMA 3.5. *We have*

$$u(a_{w^{-1}w_0,u^{-1}w_0}^{-\lambda}) = b_{u,w}^{-\lambda} \frac{\mathbf{g}}{(w^{-1}w_0)^{\mathrm{d}}\mathbf{h}}.$$

PROOF. We have

$$\begin{aligned} \mathcal{E}_w^{-\lambda} = T_{w^{-1}w_0}^{-\lambda} \bullet f_{w_0} &= \sum_{v \leq w^{-1}w_0} a_{w^{-1}w_0,v}^{-\lambda} \delta_{w^{-1}w_0}^{\mathrm{d}} \delta_v \bullet f_{w_0} \\ &= \sum_{v \leq w^{-1}w_0} f_{w_0 v^{-1}} \cdot {}^{w_0 v^{-1}} a_{w^{-1}w_0,v}^{-\lambda}. \end{aligned}$$

Letting $u = w_0 v^{-1}$, and comparing with (3.2), we obtain the result. □

3.3. The Poincaré pairing. Define

$$Y'_\Pi = \sum_{w\in W} \delta_w \prod_{\alpha>0} \frac{\theta(z_\alpha)}{\theta(\hbar - z_\alpha)} = \sum_{w\in W} \delta_w \frac{1}{\mathbf{g}}. \tag{3.4}$$

The $\bullet$-action of Y'_Π provides the following duality:

LEMMA 3.6. *We have*

$$Y'_\Pi \bullet \left(\mathbf{E}^\lambda_v \cdot (T^{-\lambda}_w)^*\right) = \mathbf{1}\frac{\delta^{\mathrm{Kr}}_{v,w}}{\mathbf{g}}.$$

PROOF. From (3.3), we have

$$\mathbf{E}^\lambda_v = T^\lambda_{v^{-1}} \bullet f_e = \sum_{w_1\in W} f_{w_1} \cdot {}^{w_1}a^\lambda_{v^{-1},w_1^{-1}}.$$

Equation (2.2) gives $(T^{-\lambda}_w)^* = \sum_v f_v b^{-\lambda}_{v,w}$, so

$$\begin{aligned}
Y'_\Pi \bullet \left(\mathbf{E}^\lambda_v \cdot (T^{-\lambda}_w)^*\right) &= Y'_\Pi \bullet \left(\sum_{w_1\in W} f_{w_1} \cdot {}^{w_1}a^\lambda_{v^{-1},w_1^{-1}} \cdot b^{-\lambda}_{w_1,w}\right) \\
&= \sum_{w_2\in W} \delta_{w_2} \frac{1}{\mathbf{g}} \bullet \left(\sum_{w_1} f_{w_1} \cdot {}^{w_1}a^\lambda_{v^{-1},w_1^{-1}} \cdot b^{-\lambda}_{w_1,w}\right) \\
&= \sum_{w_2,w_1\in W} f_{w_1w_2^{-1}} \frac{{}^{w_1}a^\lambda_{v^{-1},w_1^{-1}} \cdot b^{-\lambda}_{w_1,w}}{{}^{w_1}\mathbf{g}} \\
&\overset{w_3:=w_1w_2^{-1}}{=} \Big(\sum_{w_3\in W} f_{w_3}\Big) \sum_{w_1\in W} \frac{{}^{w_1}a^\lambda_{v^{-1},w_1^{-1}}}{{}^{w_1}\mathbf{g}} b^{-\lambda}_{w_1,w} \\
&\overset{\text{Lem. } 1.5}{=} \mathbf{1} \sum_{w_1\in W} \frac{{}^{(v^{-1})^{\mathrm{d}}}a^{-\lambda}_{v,w_1} \cdot b^{-\lambda}_{w_1,w}}{\mathbf{g}} \\
&\overset{(*)}{=} \mathbf{1}\frac{\delta^{\mathrm{Kr}}_{v,w}}{\mathbf{g}}.
\end{aligned}$$

Here equality $(*)$ follows from applying $(w^{-1})^{\mathrm{d}}$ on the second identity of (1.5). □

To work with the opposite Schubert classes, we need to consider the following element in $\mathcal{Q}_{W^{\mathrm{d}}\times W}$:

$$Y_\Pi = \sum_{v,w\in W} \delta_w\delta^{\mathrm{d}}_v \prod_{\alpha>0} \frac{\theta(z_\alpha)}{\theta(\hbar - z_\alpha)} = \sum_{v\in W} \delta^{\mathrm{d}}_v \sum_{w\in W} \delta_w \frac{1}{\mathbf{g}} = \Big(\sum_{v\in W} \delta^{\mathrm{d}}_v\Big) Y'_\Pi. \tag{3.5}$$

The map

$$Y_\Pi \bullet \_ : \mathcal{Q}^*_W \to \mathcal{Q}^*_W, f \mapsto Y_\Pi \bullet f$$

is the algebraic model for the push-forward to the base point. So $\langle \_, \_\rangle := Y_\Pi \bullet (\_ \cdot \_)$ defines the Poincaré pairing.

COROLLARY 3.7. *We have*

$$Y_\Pi \bullet (\mathbf{E}^\lambda_w \cdot (T^{-\lambda}_v)^*) = \mathbf{1}\frac{\delta^{\mathrm{Kr}}_{w,v}|W|}{\mathbf{g}}.$$

PROOF. This follows from Lemma 3.6 and the identity $Y_\Pi = (\sum_{v\in W}\delta^{\mathrm{d}}_v)Y'_\Pi$. $\square$

LEMMA 3.8. *We have the following adjoint property*

$$Y_\Pi \bullet ((x\bullet f)\cdot f') = Y_\Pi \bullet (f\cdot(\iota(x)\bullet f')), \quad f,f'\in \mathcal{Q}_W^*, x\in \mathcal{Q}_{W^{\mathrm{d}}\times W}.$$

PROOF. Note that $\mathbf{g}$ is invariant under the dynamical action. Let $f = f_v a$, $f' = f_{v'}a'$, $x = b\delta^{\mathrm{d}}_{w_1}\delta_{w_2}$ with $a,a',b\in\mathcal{Q}$, then by definition, $\iota(x) = \delta^{\mathrm{d}}_{w_1^{-1}}\delta_{w_2^{-1}}b\frac{{}^{w_2}\mathbf{g}}{\mathbf{g}}$. So

$$\begin{aligned}
Y_\Pi\bullet((x\bullet f)\cdot f') &= Y_\Pi\bullet\left((b\delta^{\mathrm{d}}_{w_1}\delta_{w_2}\bullet f_v a)\cdot f_{v'}a'\right)\\
&= Y_\Pi\bullet\left((f_{vw_2^{-1}}\cdot {}^{vw_2^{-1}}b\cdot {}^{w_1^{\mathrm{d}}}a)\cdot(f_{v'}a')\right)\\
&= \sum_{w_3,w_4\in W}\delta^{\mathrm{d}}_{w_3}\delta_{w_4}\frac{1}{\mathbf{g}}\bullet\left(f_{v'}\delta^{\mathrm{Kr}}_{v',vw_2^{-1}}\,{}^{vw_2^{-1}}b\cdot {}^{w_1^{\mathrm{d}}}a\cdot a'\right)\\
&= \sum_{w_3,w_4\in W} f_{v'w_4^{-1}}\frac{\delta^{\mathrm{Kr}}_{v',vw_2^{-1}}\cdot {}^{w_3^{\mathrm{d}}vw_2^{-1}}b\cdot {}^{(w_3w_1)^{\mathrm{d}}}a\cdot {}^{w_3^{\mathrm{d}}}a'}{{}^{v'}\mathbf{g}}\\
&= (\sum_{w_4\in W} f_{vw_4^{-1}})\sum_{w_3\in W}\frac{\delta^{\mathrm{Kr}}_{v',vw_2^{-1}}\cdot {}^{w_3^{\mathrm{d}}vw_2^{-1}}b\cdot {}^{(w_3w_1)^{\mathrm{d}}}a\cdot {}^{w_3^{\mathrm{d}}}a'}{{}^{v'}\mathbf{g}}\\
&= \mathbf{1}\sum_{w_3\in W}\frac{\delta^{\mathrm{Kr}}_{v',vw_2^{-1}}\cdot {}^{w_3^{\mathrm{d}}vw_2^{-1}}b\cdot {}^{(w_3w_1)^{\mathrm{d}}}a\cdot {}^{w_3^{\mathrm{d}}}a'}{{}^{v'}\mathbf{g}}.
\end{aligned}$$

On the other hand, we have

$$\begin{aligned}
Y_\Pi\bullet(f\cdot(\iota(x)\bullet f')) &= Y_\Pi\bullet\left(f_v a\cdot(\delta^{\mathrm{d}}_{w_1^{-1}}\delta_{w_2^{-1}}b\frac{{}^{w_2}\mathbf{g}}{\mathbf{g}}\bullet f_{v'}a')\right)\\
&= Y_\Pi\bullet\left(f_v a\cdot[f_{v'w_2}\cdot {}^{v'(w_1^{-1})^{\mathrm{d}}}b\cdot\frac{{}^{v'w_2}\mathbf{g}}{{}^{v'}\mathbf{g}}\cdot {}^{(w_1^{-1})^{\mathrm{d}}}a']\right)\\
&= \sum_{w_3,w_4\in W}\delta^{\mathrm{d}}_{w_3}\delta_{w_4}\frac{1}{\mathbf{g}}\bullet\left(f_v\delta^{\mathrm{Kr}}_{v,v'w_2}\cdot a\cdot {}^{v'(w_1^{-1})^{\mathrm{d}}}b\cdot\frac{{}^{v'w_2}\mathbf{g}}{{}^{v'}\mathbf{g}}\cdot {}^{(w_1^{-1})^{\mathrm{d}}}a'\right)\\
&= \sum_{w_3,w_4\in W} f_{vw_4^{-1}}\delta^{\mathrm{Kr}}_{v,v'w_2}\cdot\frac{1}{{}^{v}\mathbf{g}}\cdot {}^{w_3^{\mathrm{d}}}a\cdot {}^{v'(w_3w_1^{-1})^{\mathrm{d}}}b\cdot\frac{{}^{v'w_2}\mathbf{g}}{{}^{v'}\mathbf{g}}\cdot {}^{(w_3w_1^{-1})^{\mathrm{d}}}a'\\
&= \mathbf{1}\sum_{w_3\in W}\delta^{\mathrm{Kr}}_{v,v'w_2}\cdot\frac{1}{{}^{v}\mathbf{g}}\cdot {}^{w_3^{\mathrm{d}}}a\cdot {}^{v'(w_3w_1^{-1})^{\mathrm{d}}}b\cdot\frac{{}^{v'w_2}\mathbf{g}}{{}^{v'}\mathbf{g}}\cdot {}^{(w_3w_1^{-1})^{\mathrm{d}}}a'.
\end{aligned}$$

Note that $v' = vw_2^{-1}$ if and only if $v'w_2 = v$. If we let $w_3' = w_3w^{-1}$, then it is straightforward to see that

$$Y_\Pi\bullet(f\cdot(\iota(x)\bullet f')) = Y_\Pi\bullet((x\bullet f)\cdot f').$$

The conclusion is proved. $\square$

Using (1.9), we get the following corollary.

COROLLARY 3.9. *We have*

$$Y_\Pi\bullet((T_i^\lambda\bullet f)\cdot f') = Y_\Pi\bullet(f\cdot(T_i^{-\lambda}\bullet f')), \quad f,f'\in\mathcal{Q}_W^*.$$

The following theorem is the algebraic model of the Poincaré duality between the elliptic Schubert classes and opposite elliptic Schubert classes:

THEOREM 3.10. *We have*

$$Y_\Pi \bullet (\mathbf{E}^\lambda_w \cdot \mathcal{E}^{-\lambda}_v) = \mathbf{1}\delta^{\mathrm{Kr}}_{w,v} \sum_{u_1\in W} u_1^{\mathrm{d}}(\frac{\theta_\Pi(\lambda)}{\theta_\Pi(\hbar-\lambda)}).$$

We provide two proofs of this theorem, which are not entirely independent but have different flavors.

PROOF 1. We have

$$\begin{aligned}
Y_\Pi \bullet (\mathbf{E}^\lambda_w \cdot \mathcal{E}^{-\lambda}_v) &\overset{(3.2)}{=} \sum_{u_1\in W} \delta^{\mathrm{d}}_{u_1} Y'_\Pi \bullet \Big(\mathbf{E}^\lambda_w \cdot (T^{-\lambda}_v)^* \frac{\mathbf{g}}{(w^{-1}w_0)^{\mathrm{d}}\mathbf{h}}\Big)\\
&= \sum_{u_1\in W} \delta^{\mathrm{d}}_{u_1} \bullet [Y'_\Pi \bullet \Big(\mathbf{E}^\lambda_w \cdot (T^{-\lambda}_v)^* \frac{\mathbf{g}}{(w^{-1}w_0)^{\mathrm{d}}\mathbf{h}}\Big)]\\
&\overset{(*)}{=} \sum_{u_1\in W} \delta^{\mathrm{d}}_{u_1} \bullet [\mathbf{1}\delta^{\mathrm{Kr}}_{w,v} \frac{\mathbf{g}}{(w^{-1}w_0)^{\mathrm{d}}\mathbf{h}}]\\
&= \mathbf{1}\delta^{\mathrm{Kr}}_{w,v}\mathbf{g}\sum_{u_1} \frac{1}{(u_1w^{-1}w_0)^{\mathrm{d}}\mathbf{h}}.
\end{aligned}$$

Here equality $(*)$ follows from (3.2) and the fact that the action of Y'_Π is $\mathcal{Q}$-linear. Rewriting u_1w^{-1} as u'_1, we obtain the result. □

PROOF 2. From Corollary 3.9 we know that

$$Y_\Pi \bullet (\mathbf{E}^\lambda_w \cdot \mathcal{E}^{-\lambda}_v) = Y_\Pi \bullet [(T^\lambda_{w^{-1}} \bullet f_e)\cdot (T^{-\lambda}_{v^{-1}w_0} \bullet f_{w_0})] = Y_\Pi \bullet [f_e \cdot (T^{-\lambda}_{wv^{-1}w_0} \bullet f_{w_0})].$$

Since $f_e f_u = \delta^{\mathrm{Kr}}_{e,u}$, and by using the formula of the $\bullet$- action (3.1), we just need to look at the coefficient of δ_{w_0} in the expansion of $T^{-\lambda}_{wv^{-1}w_0} = \sum_u a^{-\lambda}_{wv^{-1}w_0,u}\delta^{\mathrm{d}}_{wv^{-1}w_0}\delta_u$, which is 0 if $w\neq v$, and is $a^{-\lambda}_{w_0,w_0}\delta^{\mathrm{d}}_{w_0} = \frac{\mathbf{g}}{w_0^{\mathrm{d}}\mathbf{h}}\delta^{\mathrm{d}}_{w_0}$ if $w=v$. So if $w\neq v$, we see that $Y_\Pi(\mathbf{E}^\lambda_w\cdot\mathcal{E}^{-\lambda}_v)=0$. If $w=v$, we have

$$\begin{aligned}
Y_\Pi(\mathbf{E}^\lambda_w \cdot \mathcal{E}^{-\lambda}_w) &= Y_\Pi \bullet (f_e \cdot (a^{-\lambda}_{w_0,w_0}\delta^{\mathrm{d}}_{w_0}\delta_{w_0} \bullet f_{w_0}))\\
&= Y_\Pi \bullet (f_e \frac{\mathbf{g}}{w_0^{\mathrm{d}}\mathbf{h}})\\
&= (\sum_{u\in W}\delta_u \sum_{u_1\in W}\delta^{\mathrm{d}}_{u_1}) \bullet f_e \frac{1}{w_0^{\mathrm{d}}\mathbf{h}}\\
&= \sum_{u\in W} f_{u^{-1}} \sum_{u_1\in W} u_1^{\mathrm{d}}(\frac{1}{w_0^{\mathrm{d}}\mathbf{h}})\\
&= \mathbf{1}\sum_{u_1\in W}\frac{1}{u_1^{\mathrm{d}}\mathbf{h}}.
\end{aligned}$$

□

Acknowledgments

The author would like to thank C. Lenart and G. Zhao for the collaboration on elliptic Schubert classes, which lead to the present paper. He would also like to thank Rui Xiong for helpful discussions, and to the Institut des Hautes Études Scientifiques for its hospitality and support during his visit in July 2024.

References

[1] Mina Aganagic and Andrei Okounkov, *Elliptic stable envelopes*, J. Amer. Math. Soc. **34** (2021), no. 1, 79–133, DOI 10.1090/jams/954. MR4188815

[2] Baptiste Calmès, Kirill Zainoulline, and Changlong Zhong, *Equivariant oriented cohomology of flag varieties*, Doc. Math. Extra vol.: Alexander S. Merkurjev's sixtieth birthday (2015), 113–144. MR3404378

[3] Bertram Kostant and Shrawan Kumar, *The nil Hecke ring and cohomology of G/P for a Kac-Moody group G*, Adv. in Math. **62** (1986), no. 3, 187–237, DOI 10.1016/0001-8708(86)90101-5. MR866159

[4] Bertram Kostant and Shrawan Kumar, *T-equivariant K-theory of generalized flag varieties*, J. Differential Geom. **32** (1990), no. 2, 549–603. MR1072919

[5] Shrawan Kumar, Richárd Rimányi, and Andrzej Weber, *Elliptic classes of Schubert varieties*, Math. Ann. **378** (2020), no. 1-2, 703–728, DOI 10.1007/s00208-020-02043-z. MR4150933

[6] C. Lenart, G. Zhao, C. Zhong, *Elliptic classes via the periodic Hecke module and its Langlands dual*, Preprint, `arXiv:2309.09140`, 2023.

[7] Richárd Rimányi and Andrzej Weber, *Elliptic classes of Schubert varieties via Bott-Samelson resolution*, J. Topol. **13** (2020), no. 3, 1139–1182, DOI 10.1112/topo.12152. MR4100129

Department of Mathematics and Statistics, State University of New York, Albany, Hudson 209, 1400 Washington Ave, Albany, New York 12222

Email address: `czhong@albany.edu`

Contemporary Mathematics
Volume **837**, 2026
https://doi.org/10.1090/conm/837/16796

Quasi-split symmetric pairs of type A and Steinberg varieties of classical type, II. Constructible functions

Yiqiang Li

ABSTRACT. We provide a geometric realization of the universal enveloping algebra of the fixed-point subalgebra in a quasi-split symmetric pair of $\mathfrak{sl}_n(\mathbb{C})$, and its idempotent version, in the convolution algebra of constructible functions on Steinberg ind-varieties of classical type. We construct semicanonical bases for the idempotent version.

CONTENTS

Introduction

A theme in representation theory actively pursued in the past 30 years is the geometric realization of the universal enveloping algebra of a Kac-Moody algebra and its quantum deformation. This leads to remarkable discoveries such as canonical bases and quiver varieties.

The geometric realization in the $\mathfrak{sl}_n(\mathbb{C})$ case is particularly rich. One can use sheaves on n-step flag varieties [**BLM90**] or homologies or constructible functions on the associated Steinberg varieties of type A ([**G91**, **CG97**], [**Li10**]). One can also use homologies or constructible functions on Nakajima varieties [**N94**, **N98**, **L03**] and this approach is applicable to any symmetric Kac-Moody algebras. A typical feature in the varieties used is that they are "locally of type A", in that we mean vaguely that locally they are related to Grassmannians or their cotangent bundles.

Recently, there is a surge of activities in extending the theme to encompass more god-given naturally exist mathematical objects. On the algebra side, Kac-Moody algebras are extended to fixed-point subalgebras in symmetric pairs (see e.g. [**K14**]).

This work was partially supported by the National Science Foundation under the grant DMS 1801915 and Simons Travel Support for Mathematicians.

On the geometry side, varieties of locally of type A is extended to varieties of locally of classical type, which involve n-step isotropic flag varieties, the associated Steinberg varieties and their generalizations: σ-quiver varieties in [**Li19**]. The canonical basis theory has been developed in [**BW18a**, **BW18b**], see also [**LW18**]. The extension of the work [**BLM90**] has been done in the works [**BKLW**, **FL15**, **LW18**].

The purpose of this paper is to extend the works [**N94**, **L03**, **Li10**] to this more general theme. Precisely, we provide a geometric construction of the fixed-point subalgebra in a quasi-split symmetric pair of $\mathfrak{sl}_n(\mathbb{C})$ (see [**Sp87**] for the terminology of quasi-split), and its idempotent version, in the convolution algebra of constructible functions on Steinberg ind-varieties of classical type. Along the way, we construct semicanonical bases for the idempotent version via the geometry of the irreducible components of the Steinberg ind-varieties. We refer to [**Li21**] for a theory leveraging Borel-Moore homologies.

Our approach also provides a new, and more natural, realization of the enveloping algebra of $\mathfrak{sl}_n(\mathbb{C})$ by using the Steinberg ind-varieties of type A. In literature, the universal enveloping algebra of $\mathfrak{sl}_n(\mathbb{C})$ is constructed inside the projective system of Schur algebras, geometrically defined as sheaves/homology of Steinberg varieties. It remains open if one can realize this algebra as a convolution algebra on an ind-varieties as indicated in [**BLM90**, 4.8] and [**CG97**, 4.4.32]. Our construction provides such a realization for the first time.

The extension to the geometry of locally of classical type is completely nontrivial. The verification of the rank-one inhomogenous Serre relations in the defining relations of the quasi-split fixed point subalgebra is totally new and reflects deep interactions between the geometries of orthogonal and symplectic groups. The nontrivial definition of the generators in the n even case also exhibits the complexity in the geometry of the Steinberg ind-varieties of classical type.

1. Geometric set up: Steinberg ind-varieties $Z_{n,\infty,\varepsilon}$ of classical type

In this section, we fix some notations and we recall the Steinberg ind-varieties of classical groups from [**Li21**]. There is no new result in this section.

1.1. Notations and Euler characteristic. Let $|V|$ denote the dimension of the vector space V. If $V \subseteq W$ and $|W| = |V| + 1$, we write $V \overset{1}{\subset} W$ or $W \overset{1}{\supset} V$. Let $\mathrm{Gr}(m, n)$ be the Grassmannian of m-dimensional subspaces in an n-dimensional ambient vector space.

Let X be an algebraic variety. Let Y be a locally closed subvariety of X, we write $\chi(Y)$ for the Euler characteristic of Y.

Let $f : X \to \mathbb{C}$ be a constructible function. If $Y \subseteq X$ is constructible, we write

$$\int_{y\in Y} f(y) = \sum_{c\in\mathbb{C}} c\chi(Y \cap f^{-1}(c)).$$

Let G be a reductive group over $\mathbb{C}$ and P a parabolic subgroup. Let W_G and W_P be the associated Weyl groups. Then by the Bruhat decomposition, the Euler characteristic of the partial flag variety G/P is given by

$$\chi(G/P) = \frac{\#W_G}{\#W_P}. \tag{1}$$

1.2. Notations taken from [Li21]. We borrow the following notations from [**Li21**]. We also recall some results from [**Li21**] in this section. Fix an integer $\varepsilon \in \{\pm 1\}$. Let V be a finite dimensional complex vector space of dimension v. We equip V with a non-degenerate bilinear form $(-,-)_V$ such that $(u, u')_V = \varepsilon(u', u)_V$ for all $u, u' \in V$. Let $\mathbf{G}_{v,\varepsilon}$ be the group of isometries on V with respect to the form $(-,-)_V$.

Let $\mathcal{N}_{v,\varepsilon}$ be the variety of nilpotent elements in $\mathrm{Lie}(\mathbf{G}_{v,\varepsilon})$. Let $\mathbf{G}_{v,\varepsilon}$ act on $\mathcal{N}_{v,\varepsilon}$ by conjugation. Let n be an integer greater than 1. We define the n-nilcone:

$$\mathcal{N}_{n,v,\varepsilon} = \{x \in \mathcal{N}_{v,\varepsilon} | x^n = 0\}.$$

We also fix the notation

$$r = \lfloor n/2 \rfloor.$$

Let e be an $n \times n$ Jordan block and J be an $n \times n$ anti-diagonal matrix as follows.

$$e = \begin{bmatrix} 0 & 1 & 0 & \cdots & 0 \\ 0 & 0 & 1 & \cdots & 0 \\ & & & \cdots & \\ 0 & 0 & 0 & \cdots & 1 \\ 0 & 0 & 0 & \cdots & 0 \end{bmatrix} \quad J = \begin{bmatrix} 0 & \cdots & 0 & 0 & 1 \\ 0 & \cdots & 0 & 1 & 0 \\ & \cdots & & & \\ 0 & \cdots & 0 & 0 & 0 \\ 1 & \cdots & 0 & 0 & 0 \end{bmatrix}$$

On $\mathbb{C}^{2n}$, we define an ε-form by the following matrix

$$M_\varepsilon = \begin{bmatrix} 0 & J \\ \varepsilon J & 0 \end{bmatrix}. \tag{2}$$

Implicitly, we fix a basis $\{a_i, b_i\}_{1 \le i \le n}$ of $\mathbb{C}^{2n}$ such that $(a_i, b_j) = \delta_{ij}$, $(a_i, a_j) = 0 = (b_i, b_j)$ for all $1 \le i, j \le n$. Let

$$e_\varepsilon = \begin{bmatrix} e & 0 \\ 0 & -e \end{bmatrix}$$

Then $e_\varepsilon \in \mathrm{Lie}(\mathbf{G}_{2n,\varepsilon})$ and e_ε is nilpotent of type n^2.

We define a form on $V \oplus \mathbb{C}^{2n}$ to be the direct sum of the forms on V and $\mathbb{C}^{2n}$. Then the assignment $x \mapsto x \oplus e_\varepsilon$ defines a closed immersion

$$i_\varepsilon : \mathcal{N}_{n,v,\varepsilon} \to \mathcal{N}_{n,v+2n,\varepsilon}, \quad x \mapsto x \oplus e_\varepsilon. \tag{3}$$

Thus we have a direct system $(\mathcal{N}_{n,v+2nk,\varepsilon}, i_\varepsilon)_{k \in \mathbb{N}}$. Let

$$\mathcal{N}_{n,v+\infty,\varepsilon} = \varinjlim_k (\mathcal{N}_{n,v+2nk,\varepsilon}, i_\varepsilon). \tag{4}$$

be the associated direct limit, which is an ind-variety. Similarly, the rule $g \mapsto g \oplus 1$ defines a group homomorphism $\mathbf{G}_{v,\varepsilon} \to \mathbf{G}_{v+2n,\varepsilon}$. We have the direct limit

$$\mathbf{G}_{v+\infty,\varepsilon} = \varinjlim_k \mathbf{G}_{v+2nk,\varepsilon},$$

which is an ind-group. There is a natural $\mathbf{G}_{v+\infty,\varepsilon}$ action on $\mathcal{N}_{n,v+\infty,\varepsilon}$.

Let $\mathcal{F}_{n,v,\varepsilon}$ be the variety of isotropic flags in V of the form

$$F = (0 \equiv F_0 \subseteq F_1 \subseteq \cdots \subseteq F_n \equiv V),$$

such that $F_i^\perp = F_{n-i} \forall 1 \le i \le n$, if $(n, v, \varepsilon) \neq (even, odd, 1)$. For the case $(n, v, \varepsilon) = (even, odd, 1)$, we define

$$\mathcal{F}_{n,v,\varepsilon} := \{F \in \mathcal{F}_{n+1,v,\varepsilon} | F_r \text{ is maximal isotropic}\}, \quad \text{if } (n, v, \varepsilon) = (even, odd, 1).$$

Recall that we fix a basis $\{a_i, b_i\}_{1\leq i\leq n}$ of $\mathbb{C}^{2n}$ and define a form whose matrix under the basis is M_ε in (2). Fix the following n-step isotropic flag in $\mathcal{F}_{n,2n,\varepsilon}$:

$$\mathbf{F}_\varepsilon = (0 \subset \langle a_1, b_n\rangle \subset \langle a_1, a_2, b_n, b_{n-1}\rangle \subset \cdots \subset \langle a_1, \cdots, a_{n-1}, b_n, \cdots, b_2\rangle \subset \mathbb{C}^{2n}).$$

If n is even, let $\tilde{\mathbf{F}}_\varepsilon$ be the $(n+1)$-step isotropic flag obtained from $\mathbf{F}_\varepsilon$ by adding an extra r-th step subspace in $\mathbf{F}_\varepsilon$.

For any flag $F \in \mathcal{F}_{n,v,\varepsilon}$, let $F \oplus \mathbf{F}_\varepsilon$ be a flag in $\mathcal{F}_{n,v+2n,\varepsilon}$ (with $V \oplus \mathbb{C}^{2n}$ as the underlying space) whose ith step is the sum of the ith steps of F and $\mathbf{F}_\varepsilon$. Similarly, we define $F \oplus \tilde{\mathbf{F}}_\varepsilon$. We define an embedding

$$(5) \qquad i_\varepsilon : \mathcal{F}_{n,v,\varepsilon} \to \mathcal{F}_{n,v+2n,\varepsilon}, \begin{cases} F \mapsto F \oplus \mathbf{F}_\varepsilon, & \text{if } (n, v, \varepsilon) \neq (\text{even, odd}, 1), \\ F \mapsto F \oplus \tilde{\mathbf{F}}_\varepsilon, & \text{if } (n, v, \varepsilon) = (\text{even, odd}, 1). \end{cases}$$

Let $\mathcal{M}_{n,v,\varepsilon}$ be the variety of all pairs $(x, F) \in \mathcal{N}_{n,v,\varepsilon} \times \mathcal{F}_{n,v,\varepsilon}$ such that $x(F_i) \subseteq F_{i-1}$ for all i. The associated first projection is

$$(6) \qquad \pi : \mathcal{M}_{n,v,\varepsilon} \to \mathcal{N}_{n,v,\varepsilon}, \quad (x, F) \mapsto x.$$

We have a cartesian diagram.

$$(7) \qquad \begin{array}{ccc} \mathcal{M}_{n,v,\varepsilon} & \xrightarrow{i_\varepsilon} & \mathcal{M}_{n,v+2n,\varepsilon} \\ \pi\downarrow & & \pi\downarrow \\ \mathcal{N}_{n,v,\varepsilon} & \xrightarrow{i_\varepsilon} & \mathcal{N}_{n,v+2n,\varepsilon} \end{array}$$

Since $(e_\varepsilon, \mathbf{F}_\varepsilon) \in \mathcal{M}_{n,2n,\varepsilon}$, the two embeddings under the same notations i_ε in (3) and (5) induce an embedding

(8)

$$i_\varepsilon : \mathcal{M}_{n,v,\varepsilon} \to \mathcal{M}_{n,v+2n,\varepsilon}, \begin{cases} (x, F) \mapsto (x \oplus e_\varepsilon, F \oplus \mathbf{F}_\varepsilon) & \text{if } (n, v, \varepsilon) \neq (\text{even, odd}, 1), \\ (x, F) \mapsto (x \oplus e_\varepsilon, F \oplus \tilde{\mathbf{F}}_\varepsilon) & \text{if } (n, v, \varepsilon) = (\text{even, odd}, 1). \end{cases}$$

Let

$$Z_{n,v,\varepsilon} = \mathcal{M}_{n,v,\varepsilon} \times_{\mathcal{N}_{n,v,\varepsilon}} \mathcal{M}_{n,v,\varepsilon} = \{(x, F, F') | (x, F), (x, F') \in \mathcal{M}_{n,v,\varepsilon}\}$$

be the Steinberg variety. The embedding i_ε in (8) induces an embedding

$$(9) \qquad i_\varepsilon : Z_{n,v,\varepsilon} \to Z_{n,v+2n,\varepsilon}.$$

By [**Li21**], we have

THEOREM 1.2.1. *The restriction of an irreducible component of $Z_{n,v+2n,\varepsilon}$ to $Z_{n,v,\varepsilon}$ under i_ε is either empty or an irreducible component of $Z_{n,v,\varepsilon}$.*

Define the direct limit of $(Z_{n,v+2nk,\varepsilon}, i_\varepsilon)_{k\in\mathbb{N}}$ by

$$Z_{n,v+\infty,\varepsilon} = \varinjlim_k (Z_{n,v+2nk,\varepsilon}, i_\varepsilon).$$

Then $Z_{n,v+\infty,\varepsilon}$ can be identified with the equivalence class $\sqcup_k Z_{n,v+2nk,\varepsilon}/\sim$, where the equivalence $\sim$ is defined by for all $(x, F, F') \in Z_{n,v+2nk,\varepsilon}$ and $(y, G, G') \in Z_{n,v+2k'n,\varepsilon}$ for $k < k'$, $(x, F, F') \sim (y, G, G')$ if and only if $i_{k',k}(x, F, F') = (y, G, G')$ where $i_{k',k} : Z_{n,v+2kn,\varepsilon} \to Z_{n,v+2k'n,\varepsilon}$ for $k < k'$ be the transfer map, induced by i_ε, in the direct system $(Z_{n,v+2nk,\varepsilon}, i_\varepsilon)_{k\in\mathbb{N}}$. Let $[x, F, F']$ be the equivalence class of (x, F, F'). Let $i_k : Z_{n,v+2kn,\varepsilon} \to Z_{n,\infty,\varepsilon}$ be the induced embedding from the direct system $(Z_{n,v+2nk,\varepsilon}, i_\varepsilon)_{k\in\mathbb{N}}$.

Following [**Li21**], we consider *the Steinberg ind-varieties of classical type*,

$$Z_{n,\infty,\varepsilon} := \begin{cases} \sqcup_{\ell=1}^{n} Z_{n,2\ell+\infty,\varepsilon}, & \text{if } \varepsilon = -1, \\ \sqcup_{\ell=1}^{n} Z_{n,2\ell+\infty,\varepsilon}, & \text{if } \varepsilon = 1, \\ \sqcup_{\ell=1}^{n} Z_{n,(2\ell-1)+\infty,\varepsilon}, & \text{if } \varepsilon = 1. \end{cases} \tag{10}$$

The direct limit $Z_{n,\infty,\varepsilon}$ admits a Zariski topology induced from those in $Z_{n,v+2kn,\varepsilon}$ for all $1 \leq v \leq 2n$, $k \in \mathbb{N}$. In a more precise way, we claim that a subset $X \subseteq Z_{n,\infty,\varepsilon}$ is open (resp. constructible) if the restriction $i_k^{-1}(X)$ is Zariski open (resp. constructible) in $Z_{n,v+2kn,\varepsilon}$ for all k and v. Then t the collection of all open sets in $Z_{n,\infty,\varepsilon}$ forms the desired Zariski topology.

Let

(11)

$$\Theta_{v+\infty} = \left\{ A = (a_{ij})_{1\leq i,j\leq n} \mid \sum_{1\leq i,j\leq n} a_{ij} \equiv v \ (\mathrm{mod} 2n), a_{ij} \in \mathbb{N}, a_{ij} = a_{n+1-i,n+1-j} \right\}.$$

The set $\Theta_{v+\infty}$ admits a partition

$$\Theta_{v+\infty} = \sqcup_{k:v+2nk>0} \Theta_{v+2nk}, \quad \Theta_{v+2nk} = \{A \in \Theta_{v+\infty} \mid \sum_{i,j} a_{ij} = v + 2kn\}.$$

We define an equivalence relation on $\Theta_{v,\varepsilon}$ by $A \sim B$ if and only if $A \equiv B \ (\mathrm{mod} 2I)$. Let

$$\overline{\Theta}_{v+\infty} = \Theta_{v+\infty} / \sim,$$

and $\overline{A}$ be the equivalence class of A. Note that the closed imbedding in (5) defines a direct system $(\mathcal{F}_{n,v+2kn,\varepsilon}, i_\varepsilon)$. Let

$$\mathcal{F}_{n,v+\infty,\varepsilon} = \varinjlim_{k} \mathcal{F}_{n,v+2kn,\varepsilon}$$

be the ind-variety of the direct system. The ind-group $\mathbf{G}_{v+\infty,\varepsilon}$ acts transitively on $\mathcal{F}_{n,v+\infty,\varepsilon}$, and induces a diagonal action on the product $\mathcal{F}_{n,v+\infty,\varepsilon} \times \mathcal{F}_{n,v+\infty,\varepsilon}$.

Given a point $[F, F'] \in \mathcal{F}_{n,v+\infty,\varepsilon} \times \mathcal{F}_{n,v+\infty,\varepsilon}$, we can define a matrix $A_{F,F'}$ whose (i,j)-th entry is the dimension of $F_i \cap F'_j/(F_{i-1} \cap F'_j + F_i \cap F'_{j-1})$. We let $\mathrm{dlt}(A_{F,F'})$ be the matrix obtained by deleting the $(r+1)$-th row and column from $A_{F,F'}$. By [**Li21**], we have

Lemma 1.2.2. *The assignment*

$$[F, F'] \mapsto \begin{cases} \overline{A}_{F,F'}, & \text{if } (n, v, \varepsilon) \neq (\text{even, odd}, 1), \\ \mathrm{dlt}(\overline{A}_{F,F'}), & \text{if } (n, v, \varepsilon) = (\text{even, odd}, 1), \end{cases}$$

defines a bijection $\mathbf{G}_{v+\infty,\varepsilon} \backslash \mathcal{F}_{n,v+\infty,\varepsilon} \times \mathcal{F}_{n,v+\infty,\varepsilon} \cong \overline{\Theta}_{v+\infty}$.

We shall denote $\mathcal{O}_{\overline{A},\varepsilon}$ the $\mathbf{G}_{v+\infty,\varepsilon}$-orbit in $\mathcal{F}_{n,v+\infty,\varepsilon} \times \mathcal{F}_{n,v+\infty,\varepsilon}$ indexed by $\overline{A}$. Let $p : Z_{n,v+\infty,\varepsilon} \to \mathcal{F}_{n,v+\infty,\varepsilon} \times \mathcal{F}_{n,v+\infty,\varepsilon}$ be the projection map $[x, F, F'] \mapsto [F, F']$. Let

$$Z_{\overline{A},\varepsilon} = p^{-1}(\mathcal{O}_{\overline{A},\varepsilon}).$$

Clearly we have

$$Z_{n,v+\infty,\varepsilon} = \sqcup_{\overline{A} \in \overline{\Theta}_{v+\infty}} Z_{\overline{A},\varepsilon}.$$

Let $\overline{Z}_{\overline{A},\varepsilon}$ be the closure of $Z_{\overline{A},\varepsilon}$ in $Z_{n,v+\infty,\varepsilon}$. By [**Li21**], we have

PROPOSITION 1.2.3.

(1) *All irreducible components of* $Z_{n,v+\infty,\varepsilon}$ *are of the form* $\overline{Z}_{\overline{A},\varepsilon}$ *for some* $\overline{A} \in \overline{\Theta}_{v+\infty}$ *if* $(v,\varepsilon) \neq (even, 1)$. *Otherwise,* $\overline{Z}_{n,v+\infty,\varepsilon}$ *is a union of finitely many irreducible components of* $Z_{n,v+\infty,\varepsilon}$ *and all irreducible components of* $Z_{n,v+\infty,\varepsilon}$ *appear this way.*

(2) *Let* $Z_{A,\varepsilon}$ *be the intersection* $Z_{\overline{A},\varepsilon} \cap Z_{n,v+2n,\varepsilon}$ *if* $A \in \Theta_{v+2n}$. *Then* $\overline{Z}_{A,\varepsilon}$ *is a (union of) irreducible component of* $Z_{n,v+2n,\varepsilon}$. *Moreover,* $i_{\varepsilon}^{-1}(\overline{Z}_{A,\varepsilon}) = \overline{Z}_{A-2I,\varepsilon}$ *if* $A - 2I \in \Theta_{v+\infty}$ *and empty otherwise.*

2. Geometric realization of $\mathrm{U}(\mathfrak{sl}_n^\theta)$

In this section, we provide geometric realization of the universal enveloping algebra $\mathrm{U}(\mathfrak{sl}_n^\theta)$, together with its idempotent version, where $(\mathfrak{sl}_n, \mathfrak{sl}_n^\theta)$ is a quasi-split symmetric pair. We have to treat n being odd or even separately.

2.1. Convolution algebra on $Z_{n,\infty,\varepsilon}$**.** We call a $\mathbb{C}$-valued function $f : Z_{n,\infty,\varepsilon} \to \mathbb{C}$ constructible if the pullback $i_k^*(f)$ is constructible for all k and $1 \leq v \leq 2n$. Let us consider the space of all constructible functions on $Z_{n,\infty,\varepsilon}$:

$$\mathscr{C}(Z_{n,\infty,\varepsilon}) = \{\text{all } \mathbb{C}\text{-valued constructible functions on } Z_{n,\infty,\varepsilon}\}.$$

Write π_k for the proper map $\pi : \mathcal{M}_{n,v+2kn,\varepsilon} \to \mathcal{N}_{n,v+2kn,\varepsilon}$ to emphasize the dependence on k. Let $\mathscr{C}(Z_{n,v+2kn,\varepsilon})$ be the space of constructible functions on $Z_{n,v+2kn,\varepsilon}$. There is a convolution product on $\mathscr{C}(Z_{n,v+2kn,\varepsilon})$ defined as follows. For all $\phi_1, \phi_2 \in \mathscr{C}(Z_{n,v+2kn,\varepsilon})$ and $(x, F, F') \in Z_{n,v+2kn,\varepsilon}$ we define

$$\phi_1 \star \phi_2(x, F, F') = \int_{\tilde{F} \in \pi_k^{-1}(x)} \phi_1(x, F, \tilde{F})\phi_2(x, \tilde{F}, F').$$

Now we define a convolution product on $\mathscr{C}(Z_{n,\infty,\varepsilon})$. For any constructible functions $\phi_1, \phi_2 \in \mathscr{C}(Z_{n,\infty,\varepsilon})$ and $z = [x, F, F'] \in Z_{n,\infty,\varepsilon}$, $(x, F, F') \in Z_{n,v+2kn,\varepsilon}$, we define

$$\phi_1 \star \phi_2(z) := i_k^*(\phi_1) \star i_k^*(\phi_2)(x, F, F'). \tag{12}$$

We must show that the convolution in (12) is well-defined. More precise, if $z = [y, G, G']$ with $(y, G, G') \in Z_{n,\varepsilon+2k'n,\varepsilon}$ for some k' such that $k < k'$, then we must have

$$i_k^*(\phi_1) \star i_k^*(\phi_2)(x, F, F') = i_{k'}^*(\phi_1) \star i_{k'}^*(\phi_2)(y, G, G')$$

Note that by assumption, there is

$$(y, G, G') = i_{k',k}(x, F, F').$$

So we must show that

$$\begin{aligned} i_k^*(\phi_1) \star i_k^*(\phi_2)(x, F, F') &= i_{k'}^*(\phi_1) \star i_{k'}^*(\phi_2) i_{k',k}(x, F, F') \\ &= i_{k',k}^*(i_{k'}^*(\phi_1) \star i_{k'}^*(\phi_2))(x, F, F'). \end{aligned}$$

It is equivalent to show that

$$i_k^*(\phi_1) \star i_k^*(\phi_2) = i_{k',k}^*(i_{k'}^*(\phi_1) \star i_{k'}^*(\phi_2)).$$

The latter is due to (7) and can be proved in a way similar to [**Li10**, Lemma 5.3]. So the convolution (12) is well-defined.

It is clear that the convolution (12) is associative because the convolution on $\mathscr{C}(Z_{n,v+2kn,\varepsilon})$ is so. The algebra $\mathscr{C}(Z_{n,\infty,\varepsilon})$ has a unit $\mathbf{1}$ defined by $\mathbf{1}(z) = 1$ if z has a representative of the form (x, F, F) and $\mathbf{1}(z) = 0$ otherwise.

The direct system $(Z_{n,v+2nk,\varepsilon}, i_\varepsilon)$ induces a projective system

$$(\mathscr{C}(Z_{n,v+2nk,\varepsilon}), i^*_{k',k})_{k,k'\in\mathbb{N}}$$

of associative algebras. One can consider the projective limit $\varprojlim_k \mathscr{C}(Z_{n,v+2nk,\varepsilon})$. The convolution algebra $\mathscr{C}(Z_{n,\infty,\varepsilon})$ provides a geometric realization of this projective limit.

Summing up the above analysis, we have

PROPOSITION 2.1.1. *The pair $(\mathscr{C}(Z_{n,\infty,\varepsilon}), \star)$ is an associative algebra, and is isomorphic to the projective limit $\varprojlim_k \mathscr{C}(Z_{n,v+2nk,\varepsilon})$.*

PROOF. By the universality, there is an algebra homomorphism

$$\mathscr{C}(Z_{n,\infty,\varepsilon}) \to \varprojlim_k \mathscr{C}(Z_{n,v+2nk,\varepsilon}).$$

Clearly, this homomorphism is bijective and the proposition follows. □

2.2. Geometric realization of $\mathrm{U}(\mathfrak{sl}_n^\theta)$ for n odd. In this section, assume that n is odd.

Let us consider the following constructible functions on $Z_{n,\infty,\varepsilon}$. For $1 \le i \le n-1$, $z = [x, F, F'] \in Z_{n,\infty,\varepsilon}$,

$$\mathbf{e}_{i,\varepsilon}(z) = \begin{cases} 1 & \text{if } F_i \overset{1}{\subset} F'_i, F_j = F'_j, \forall j \neq i, n-i, \\ 0 & \text{o.w.} \end{cases} \tag{13}$$

$$\mathbf{h}_{i,\varepsilon}(z) = \delta_{F,F'}(|F_{i+1}/F_i| - |F_i/F_{i-1}| + \delta_{1,\varepsilon}\delta_{-1,(-1)^v}(\delta_{i,r+1} - \delta_{i,r})). \tag{14}$$

One can show that $\mathbf{h}_{i,\varepsilon}$ is well-defined. Let $\mathrm{U}_{n,\varepsilon}$ be the subalgebra of $\mathscr{C}(Z_{n,\infty,\varepsilon})$ generated by the elements $\mathbf{e}_{i,\varepsilon}$ and $\mathbf{h}_{i,\varepsilon}$ for all $1 \le i \le n-1$.

Let θ be the bijection on the set $\{1, \cdots, n-1\}$ defined by $i \mapsto n-i$. The bijection defines an involution, still denoted by the same notation, on $\mathfrak{sl}_n \equiv \mathfrak{sl}_n(\mathbb{C})$ by $e_i \mapsto f_{\theta(i)}$, $f_i \mapsto e_{\theta(i)}$ and $h_i \mapsto -h_{\theta(i)}$ for $1 \le i \le n-1$. Let $\mathfrak{sl}_n^\theta$ be the fixed-point subalgebra of $\mathfrak{sl}_n$ under the involution θ. Let $\mathrm{U}(\mathfrak{sl}_n^\theta)$ be the enveloping algebra of $\mathfrak{sl}_n^\theta$. The pair $(\mathfrak{sl}_n, \mathfrak{sl}_n^\theta)$ is an infinitesimal quasi-split symmetric pair. It is known that $\mathrm{U}(\mathfrak{sl}_n^\theta)$ is generated by $e_{i,\theta} = e_i + f_{\theta(i)}$ and $h_{i,\theta} = h_i - h_{\theta(i)}$, for all $1 \le i \le n-1$, and subject to the following defining relations (setting $a_{ij} = 2\delta_{ij} - \delta_{i,j+1} - \delta_{i,j-1}$).

$$\begin{cases} h_{i,\theta} + h_{\theta(i),\theta} = 0 & \forall 1 \le i \le n-1. \\ [h_{i,\theta}, e_{j,\theta}] = (a_{ij} - a_{\theta(i),j})e_{j,\theta} & \forall 1 \le i,j \le n-1. \\ [e_{i,\theta}, e_{j,\theta}] = \delta_{i,\theta(j)}h_{i,\theta} & \text{if } a_{ij} = 0. \\ e_{i,\theta}e_{i,\theta}e_{j,\theta} - 2e_{i,\theta}e_{j,\theta}e_{i,\theta} + e_{j,\theta}e_{i,\theta}e_{i,\theta} = 0 & \text{if } a_{ij} = -1, i \neq \theta(j). \\ e_{i,\theta}e_{i,\theta}e_{j,\theta} - 2e_{i,\theta}e_{j,\theta}e_{i,\theta} + e_{j,\theta}e_{i,\theta}e_{i,\theta} = -4e_{i,\theta} & \text{if } a_{ij} = -1, i = \theta(j). \end{cases} \tag{15}$$

We are ready to state a geometric/topological realization of $\mathrm{U}(\mathfrak{sl}_n^\theta)$ for n being odd.

THEOREM 2.2.1. *There is a unique algebra isomorphism defined by*

$$\mathrm{U}(\mathfrak{sl}_n^\theta) \to \mathrm{U}_{n,\varepsilon}, e_{i,\theta} \mapsto \mathbf{e}_{i,\varepsilon}, h_{i,\theta} \mapsto \mathbf{h}_{i,\varepsilon}, \quad \forall 1 \le i \le n-1.$$

PROOF. Recall the embedding $i_k : Z_{n,v+2kn,\varepsilon} \to Z_{n,\infty,\varepsilon}$. Let $\mathbf{S}_{n,v+2nk,\varepsilon}$ be the pullback of $\mathrm{U}_{n,\varepsilon}$ under i_k. Let $\mathbf{e}_i^{\mathfrak{c}}$ and $\mathbf{h}_i^{\mathfrak{c}}$ denote the image of the respective generators in the same notation in $\mathrm{U}_{n,\varepsilon}$ under i_k^*. Then there exists a surjective algebra homomorphism $\mathrm{U}(\mathfrak{sl}_n^\theta) \to \mathbf{S}_{n,v+2nk,\varepsilon}$ defined by sending the generators $e_{i,\theta}, h_{i,\theta}$ to $\mathbf{e}_i^{\mathfrak{c}}$, $\mathbf{h}_i^{\mathfrak{c}}$, respectively. The nontrivial proof is given in Proposition 4.3.6. By the universality of projective limits, this induces a surjective algebra homomorphism from $\mathrm{U}(\mathfrak{sl}_n^\theta)$ to $\mathrm{U}_{n,\varepsilon}$, as a subalgebra of the projective limit of $\mathbf{S}_{n,v+2nk,\varepsilon}$.

It remains to show that this homomorphism is injective. We consider the projective system $(\mathscr{C}_{\mathbf{G}_{v+2kn,\varepsilon}}(\mathcal{F}_{n,v+2kn,\varepsilon} \times \mathcal{F}_{n,v+2kn,\varepsilon}))_{k\in\mathbb{N}}$ of convolution algebras of $\mathbf{G}_{v+2kn,\varepsilon}$-invariant functions. It is known from the work [**LW18**], after a specialization of quantum parameter to 1, that $\mathrm{U}(\mathfrak{sl}_n^\theta)$ is naturally a subalgebra of the projective limit of $(\mathscr{C}_{\mathbf{G}_{v+2kn,\varepsilon}}(\mathcal{F}_{n,v+2kn,\varepsilon} \times \mathcal{F}_{n,v+2kn,\varepsilon}))_{k\in\mathbb{N}}$.

Now regarding $\mathcal{F}_{n,v+2kn,\varepsilon} \times \mathcal{F}_{n,v+2kn,\varepsilon}$ as the zero section of $Z_{n,v+2nk,\varepsilon}$ by $(F, F') \mapsto (0, F, F')$. Taking restriction defines a map from the projective system $(\mathscr{C}(Z_{n,v+2nk,\varepsilon}))_k$ to the system $(\mathscr{C}_{\mathbf{G}_{v+2kn,\varepsilon}}(\mathcal{F}_{n,v+2kn,\varepsilon} \times \mathcal{F}_{n,v+2kn,\varepsilon}))_{k\in\mathbb{N}}$. By universality, this induces a surjective morphism from $\mathrm{U}_{n,\varepsilon}$ to $\mathrm{U}(\mathfrak{sl}_n^\theta)$ defined by sending the generators to their counterparts. This is an inverse to the previously defined algebra homomorphism, and so the algebra homomorphism $\mathrm{U}(\mathfrak{sl}_n^\theta) \to \mathrm{U}_{n,\varepsilon}$ is injective. The proof is finished. □

Before we proceed to the n =even case, we make the following

REMARK 2.2.2. Applying same argument in the proof of the above theorem to the $\mathbf{G}_v = \mathrm{GL}_v(\mathbb{C})$ version, provides a geometric realization of $\mathrm{U}(\mathfrak{sl}_n)$ itself.

2.3. Geometric realization of $\mathrm{U}(\mathfrak{sl}_n^\theta)$ for n even. In this section, we assume that n is even, and recall that $r = n/2$. Similar to the case when n is odd in Section 2.2, we provide a geometric realization of $\mathrm{U}(\mathfrak{sl}_n^\theta)$ inside the convolution algebra $\mathscr{C}(Z_{n,\infty,\varepsilon})$.

On the algebraic side, the fixed-point subalgebra $\mathrm{U}(\mathfrak{sl}_n^\theta)$ in Section 2.2 is still well-defined when n is odd. Moreover, it admits a presentation as follows.

$$
(16)\qquad \begin{cases}
h_{i,\theta} + h_{\theta(i),\theta} = 0 & \forall 1 \le i \le n-1.\\
[h_{i,\theta}, e_{j,\theta}] = (a_{ij} - a_{\theta(i),j})e_{j,\theta} & \forall 1 \le i,j \le n-1.\\
[e_{i,\theta}, e_{j,\theta}] = \delta_{i,\theta(j)} h_{i,\theta} & \text{if } a_{ij} = 0.\\
e_{i,\theta}e_{i,\theta}e_{j,\theta} - 2e_{i,\theta}e_{j,\theta}e_{i,\theta} + e_{j,\theta}e_{i,\theta}e_{i,\theta} = 0 & \text{if } a_{ij} = -1, i \ne \theta(i).\\
e_{i,\theta}e_{i,\theta}e_{j,\theta} - 2e_{i,\theta}e_{j,\theta}e_{i,\theta} + e_{j,\theta}e_{i,\theta}e_{i,\theta} = e_{j,\theta} & \text{if } a_{ij} = -1, i = \theta(i).
\end{cases}
$$

On the geometric side, the generators $\mathbf{e}_{i,\varepsilon}$ and $\mathbf{h}_{i,\varepsilon}$ are still well-defined, except $\mathbf{e}_{r,\varepsilon}$. Instead, for all $z \in Z_{n,\infty,\varepsilon}$, we define

$$
\mathbf{e}_{r,\varepsilon}(z) = \begin{cases}
1 & \text{if } z = [x, F, F'], |F_j \cap F'_j| = |F_j| - \delta_{j,r}, \forall 1 \le j \le n,\\
-1 & \text{if } z = [x, F, F], \mathrm{rank}(x|_{F_{r+1}/F_{r-1}}) \text{ is odd}, \varepsilon = -1,\\
0 & \text{o.w.}
\end{cases}
$$

It is straightforward to check that $\mathbf{e}_{r,\varepsilon}$ is well-defined. Note that $\mathbf{h}_{r,\varepsilon} \equiv 0$. Finally, let $\mathrm{U}_{n,\varepsilon}$ be the subalgebra of $\mathscr{C}(Z_{n,\infty,\varepsilon})$ generated by the generators $\mathbf{e}_{i,\varepsilon}$ and $\mathbf{h}_{i,\varepsilon}$.

THEOREM 2.3.1. *The statement in Theorem 2.2.1 still holds for n being odd.*

PROOF. The existence of an algebra homomorphism from $\mathrm{U}(\mathfrak{sl}_n^\theta)$ to $\mathrm{U}_{n,\varepsilon}$ is due to Proposition 4.4.3. The argument in the proof of Theorem 2.2.1 applies here to finish the proof. □

2.4. Idempotent version. Let

$$\Lambda_{v+\infty} = \left\{ \lambda = (\lambda_i)_{1\leq i \lneq n} \in \mathbb{N}^n | \sum_{i=1}^{n} \lambda_i \equiv v (\mathrm{mod} 2n), \lambda_i = \lambda_{n+1-i} \right\}.$$

The set $\Lambda_{v+\infty}$ admits a partition

$$\Lambda_{v+\infty} = \sqcup_{k:v+2kn>0} \Lambda_{v+2kn}, \Lambda_{v+2kn} = \{\lambda \in \Lambda_{v+\infty} | \sum_i \lambda_i = v+2kn\}.$$

We define an equivalence relation $\sim$ on $\Lambda_{v+\infty}$ by

$$\lambda \sim \mu \text{ if and only if } \lambda - \mu \equiv 0 \ (\mathrm{mod} 2).$$

Let $\bar{\Lambda}_{v+\infty} = \Lambda_{v+\infty}/\sim$ and $\bar{\lambda}$ be the equivalence class of λ.

Let $(\delta_i)_{1\leq i\leq n}$ be the standard basis of $\mathbb{N}^n$. If n is odd, we define a map

$$\mathrm{add}: \Lambda_{v+\infty} \to \Lambda_{(v+1)+\infty}, \quad \lambda \mapsto \lambda + \delta_r.$$

It clearly descends to a map $\mathrm{add}: \bar{\Lambda}_{v+\infty} \to \bar{\Lambda}_{(v+1)+\infty}$.

We define two maps

$$\mathrm{ro}, \mathrm{co}: \Theta_{v+\infty} \to \Lambda_{v+\infty},$$

where the i-th entry of $\mathrm{ro}(A)$ (resp. $\mathrm{co}(A)$) is equal to $\sum_l a_{il}$ (resp. $\sum_k a_{ki}$). Clearly these maps induce maps $\mathrm{ro}, \mathrm{co}: \bar{\Theta}_{v+\infty} \to \bar{\Lambda}_{v+\infty}$.

For each $\bar{\lambda} \in \bar{\Lambda}_{v+\infty}$, we define a function $\mathbf{1}_{\bar{\lambda},\varepsilon}$ on $Z_{n,\infty,\varepsilon}$ as follows. For any $z = [x, F, F'] \in Z_{n,\infty,\varepsilon}$, we set

$$\mathbf{1}_{\bar{\lambda},\varepsilon}(z) = \begin{cases} \delta_{F,F'}\delta_{\bar{\lambda},\mathrm{ro}(A_{F,F'})} & \text{if } (n,v,\varepsilon) \neq (\text{even, odd}, 1), \\ \delta_{F,F'}\delta_{\bar{\lambda},\mathrm{ro}(\mathrm{dlt}A_{F,F'})} & \text{if } (n,v,\varepsilon) = (\text{even, odd}, 1). \end{cases}$$

Let $\dot{\mathrm{U}}_{n,\varepsilon}$ be the $\mathrm{U}_{n,\varepsilon}$-bimodule generated by $\mathbf{1}_{\bar{\lambda},\varepsilon}$ for all $\bar{\lambda} \in \bar{\Lambda}_{v+\infty}$ and $1 \leq v \leq 2n$. It is clear that

$$\begin{aligned} &\mathbf{1}_{\bar{\lambda},\varepsilon}\mathbf{1}_{\bar{\mu},\varepsilon} = \delta_{\bar{\lambda},\bar{\mu}}\mathbf{1}_{\bar{\lambda},\varepsilon}, \ \forall \bar{\lambda}, \bar{\mu}. \\ &\mathbf{e}_{i,\varepsilon}\mathbf{1}_{\bar{\lambda},\varepsilon} = \mathbf{1}_{\bar{\lambda}-\bar{\delta}_i+\bar{\delta}_{i+1}+\bar{\delta}_{\theta(i)}-\bar{\delta}_{\theta(i)+1},\varepsilon}\mathbf{e}_{i,\varepsilon}, \ \forall i, \bar{\lambda}. \end{aligned} \tag{17}$$

So $\dot{\mathrm{U}}_{n,\varepsilon}$, together with the convolution, is naturally an associative algebra without unit.

Let $\dot{\mathrm{U}}(\mathfrak{sl}_n^\theta)$ be the idempotent form of $\mathrm{U}(\mathfrak{sl}_n^\theta)$. This is an associative algebra with generators $1_{\bar{\lambda}}$ and $e_{i,\theta}1_{\bar{\lambda}}$, $\forall \bar{\lambda} \in \bar{\Lambda}_{v+\infty}, v = 2, 4, \cdots, 2n$ and as $\mathrm{U}(\mathfrak{sl}_n^\theta)$-bimodule generated by $1_{\bar{\lambda}}$ and such that

$$1_{\bar{\lambda}} h_{i,\theta} = h_{i,\theta} 1_{\bar{\lambda}} = (\lambda_{i+1} - \lambda_i) 1_{\bar{\lambda}}, \quad \forall i, \bar{\lambda}.$$

The following is an idempotent version of Theorems 2.2.1 and 2.3.1, with a similar proof.

THEOREM 2.4.1. *There exists an algebra isomorphism $\dot{\mathrm{U}}(\mathfrak{sl}_n^\theta) \to \dot{\mathrm{U}}_{n,\varepsilon}$ defined by*

$$e_{i,\theta}^\ell 1_{\bar{\lambda}} \mapsto \begin{cases} \mathbf{e}_{i,\varepsilon}^\ell \mathbf{1}_{\mathrm{add}(\bar{\lambda}),\varepsilon} & \text{if } (n,v,\varepsilon) = (\text{odd, odd}, 1), \\ \mathbf{e}_{i,\varepsilon}^\ell \mathbf{1}_{\bar{\lambda},\varepsilon} & \text{o.w.} \end{cases} \qquad \forall \ell = 0, 1$$

REMARK 2.4.2. Applying a similar argument of the proof of Theorem 2.4.1 yields a geometric realization of the idempotent enveloping algebra $\dot{\mathrm{U}}(\mathfrak{sl}_n)$.

3. Semicanonical bases

In this section, we shall construct semicanonical bases for the algebra $\dot{\mathrm{U}}(\mathfrak{sl}_n^\theta)$. We treat the n=odd case first, and then the harder n=even case. For the latter case, we have to use the results in n=odd case to establish the desired result.

Recall $\Theta_{v+\infty}$ from (11). Following [**Li21**], we define a partial order $\preceq$ on $\Theta_{v+\infty}$ by

$$A \preceq B \text{ iff } \sum_{r\leq i; s\geq j} a_{rs} \leq \sum_{r\leq i; s\geq j} b_{rs}, \quad \forall i<j.$$

Following [**Li21**], we define a refinement $\sqsubseteq$ by

$$A \sqsubseteq B \text{ iff } A \preceq B, \mathrm{ro}(A) = \mathrm{ro}(B), \mathrm{co}(A) = \mathrm{co}(B).$$

3.1. The odd case. In this section, we assume that n is odd. For any $\overline{C} \in \overline{\Theta}_{v+\infty}$, let $\xi_{\overline{C},\varepsilon}$ be the characteristic function of $Z_{\overline{C},\varepsilon}$. Let $E^\theta_{h,h+1} = E_{h,h+1} + E_{n+1-h,n-h}$. For $h, j_0 \in [1, n-1]$, $c \in \mathbb{N}$ and $A \in \Theta_{v+\infty}$, we set

$$A_{h,c;j_0} = A - c(E^\theta_{h,j_0} - E^\theta_{h+1,j_0}).$$

For any function f in $\mathscr{C}(Z_{n,\infty,\varepsilon})$, we write $\mathrm{supp}(f)$ to be the support of f, i.e., the closure of all points z such that $f(z) \neq 0$. The following is an analogue of [**Li21**, Lemma 3.1.1].

LEMMA 3.1.1. *Assume that* $C, A \in \Theta_{v+\infty}$. *Assume further that* $\mathrm{co}(C) = \mathrm{ro}(A)$ *and* $C - cE^\theta_{h,h+1}$ *is diagonal for some* $c \in \mathbb{N}$, $1 \leq h \leq n-1$. *If there is an integer* j_0 *such that*

$$\begin{cases} j_0 \in [1,n] & \text{if } h \neq r, r+1, \\ j_0 \in [1,r) & \text{if } h = r, \\ j_0 \in (r,n] & \text{if } h = r+1, \end{cases} \quad \text{and} \quad \begin{cases} a_{h,j} = 0 & \forall 1 \leq j \leq j_0 - 1, \\ a_{h,j_0} = c & \\ a_{h+1,j} = 0 & \forall 1 \leq j \leq j_0, \end{cases}$$

then we have

$$\xi_{\overline{C},\varepsilon} * \xi_{\overline{A},\varepsilon}|_{Z_{\overline{A}_{h,c;j_0},\varepsilon}} \equiv 1 \text{ and } \mathrm{supp}(\xi_{\overline{C},\varepsilon} * \xi_{\overline{A},\varepsilon}) \subseteq p^{-1}(\overline{\mathcal{O}}_{\overline{A}_{h,c;j_0},\varepsilon}). \tag{18}$$

PROOF. We shall provide the proof under the assumption that $\varepsilon = -1$. The case when $\varepsilon = 1$ can be proved similarly.

From the analysis of Lemma 3.9 in [**BKLW**], we only need to show that the restriction of $\xi_{\overline{C},\varepsilon} * \xi_{\overline{A},\varepsilon}$ to $Z_{\overline{A}_{h,c;j_0},\varepsilon}$ is 1. Assume that $z = [x, F, F']$ with $(x, F, F') \in Z_{n,v,\varepsilon}$ such that $(F, F') \in \mathcal{O}_{A_{h,c;j_0},\varepsilon}$. Then the value of $\xi_{\overline{C},\varepsilon} * \xi_{\overline{A},\varepsilon}$ at z is equal to the Euler characteristic of

$$X = \{F'' \in \mathcal{F}_{n,v,\varepsilon} | (x, F'') \in \mathcal{M}_{n,v,\varepsilon}, (F, F'') \in \mathcal{O}_{C,\varepsilon}, (F'', F') \in \mathcal{O}_{A,\varepsilon}\}.$$

It is reduced to show that X consists of a single point.

Assume now that $h \neq r, r+1$. Pick an element $F'' \in X$. The condition $(F, F'') \in \mathcal{O}_{C,\varepsilon}$ implies that the flag F'' satisfies $F''_i = F_i$ for all $i \neq h, n-h$, $F_h \subseteq F''_h \subseteq F_{h+1}$ and $\dim F''_h/F_h = c$. The condition $(F'', F') \in \mathcal{O}_{A,\varepsilon}$ implies that

$$F''_h = F''_h + F_{h+1} \cap F'_1 = \cdots = F''_h + F_{h+1} \cap F'_{j_0}.$$

In particular, we have $F_{h+1} \cap F'_{j_0} \subseteq F''_h$. So $F_h + F_{h+1} \cap F'_{j_0} \subseteq F''_h$. Now we compute

$$|F_h + F_{h+1} \cap F'_{j_0} / F_h| = |F_{h+1} \cap F'_{j_0}| - |F_h \cap F'_{j_0}| = c,$$

where the last equality is due to the condition $(F, F') \in \mathcal{O}_{A_{h,c;j_0},\varepsilon}$. This leads to

$$F''_h = F_h + F_{h+1} \cap F'_{j_0}. \tag{19}$$

Hence, there is at most one point in X, whose h-component is given by (19).

Let F'' be the flag such that $F''_i = F_i$ for all $i \neq h, n-h$ and satisfies (19). To show that X consists of one point, it is enough to show that

$$(x, F'') \in \mathcal{M}_{n,v,\varepsilon}. \tag{20}$$

This is reduced to prove that $x(F''_{h+1}) \subseteq F''_h$ and $x(F''_h) \subseteq F''_{h-1}$. For the first one, we have $x(F''_{h+1}) = x(F_{h+1}) \subseteq F_h \subseteq F''_h$. For the second one, it is given by

$$x(F''_h) = x(F_h + F_{h+1} \cap F'_{j_0}) \subseteq F_{h-1} + F_h \cap F'_{j_0-1} = F_{h-1} = F''_{h-1},$$

where we use the condition that $(F, F') \in \mathcal{O}_{A_{h,c;j_0},\varepsilon}$ for the second equality. The property (20) is thus proved, and so the lemma holds for $h \neq r, r+1$.

The proof of the case when $h = r$ or $r+1$ is a refinement of the above analysis. Let us deal with the $h = r$ case. In this case, the flag F'' in X has to be of the form

$$0 \subseteq F_1 \subseteq \cdots \subseteq F_{r-1} \subseteq F''_r \subseteq (F''_r)^\perp \subseteq F_{r+2} \subseteq \cdots,$$

$F_r \subseteq F''_r \subseteq (F''_r)^\perp \subseteq F_{r+1}$ and $|F''_r/F_r| = c$. In particular, we have $(F''_r)^\perp \cap F'_{j_0} \subseteq F_{r+1} \cap F'_{j_0}$. By the conditions $(F'', F') \in \mathcal{O}_{A,\varepsilon}$ and $(F, F') \in \mathcal{O}_{A_{h,c;j_0},\varepsilon}$, we see that the dimensions of $(F''_r)^\perp \cap F'_{j_0}$ and $F_{r+1} \cap F'_{j_0}$ are the same, and hence we have

$$(F''_r)^\perp \cap F'_{j_0} = F_{r+1} \cap F'_{j_0}. \tag{21}$$

The condition that $(F'', F') \in \mathcal{O}_{A,\varepsilon}$ implies that $F''_r = F''_r + (F''_r)^\perp \cap F'_{j_0}$, so we have $F_r + (F''_r)^\perp \cap F'_{j_0} = F_r + F_{r+1} \cap F'_{j_0} \subseteq F''_r$. By comparing the dimension, we have

$$F''_r = F_r + F_{r+1} \cap F'_{j_0},$$

similar to (19). Clearly, $(F''_r)^\perp = F_r + F_{r+1} \cap F'_{n-j_0} \supseteq F''_r$. It remains to show that this F'' satisfies (20). It is enough to show that $x(F_{r+1}) \subseteq (F''_r)^\perp$, $x((F''_r)^\perp) \subseteq F''_r$ and $x(F''_r) \subseteq F_{r-1}$. The second one is followed by the argument:

$$x((F''_r)^\perp) \subseteq x(F_{r+1}) \subseteq F_r \subseteq F''_r.$$

Since $x \in \mathcal{N}_{n,v,\varepsilon}$, we see that the third one implies the first one. Finally, for the third condition, we have

$$x(F''_r) = x(F_r + (F''_r)^\perp \cap F'_{j_0}) \subseteq F_{r-1} + F''_r \cap F'_{j_0-1} = F_{r-1},$$

where the first equality is by (21) and the last equality is due to the condition $(F'', F') \in \mathcal{O}_{A,\varepsilon}$. Therefore, the lemma holds for $h = r$.

Finally, we consider $h = r+1$. If $F'' \in X$, then the condition $(F, F'') \in \mathcal{O}_{C,\varepsilon}$ implies that the flag F'' is of the form

$$0 \subseteq F_1 \subseteq \cdots \subseteq F_{r-1} \subseteq F''_r \subseteq (F''_r)^\perp \subseteq F_{r+2} \subseteq \cdots,$$

$F''_r \subseteq F_r \subseteq F_{r+1} \subseteq (F''_r)^\perp$ and $|F_r/F''_r| = c$. The conditions $(F'', F') \in \mathcal{O}_{A,\varepsilon}$ and $(F, F') \in \mathcal{O}_{A_{r+1,c;j_0},\varepsilon}$ imply that $F''_r \cap F'_{j_0-1} = F_r \cap F'_{j_0-1}$. This implies that $F''_r = F_{r-1} + F_r \cap F'_{j_0-1}$ since $F''_r \supseteq F_{r-1} + F_r \cap F'_{j_0-1}$ and both have the same dimension. The conditions $x(F''_r) \subseteq F_{r-1}$ and $F''_r \subseteq (F''_r)^\perp$ are easily checked.

The last condition $x((F''_r)^{\perp}) \subseteq F''_r$ can be proved as follows. We have $(F''_r)^{\perp} = (F_{r-1} + F_r \cap F'_{j_0-1})^{\perp} = F_{r+1} + F_{r+2} \cap F'_{n+1-j_0}$, so $x((F''_r)^{\perp}) \subseteq F_r + F_{r+1} \cap F'_{n-j_0}$. On the other hand, $(F''_r)^{\perp} = F''_r + (F''_r)^{\perp} \cap F'_{j_0}$, so $x((F''_r)^{\perp}) \subseteq F_{r-1} + F'_{j_0-1}$. Combining the above two inclusions, we have

$$x((F''_r)^{\perp}) \subseteq (F_r + F_{r+1} \cap F'_{n-j_0}) \cap (F_{r-1} + F'_{j_0-1}).$$

Observe that

$$\begin{aligned}(F_r + F_{r+1} \cap F'_{n-j_0}) \cap (F_{r-1} + F'_{j_0-1}) &= F_{r-1} + F'_{j_0-1} \cap (F_r + F_{r+1} \cap F'_{n-j_0})\\ = F_{r-1} + F_{r+1} \cap F'_{n-j_0} + F_r \cap F'_{j_0-1} &= F_{r-1} + F_r \cap F'_{j_0-1} = F''_r,\end{aligned}$$

where we use $F_{r+1} \cap F'_{n-j_0} = F_r \cap F'_{n-j_0} \subseteq F_r \cap F'_{j_0-1}$ by $(F, F') \in \mathcal{O}_{A_{r+1,c:j_0},\varepsilon}$. Hence we have $x((F''_r)^{\perp}) \subseteq F''_r$, and so the $h = r$ case is done. The lemma is thus proved. □

Following [**Li21**], we define a total order on the set $\{(i,j) | 1 \leq j < i \leq n\}$ by setting $(i,j) \lhd (k,l)$ if and only if $i - j > k - l$ or $i - j = k - l$ and $i < k$. For all $A \in \Theta_{v+\infty}$ with $a_{ii} >> 0$, we define inductively with respect to the order $\lhd$ the $\frac{n(n-1)}{2}$ matrices $G_{ij} \equiv G^A_{ij} \in \Theta_{v+\infty}$ for $i > j$ as follows. $G_{n,1}$ is defined by the condition $G_{n,1} - a_{n,n-1}E^{\theta}_{n,n-1}$ is diagonal and $\mathrm{ro}(G_{n,1}) = \mathrm{ro}(A)$. Put $G_{i-1,j-1} = G_{n,n-i}$ if $j = 1$. Assume that $G_{i-1,j-1}$ is defined, we define G_{ij} by the conditions that $\mathrm{ro}(G_{i,j}) = \mathrm{co}(G_{i-1,j-1})$ and $G_{ij} - \sum_{k \geq i} a_{kj} E^{\theta}_{i,i+1}$ is diagonal. In particular, for all $2 \leq i \leq n$, we have

$$\mathrm{ro}(G_{n,1}) = \mathrm{ro}(A), \mathrm{co}(G_{n,n-i}) = \mathrm{ro}(G_{i,1}),$$
$$\mathrm{co}(G_{ij}) = \mathrm{ro}(G_{i+1,j+1}), \mathrm{co}(G_{n,n-1}) = \mathrm{co}(A).$$

Similar to [**Li21**, (41)], we define

$$m_{\overline{A},\varepsilon} = \xi_{\overline{G}_{n,1},\varepsilon}(\xi_{\overline{G}_{n-1,1},\varepsilon}\xi_{\overline{G}_{n,2},\varepsilon}) \cdots (\xi_{\overline{G}_{i,1},\varepsilon}\xi_{\overline{G}_{i+1,2},\varepsilon} \cdots \xi_{\overline{G}_{n,n-i+1},\varepsilon}) \cdots (\xi_{\overline{G}_{21},\varepsilon} \cdots \xi_{\overline{G}_{n,n-1},\varepsilon}), \tag{22}$$

where the product is taken with respect to the order $\lhd$. If a function f satisfies conditions in (18) with respect to a matrix M, we shall write it as "$f = \xi_{Z_{\overline{M}},\varepsilon}$ + lower term." By applying inductively Lemma 3.1.1, we have

PROPOSITION 3.1.2. *For all $A \in \Theta_{v+\infty}$, we have $m_{\overline{A},\varepsilon} = \xi_{\overline{A},\varepsilon}$ + lower term.*

Since the set $\{\xi_{\overline{A},\varepsilon} | A \in \Theta_{v+\infty}\}$ is linear independent in $\mathscr{C}(Z_{n,\infty,\varepsilon})$, we have

PROPOSITION 3.1.3. *The set $\{m_{\overline{A},\varepsilon} | \overline{A} \in \overline{\Theta}_{v+\infty}, 1 \leq v \leq 2n\}$ is linearly independent in $\dot{\mathrm{U}}_{n,\varepsilon}$.*

Moreover, there is

PROPOSITION 3.1.4. *For all $A \in \Theta_{v+\infty}$, there exists an element $s_{\overline{A},\varepsilon} \in \dot{\mathrm{U}}_{n,\varepsilon}$ satisfying the following conditions.*

$$\begin{cases} \text{(a)} & s_{\overline{A},\varepsilon}|_{Z_{\overline{A},\varepsilon}} \equiv 1, \\ \text{(b)} & s_{\overline{A},\varepsilon}|_U \equiv 0, \textit{for some open dense subset } U \textit{ in } Z_{\overline{B},\varepsilon} \textit{ if } B \sqsubset A, \\ \text{(c)} & s_{\overline{A},\varepsilon}|_{Z_{\overline{A'},\varepsilon}} \equiv 0 \textit{ for any } A' \not\sqsubseteq A. \end{cases} \tag{23}$$

PROOF. One can produce a proof entirely similar to the proof of Proposition 4.4 in [**Li10**]. Here we provide a refinement using the element $m_{\overline{A},\varepsilon}$. It is clear that the statement is true for $s_{\overline{A},\varepsilon} := m_{\overline{A},\varepsilon}$ when $A - cE^{\theta}_{h,h+1}$ is diagonal for some $c \in \mathbb{N}$. For any $B \sqsubset A$, we assume that there is an element $s_{\overline{B},\varepsilon}$ in $\dot{\mathrm{U}}_{n,\varepsilon}$ satisfying the conditions in the proposition. Consider the monomial $m_{\overline{A},\varepsilon}$. The restriction $m_{\overline{A},\varepsilon}|_{Z_{\overline{B},\varepsilon}}$ to $Z_{\overline{B},\varepsilon}$ is constructible, and hence there is a dense open subvariety, say $\tilde{U}_B$, of $Z_{\overline{B},\varepsilon}$ and an integer $c_{B,A}$ such that $m_{\overline{A},\varepsilon}|_{\tilde{U}_B} \equiv c_{B,A}$. We set

$$s_{\overline{A},\varepsilon} = m_{\overline{A},\varepsilon} - \sum_{B \sqsubset A} c_{B,A} s_{\overline{B},\varepsilon}. \tag{24}$$

By Proposition 3.1.2 and the construction, we see that $s_{\overline{A},\varepsilon}$ satisfies the first and third conditions. It remains to show that $s_{\overline{A},\varepsilon}$ satisfies the second condition. For B' such that $B \sqsubset B' \sqsubset A$, we can find an open dense subset $U_{B',B}$ such that $s_{B'}|_{U_{B',B}} \equiv 0$. Let $U = \tilde{U}_B \cap \cap_{B':B \sqsubset B' \sqsubset A} U_{B',B}$. It is a finite intersection and hence is still open dense in $Z_{\overline{B},\varepsilon}$. By induction, it is clear that $s_{\overline{A},\varepsilon}|_U \equiv 0$. Hence the second condition holds for $s_{\overline{A},\varepsilon}$. The proposition is thus proved. □

THEOREM 3.1.5. *The sets* $\{m_{\overline{A},\varepsilon} | A \in \Theta_{v+\infty}, 1 \le v \le 2n\}$ *and* $\{s_{\overline{A},\varepsilon} | A \in \Theta_{v+\infty}, 1 \le v \le 2n\}$ *are two bases of* $\dot{\mathrm{U}}_{n,\varepsilon}$, *called the monomial and semicanonical basis, respectively.*

PROOF. That the set $\{m_{\overline{A},\varepsilon}\}$ is a basis of $\dot{\mathrm{U}}_{n,\varepsilon}$ is known in light of [**BKLW**]. Since the transition matrix of $\{m_{\overline{A},\varepsilon}\}$ and $\{s_{\overline{A},\varepsilon}\}$ is upper triangular, the set $\{s_{\overline{A},\varepsilon}\}$ is also a basis of $\dot{\mathrm{U}}_{n,\varepsilon}$. □

REMARK 3.1.6.

(1) Via the identification in Theorem 2.4.1, $\{m_{\overline{A},\varepsilon}\}$ and $\{s_{\overline{A},\varepsilon}\}$ are bases of $\dot{\mathrm{U}}(\mathfrak{sl}^{\theta}_n)$. The basis $\{m_{\overline{A},\varepsilon}\}$ is independent of the choice of ε, but not clear if it is the case for the basis $s_{\overline{A},\varepsilon}$. See Example 4.4.5.

(2) Apply the same argument in the proof of Theorem 3.1.5 to the $\mathbf{G}_v$ case, we obtain a semicanonical basis for the idempotent enveloping algebra $\dot{\mathrm{U}}(\mathfrak{sl}_n)$.

In light of Theorem 3.1.5, we have the following by a standard argument (see [**Li10**, 4.7]).

THEOREM 3.1.7. *The* $s_{\overline{A},\varepsilon}$ *in* $\dot{\mathrm{U}}_{n,\varepsilon}$ *is the unique element satisfying the conditions* (23).

Let ${}_{\mathbb{Z}}\dot{\mathrm{U}}_{n,\varepsilon}$ be the $\mathbb{Z}$-subalgebra of $\dot{\mathrm{U}}_{n,\varepsilon}$ spanned by all monomials in $\xi_{A,\varepsilon}$ where $A - cE^{\theta}_{h,h+1}$ is diagonal. Due to (24), we have

PROPOSITION 3.1.8. *The set* $\{s_{\overline{A},\varepsilon} | A \in \Theta_{v+\infty}, 1 \le v \le 2n\}$ *is a basis of* ${}_{\mathbb{Z}}\dot{\mathrm{U}}_{n,\varepsilon}$.

For the remaining part of this section, we construct the semicanonical basis on the Schur algebra level. Recall there is an embedding $i_k : Z_{n,v+2nk,\varepsilon} \to Z_{n,v+\infty,\varepsilon}$, the pullback of i_k defines a surjective algebra homomorphism

$$i_k^* : \dot{\mathrm{U}}_{n,\varepsilon} \to \mathbf{S}_{n,v+2nk,\varepsilon},$$

where $\mathbf{S}_{n,v+2kn,\varepsilon}$ is the pullback of $\dot{\mathrm{U}}_{n,\varepsilon}$ (and $\mathrm{U}_{n,\varepsilon}$) under i_k^*. Note that

$$i_k^{-1}(\overline{Z}_{\overline{A}}) = \begin{cases} \overline{Z}_{A+2kI} & \text{if } A+2k'I \in \Theta_{v+2kn}, \text{ for some } k', \\ \emptyset & \text{o.w.} \end{cases}$$

This implies that

$$i_k^*(s_{\overline{A},\varepsilon}) = 0, \text{if there exists no } k' \text{ such that } A + 2k'I \in \Theta_{v+2kn}.$$

We set

$$s_{A,\varepsilon} := i_k^*(s_{\overline{A},\varepsilon}), \quad A \in \Theta_{v+2kn}.$$

Then we have

LEMMA 3.1.9. *The set $\{s_{A,\varepsilon} | A \in \Theta_{v+2nk}\}$ is a basis of $\mathbf{S}_{n,v+2nk,\varepsilon}$. Moreover, $s_{A,\varepsilon}$ is the unique element in $\mathbf{S}_{n,v+2kn,\varepsilon}$ subject to the conditions* (23) *(with the bars removed).*

PROOF. That $s_{A,\varepsilon}$ is a spanning set of $\mathbf{S}_{n,v+2nk,\varepsilon}$ is clear. Moreover, by restriction to various $Z_{A,\varepsilon}$, one see that the $s_{A,\varepsilon}$'s are linearly independent. Hence $\{s_{A,\varepsilon} | A \in \Theta_{v+2kn}\}$ forms a basis for $\mathbf{S}_{n,v+2nk,\varepsilon}$. It is clear that $s_{A,\varepsilon}$ satisfies the conditions in (23) and the uniqueness follows from a standard argument as before. The lemma is proved. □

3.2. The even case. In this section, assume that n is even. Due to the fact that $\dot{\mathrm{U}}(\mathfrak{sl}_n^\theta)$ is a subquotient of $\dot{\mathrm{U}}(\mathfrak{sl}_{n+1}^\theta)$, we can use results in previous section to obtain the following result.

THEOREM 3.2.1. *Assume that n is even. There is a basis $\{s_{\overline{A},\varepsilon} | A \in \Theta_{v+\infty}, 1 \leq v \leq 2n\}$ of $\dot{\mathrm{U}}_{n,\varepsilon}$ where $s_{\overline{A},\varepsilon}$ is the unique element in $\dot{\mathrm{U}}_{n,\varepsilon}$ subject to the three conditions in* (23).

The remaining part of this section is devoted to the proof of this theorem.

Write $\Theta_{n,v}$ (resp. $\Lambda_{n,v}$) for Θ_v (resp. Λ_v) to emphasize the dependence on n. Recall from Lemma 3.1.9, there is a basis $\{s_{A,\varepsilon} | A \in \Theta_{n+1,v}\}$ for the Schur algebra $\mathbf{S}_{n+1,v,\varepsilon}$. For each $\lambda \in \Lambda_{n+1,v}$ we set $\mathbf{1}_{\lambda,\varepsilon} = i_\varepsilon^*(\mathbf{1}_{\overline{\lambda},\varepsilon}) \in \mathbf{S}_{n+1,v,\varepsilon}$. Consider the following element in $\mathbf{S}_{n+1,v,\varepsilon}$.

$$\breve{\mathbf{j}}_{r,\varepsilon} = \begin{cases} \sum_{\lambda:\lambda_{r+1}=1} \mathbf{1}_{\lambda,\varepsilon} & \text{if } (v,\varepsilon) = (\text{odd}, 1), \\ \sum_{\lambda:\lambda_{r+1}=0} \mathbf{1}_{\lambda,\varepsilon} & \text{o.w.} \end{cases} \tag{25}$$

It is then clear that

$$\mathbf{S}_{n,v,\varepsilon} \cong \breve{\mathbf{j}}_{r,\varepsilon} \mathbf{S}_{n+1,v,\varepsilon} \breve{\mathbf{j}}_{r,\varepsilon}. \tag{26}$$

From this observation, there is

LEMMA 3.2.2. *The statement in Lemma* 3.1.9 *holds for n being even.*

Again from the observation (25), there is

LEMMA 3.2.3. *The statement in Lemma* 3.1.1 *remain holds for n being even.*

PROOF. The only new case from Lemma 3.1.1 is when $C - cE^\theta_{r,r+1}$ is diagonal. But in light of the imbedding $\mathbf{S}_{n,v,\varepsilon} \subseteq \mathbf{S}_{n+1,v,\varepsilon}$ via (26), this case can be proved by applying Lemma 3.1.1 twice (c.f. [**FLLLW**, Lemma 7.3.2]). □

Recall that there is an algebra homomorphism

$$i_{k',k}^{*}: \mathbf{S}_{n,v+2nk,\varepsilon} \to \mathbf{S}_{n,v+2nk',\varepsilon}.$$

LEMMA 3.2.4. *For any $A \in \Theta_{n,v+2nk}$,*

$$i_{k',k}^{*}(s_{A,\varepsilon}) = \begin{cases} s_{A-2(k-k')n,\varepsilon} & \text{if } A-2(k-k')n \in \Theta_{n,v+2nk'}, \\ 0 & o.w. \end{cases}$$

PROOF. This is due to Theorem 1.2.1, Proposition 1.2.3 (2) and the uniqueness of semicannonical basis with respect to the conditions in (23). □

Now we are ready to prove Theorem 3.2.1.

By Lemmas 3.2.3 and 3.2.4, we see, via an induction with respect to the partial order $\sqsubseteq$, that there is an element $s_{\overline{A},\varepsilon}$ for all $\overline{A} \in \overline{\Theta}_{v+\infty}$ in $\dot{\mathrm{U}}_{n,\varepsilon}$, for $n=2$, such that it satisfies the conditions in (23). But then the various $s_{\overline{A},\varepsilon}$ must span $\dot{\mathrm{U}}_{n,\varepsilon}$, for $n=2$, and hence must be the unique element in $\mathrm{U}_{n,\varepsilon}$ subject to the conditions in (23). The proof of Theorem 3.2.1 for $n=2$ is thus finished.

For general n case, we obtain a hybrid monomial basis, say $h_{\overline{A},\varepsilon}$, by substitute the element $\xi_{\overline{C},\varepsilon}$ in m_A in (22) by the associated semincanonical basis element $s_{\overline{C},\varepsilon}$ if $C - cE^{\theta}_{r,r+1}$ is diagonal. Note that m_A may not be in $\dot{\mathrm{U}}_{n,\varepsilon}$ but $h_{\overline{A},\varepsilon}$ is so. Now applying the argument in the proof of Proposition 3.1.4 to prove the existence of the basis $s_{\overline{A},\varepsilon}$ and the uniqueness property with respect to the conditions in (23) can be proved by a standard argument. This finishes the proof.

4. Defining relations in Schur algebras

In this section, we shall verify defining relations in (15)–(16) on Schur algebra level.

4.1. Rank one case in symplectic setting. In this section, we preserve the setting in Section 1.2. We further assume that $\varepsilon = -1$, so that the dimension v of the space V is even. Let $\mathrm{Gr}_{\mathrm{Sp}}(m,v)$ be the isotropic Grassmannian of m-dimensional isotropic subspace in a v-dimensional ambient vector space, equipped with a symplectic form. Let $T^*\mathrm{Gr}_{\mathrm{Sp}}(m,v)$ be its cotangent bundle, which is a connected component of $\mathcal{M}_{2,v,\varepsilon}$.

We need the following auxiliary lemmas.

LEMMA 4.1.1. *Let $(x,F) \in T^*\mathrm{Gr}_{\mathrm{Sp}}(|F|,v)$ and set*

$$A_{x,F} = \{\tilde{F} \in \mathrm{Gr}_{\mathrm{Sp}}(|F|+1,v) | F \subset \tilde{F}, (x,\tilde{F}) \in T^*\mathrm{Gr}_{\mathrm{Sp}}(|F|+1,v)\}$$

Then $\chi(A_{x,F}) = |F^{\perp}/F| - |x(F^{\perp})|$.

PROOF. The condition $(x,\tilde{F}) \in T^*\mathrm{Gr}_{\mathrm{Sp}}(|F|+1,v)$ is equivalient to $x(\tilde{F}) = 0$. Then $A_{x,F}$ consists of all isotropic subspaces $\tilde{F}$ such that $F \subset \tilde{F} \subseteq \ker(x|_{F^{\perp}})$ and $|\tilde{F}/F| = 1$. Hence

$$A_{x,F} \cong \mathrm{Gr}(1, |\ker(x|_{F^{\perp}})/F|).$$

Note that $|\ker(x|_{F^{\perp}})| = |F^{\perp}| - |x(F^{\perp})|$, so

$$\chi(A_{x,F}) = |F^{\perp}| - |x(F^{\perp})| - |F| = |F^{\perp}/F| - |x(F^{\perp})|.$$

The lemma is proved. □

This following lemma is an exercise in linear algebras.

LEMMA 4.1.2. *Let $x \in \mathrm{End}(W)$ and suppose that $\tilde{W}$ is a codimension one subspace in W such that*

- $x(\tilde{W}) \subseteq \tilde{W}$,
- *there exists $\tilde{w} \in \tilde{W}$ such that $x(\tilde{w}) \in \ker(x)$, and*
- *there exists $w \in W\backslash\tilde{W}$ such that $x(w) \notin x(\tilde{W})$,*

then $\mathrm{Rank}(x|_{\tilde{W}/\langle x(\tilde{w})\rangle}) = \mathrm{Rank}(x) - 2$.

We shall need the following lemma.

LEMMA 4.1.3. *Let $(x, F) \in T^*\mathrm{Gr}_{\mathrm{Sp}}(|F|, v)$ and $(x, F') \in T^*\mathrm{Gr}_{\mathrm{Sp}}(|F'|, v)$ and $F \overset{1}{\subset} F'$. If $x(F'^{\perp}) \subsetneqq x(F^{\perp})$, then*

$$\mathrm{Rank}(x|_{x(F^{\perp})^{\perp}/x(F^{\perp})}) = \mathrm{Rank}(x|_{x(F'^{\perp})^{\perp}/x(F'^{\perp})}) + 2.$$

PROOF. By assumption, there is an element

$$\beta \in F^{\perp}\backslash F'^{\perp} \quad \text{and} \quad \alpha \in x(F^{\perp})\backslash x(F'^{\perp})$$

such that $x(\beta) = \alpha$. On the other hand, there exists a vector $\gamma \in x(F'^{\perp})^{\perp}\backslash x(F^{\perp})^{\perp}$ such that $(\alpha, \gamma) = 1$. Obviously, we have $x(\gamma) \in F'$. We claim that $x(\gamma) \notin F$. Otherwise we would have the following contradiction.

$$1 = (\alpha, \gamma) = (x(\beta), \gamma) = -(\beta, x(\gamma)) \in (F^{\perp}, F) = \{0\}.$$

Now apply Lemma 4.1.2 with $W = x(F'^{\perp})^{\perp}/x(F'^{\perp})$ to finish the proof. $\square$

LEMMA 4.1.4. *Let $(x, F) \in T^*\mathrm{Gr}_{\mathrm{Sp}}(|F|, v)$. Assume further that $x(F^{\perp}) = 0$. Set*

$$\tilde{B}_{x,F} = \{U \overset{1}{\subset} F | x(U^{\perp}) \subseteq U\}.$$

Then $\chi(\tilde{B}_{x,F}) = |F| + \frac{(-1)^{\mathrm{Rank}(x)} - 1}{2}$.

PROOF. Recall that V is the v-dimensional underlying symplectic space. We fix a decomposition $V = (F^{\perp}/F) \oplus F \oplus F^{\natural}$ such that

- $F^{\perp} = (F^{\perp}/F) \oplus F$,
- the restriction of the symplectic form to $F^{\perp}/F$ and $F \oplus F^{\natural}$ are non-degenerate,
- $F^{\perp}/F \perp F \oplus F^{\natural}$,
- $F, F^{\natural}$ are maximal isotropic with respect to the symplectic form on $F \oplus F^{\natural}$,
- $F, F^{\natural}$ are dual to each other.

Let $\pi_{F^{\natural}} : V \to F^{\natural}$ be the canonical projection with respect to the decomposition. Let

$$\tilde{B}^{\natural}_{x,F} = \{L \in \mathrm{Gr}(1, F^{\natural}) | (x(L), L) = 0\}.$$

The assignment $U \mapsto \pi_{F^{\natural}}(U^{\perp})$ defines a bijection

$$\tilde{B}_{x,F} \cong \tilde{B}^{\natural}_{x,F}.$$

Its inverse is given by $L \mapsto \pi_{F^{\natural}}^{-1}(L)^{\perp}$. Note that $\pi_{F^{\natural}}^{-1}(L) = F^{\perp} \oplus L$. We have

$$(x(L), F^{\perp} \oplus L) = -(L, x(F^{\perp}) \oplus x(L) = (L, x(L)) = 0,$$

which implies that $x(\pi_{F^{\natural}}^{-1}(L)) = x(L) \subseteq \pi_{F^{\natural}}^{-1}(L)^{\perp}$. So the inverse map is well-defined.

Now we characterize $\tilde{B}^{\natural}_{x,F}$. Fix a basis on F and $F^{\natural}$ so that the anti-symmetric matrix of the symplectic form is of the following form with respect to the decomposition of V.

$$J = \begin{bmatrix} J_0 & 0 & 0 \\ 0 & 0 & 1 \\ 0 & -1 & 0 \end{bmatrix}$$

where J_0 is a certain anti-symmetric matrix unrelated to the discussion. The assumption $x(F^{\perp}) = 0$ forces x to be of the form

$$x = \begin{bmatrix} 0 & 0 & x_1 \\ 0 & 0 & x_2 \\ 0 & 0 & x_3 \end{bmatrix}.$$

The fact that $x \in \mathcal{N}_{v,\varepsilon}$ is equivalent to ${}^t xJ + Jx = 0$ which is equivalent to

$$x = \begin{bmatrix} 0 & 0 & 0 \\ 0 & 0 & x_2 \\ 0 & 0 & 0 \end{bmatrix}, \quad {}^t x_2 = x_2.$$

By identifying F with $F^{\natural}$ via the symplectic form on V, we see that x_2, being symmetric, defines a possibly singular symmetric bilinear form on $F^{\natural}$. Then the set $\tilde{B}^{\natural}_{x,F}$ can be identified with $\mathrm{Gr}_{\mathrm{O}}(1, F^{\natural})$, the variety of isotropic lines in $F^{\natural}$ with respect to the form defined by x_2. Fix a decomposition $F^{\natural} = F^{\natural}_1 \oplus F^{\natural}_2$ where $F^{\natural}_1 = x_2(F^{\natural})$ and $F^{\natural}_2 = F^{\natural}/F^{\natural}_1$. Note that the restriction of the form to $F^{\natural}_1$ is non-degenerate, and 0 on $F^{\natural}_2$. Note also that $|F^{\natural}_1| = \mathrm{Rank}(x)$. Let $S_2 = \{L \in \mathrm{Gr}_{\mathrm{O}}(1, F^{\natural}) | L \subseteq F^{\natural}_2\}$ and $S_1 = \mathrm{Gr}_{\mathrm{O}}(1, F^{\natural}) \backslash S_2$. The assignment $L \mapsto L + F^{\natural}_1/F^{\natural}_1$ defines a vector bundle $S_1 \to \mathrm{Gr}_{\mathrm{O}}(1, F^{\natural}_1)$ with fiber isomorphic to $F^{\natural}_2$. Therefore, we have

$$\begin{aligned} \chi(\tilde{B}_{x,F}) &= \chi(\tilde{B}^{\natural}_F) = \chi(\mathrm{Gr}_{\mathrm{O}}(1, F^{\natural})) \\ &= \chi(S_1) + \chi(S_2) = \chi(F^{\natural}_2)\chi(\mathrm{Gr}_{\mathrm{O}}(1, F^{\natural}_1)) + \chi(\mathrm{Gr}(1, F^{\natural}_2)) \\ &= |F^{\natural}_1| + \frac{(-1)^{|F^{\natural}_1|} - 1}{2} + |F^{\natural}_2| = |F| + \frac{(-1)^{\mathrm{Rank}(x)} - 1}{2}. \end{aligned}$$

This finishes the proof. □

We will need a stronger lemma than Lemma 4.1.4 as follows.

LEMMA 4.1.5. *Let* $(x, F) \in T^*\mathrm{Gr}_{\mathrm{Sp}}(|F|, v)$. *Set*

$$B_{x,F} = \{U \in \mathrm{Gr}(|F| - 1, v) | U \subset F, x(U^{\perp}) \subseteq U\}.$$

Then $\chi(B_{x,F}) = |F/x(F^{\perp})| + \frac{(-1)^{r_{x,F}} - 1}{2}$, *with* $r_{x,F} = \mathrm{Rank}(x|_{x(F^{\perp})^{\perp}/x(F^{\perp})})$.

PROOF. Let $\tilde{B}_{x|_{x(F^{\perp})^{\perp}/x(F^{\perp})}, F/x(F^{\perp})}$ be the variety defined in Lemma 4.1.4 with the underlying symplectic space being $(x(F^{\perp})^{\perp}/x(F^{\perp})$. The assignment $U \mapsto U/x(F^{\perp})$ defines an isomorphism $B_{x,F} \cong \tilde{B}_{x|_{x(F^{\perp})^{\perp}/x(F^{\perp})}, F/x(F^{\perp})}$. Then the lemma follows by applying Lemma 4.1.4 to $\tilde{B}_{x|_{x(F^{\perp})^{\perp}/x(F^{\perp})}, F/x(F^{\perp})}$. □

4.2. Rank one case in orthogonal setting. In this section, we preserve the setting in Section 1.2. We further assume that $\varepsilon = 1$. Let $\mathrm{Gr}_{\mathrm{O}}(m, v)$ and $T^*\mathrm{Gr}_{\mathrm{O}}(m, v)$ be the orthogonal analogues of $\mathrm{Gr}_{\mathrm{Sp}}(m, v)$ and $T^*\mathrm{Gr}_{\mathrm{Sp}}(m, v)$, respectively.

Let $(x, F) \in T^*\mathrm{Gr}_{\mathrm{O}}(|F|, v)$. The underlying symmetric bilinear form descends to a non-degenerate symmetric bilinear form on $F^{\perp}/F$. It thus descends to a possibly singular symmetric bilinear form on $\ker(x|_{F^{\perp}})/F$. Let $M_{x,F}$ be the associated symmetric matrix with respect to a fixed basis in $\ker(x|_{F^{\perp}})/F$. Let

$$d_{x,F} = \mathrm{Rank}(M_{x,F}). \tag{27}$$

Of course, $d_{x,F}$ is invariant under the choices of bases in $\ker(x|_{F^{\perp}})/F$.

LEMMA 4.2.1. *Let $(x, F) \in T^*\mathrm{Gr}_{\mathrm{O}}(|F|, v)$ and $(x, F') \in T^*\mathrm{Gr}_{\mathrm{O}}(|F'|, v)$. Suppose that $F \overset{1}{\subset} F'$. If $x(F^{\perp}) = x(F'^{\perp})$, then $d_{x,F} = d_{x,F'} + 2$. If $x(F^{\perp}) \neq x(F'^{\perp})$, then $d_{x,F} = d_{x,F'}$.*

PROOF. Let us fix a decomposition $F^{\perp} = F'^{\perp} \otimes \langle \beta \rangle$ and $F' = F \oplus \langle \alpha \rangle$ such that $(\alpha, \beta) = 1$. If $x(F^{\perp}) = x(F'^{\perp})$, then $\ker(x|_{F^{\perp}}) \overset{1}{\supset} \ker(x|_{F'^{\perp}})$ by dimension comparison. And there exists $\delta \in F'^{\perp}$ such that $x(\beta) = x(\delta)$. In this setup, we have

$$\ker(x|_{F^{\perp}}) = \ker(x|_{F'^{\perp}}) \oplus \langle \beta - \delta \rangle.$$

Since the induced symmetric form on $F^{\perp}/F$ is nondegenerate, there is $\delta^* \in F^{\perp}/F$ such that

$$(\delta, \delta^*) = 1, (\delta, \alpha) = 0, (\beta, \delta^*) = 0.$$

(The first two conditions are easily satisfied and if the last one is not then we replace δ^* by $\delta^* - (\beta, \delta^*)\alpha$.) Fix a basis, say B, of $F'^{\perp}$ containing α, $\alpha + \delta^*$ as the first two terms such that the basis elements are orthogonal to $\beta - \delta$ via Gram-Schmidt. Then the basis B together with $\beta - \delta$ is a basis of $F'^{\perp}$. Then the matrices $M_{x,F}$ and $M_{x,F'}$ with respect to the basis $\beta - \delta \sqcup B$ is of the form

$$M_{x,F} = \left[\begin{array}{cc|c} * & 1 & 0 \\ 1 & 0 & 0 \\ \hline 0 & 0 & M_{x,F'} \end{array}\right]$$

So we have $d_{x,F} = d_{x,F'} + 2$ readily.

If $x(F^{\perp}) \neq x(F'^{\perp})$, then $\ker(x|_{F^{\perp}}) = \ker(x|_{F'^{\perp}})$ and there exists $\gamma \in x(F^{\perp}) - x(F'^{\perp})$ such that $x(\beta) = \gamma$. This implies that $\alpha \perp \ker(x|_{F^{\perp}})$, and hence we have $d_{x,F} = d_{x,F'}$. The lemma is thus proved. □

LEMMA 4.2.2. *Let $(x, F) \in T^*\mathrm{Gr}_{\mathrm{O}}(|F|, v)$. Set*

$$A^{\varnothing}_{x,F} = \{\tilde{F} \in \mathrm{Gr}_{\mathrm{O}}(|F| + 1, v) | F \subset \tilde{F}, x(\tilde{F}) = 0\}.$$

Then $\chi(A^{\varnothing}_{x,F}) = |F^{\perp}/F| - |x(F^{\perp})| + \frac{(-1)^{d_{x,F}} - 1}{2}$.

PROOF. The proof goes in exactly the same as the proof of Lemma 4.1.1. We have

$$A^{\varnothing}_{x,F} \cong \mathrm{Gr}_{\mathrm{O}}(1, \ker(x|_{F^{\perp}})/F).$$

Fix a decomposition $\ker(x|_{F^\perp})/F) = W_1 \oplus W_2$ such that the restriction of the form on $\ker(x|_{F^\perp})/F$ to W_1 is non-degenerate and trivial on W_2. Note that $|W_1| = d_{x,F}$ in (27). The orthogonal Grassmannian $\mathrm{Gr}_{\mathrm{O}}(1, \ker(x|_{F^\perp})/F)$ admits a partition

$$\mathrm{Gr}_{\mathrm{O}}(1, \ker(x|_{F^\perp})/F) = X_1 \sqcup X_2,\ X_1 = \{L \in \mathrm{Gr}_{\mathrm{O}}(1, \ker(x|_{F^\perp})/F) | L \in W_2\}.$$

Let $\pi_{W_1} : \ker(x|_{F^\perp})/F \to W_1$ be the canonical projection. Then the assignment $L \mapsto \pi_{W_1}(L)$ defines a vector bundle $X_2 \to \mathrm{Gr}_{\mathrm{O}}(1, W_1)$ of fiber isomorphic to W_2. So the Euler characteristic of $\mathrm{Gr}_{\mathrm{O}}(1, \ker(x|_{F^\perp})/F)$ can be computed as follows.

$$\begin{aligned}\chi(\mathrm{Gr}_{\mathrm{O}}(1, \ker(x|_{F^\perp})/F)) &= \chi(X_1) + \chi(X_2) = |W_2| + \chi(W_2)\chi(\mathrm{Gr}_{\mathrm{O}}(1, W_1)) \\ &\overset{(1)}{=} |W_2| + |W_1| + \frac{(-1)^{|W_1|} - 1}{2} \\ &= |\ker(x|_{F^\perp})/F| + \frac{(-1)^{d_{x,F}} - 1}{2} \\ &= |F^\perp/F| - |x(F^\perp)| + \frac{(-1)^{d_{x,F}} - 1}{2}.\end{aligned} \tag{28}$$

The lemma is thus proved. □

LEMMA 4.2.3. *Let* $(x, F) \in T^*\mathrm{Gr}_{\mathrm{O}}(|F|, v)$. *Set*

$$B^{\varnothing}_{x,F} = \{U \in \mathrm{Gr}(|F| - 1, v) | U \subset F, x(U^\perp) \subseteq U\}.$$

Then $\chi(B^{\varnothing}_{x,F}) = |F/x(F^\perp)|$.

PROOF. The proof is the same as the proof of Lemma 4.1.5, and Lemma 4.1.4. Instead of x_2 be symmetric in the proof of Lemma 4.1.4, we have x_2 being anti-symmetric in this case. So we have the formula. □

4.3. Defining relations in the n =odd case. In this section, the integer n is odd. Recall that we define the $\mathbf{G}_{v,\varepsilon}$-invariant functions $\mathbf{e}_i^{\mathfrak{c}}$ and $\mathbf{h}_i^{\mathfrak{c}}$, for all $1 \le i \le n-1$, on $Z_{n,v,\varepsilon}$ to be the respective restrictions of $\mathbf{e}_{i,\varepsilon}$ and $\mathbf{h}_{i,\varepsilon}$ to $Z_{n,v,\varepsilon}$. Recall $\mathbf{S}_{n,v,\varepsilon} = \langle \mathbf{e}_i^{\mathfrak{c}}, \mathbf{h}_i^{\mathfrak{c}} | 1 \le i \le n-1\rangle$ is the subalgebra of $\mathscr{C}_{\mathbf{G}_{v,\varepsilon}}(Z_{n,v,\varepsilon})$ generated by $\mathbf{e}_i^{\mathfrak{c}}, \mathbf{h}_i^{\mathfrak{c}}$.

We now verify that the generators $\mathbf{e}_i^{\mathfrak{c}}$ and $\mathbf{h}_i^{\mathfrak{c}}$ satisfy the relations in (15). By using the condition $|F_i| + |F_{\theta(i)}| = |V|$, we clearly have

LEMMA 4.3.1. $\mathbf{h}_i^{\mathfrak{c}} + \mathbf{h}_{\theta(i)}^{\mathfrak{c}} = 0$, *for all* $1 \le i \le n-1$.

Recall $(a_{ij})_{1 \le i,j \le n-1}$ is the Cartan matrix of type A_{n-1} defined by $a_{ij} = 2\delta_{ij} - \delta_{i,j+1} - \delta_{i,j-1}$.

LEMMA 4.3.2. *We have* $[\mathbf{h}_i^{\mathfrak{c}}, \mathbf{e}_j^{\mathfrak{c}}] = (a_{ij} - a_{\theta(i),j})\mathbf{e}_j^{\mathfrak{c}}$ *for all* $1 \le i, j \le n-1$.

PROOF. We assume that $\varepsilon = -1$. By definition, we have

$$[\mathbf{h}_i^{\mathfrak{c}}, \mathbf{e}_j^{\mathfrak{c}}](x, F, F') = \big((|F_{i+1}| - |F'_{i+1}|) - 2(|F_i| - |F'_i|) + (|F_{i-1}| - |F'_{i-1}|)\big)\, \mathbf{e}_j^{\mathfrak{c}}(x, F, F'),$$

for all $(x, F, F') \in Z_{n,v,\varepsilon}$. It is thus enough to show that if $\mathbf{e}_j^{\mathfrak{c}}(x, F, F') = 1$, then

$$(|F_{i+1}| - |F'_{i+1}|) - 2(|F_i| - |F'_i|) + (|F_{i-1}| - |F'_{i-1}|) = a_{ij} - a_{i,\theta(j)}. \tag{29}$$

The equality (29) can further be verified in the following 6 subcases.

(30) $$i = j, i \neq r, r+1,$$

(31) $$i = j = r \text{ or } r+1,$$

(32) $$i = \theta(j), i \neq r, r+1,$$

(33) $$i = r, j = r+1 \text{ or } i = r+1, j = r,$$

(34) $$i \neq j, \theta(j), i = j \pm 1 \text{ or } \theta(j) \pm 1,$$

(35) $$i \neq j, \theta(j), j \pm 1, \theta(j) \pm 1.$$

For subcase (30), we have $a_{ij} = 2$ and $a_{i,\theta(j)} = 0$. On the other hand, if $\mathbf{e}^{\mathfrak{c}}_j(x, F, F') = 1$, then $|F_i| - |F'_i| = -1$, $|F_{i+1}| - |F'_{i+1}| = |F_{i-1}| - |F'_{i-1}| = 0$. This yields (29). For subcase (31), we have $a_{ij} = 2$ and $a_{i,\theta(j)} = -1$. If $i = j = r$ and $\mathbf{e}^{\mathfrak{c}}_j(x, F, F') = 1$, then $|F_i| - |F'_i| = -1$, $|F_{i+1}| - |F'_{i+1}| = 1$ and $|F_{i-1}| - |F'_{i-1}| = 0$. If $i = j = r+1$ and $\mathbf{e}^{\mathfrak{c}}_j(x, F, F') = 1$, then $|F_i| - |F'_i| = -1$, $|F_{i+1}| - |F'_{i+1}| = 0$ and $|F_{i-1}| - |F'_{i-1}| = 1$. Thus the equality (29) holds in this subcase. For subcase (32), we get $a_{ij} = 0$ and $a_{i,\theta(j)} = 2$. If $\mathbf{e}^{\mathfrak{c}}_j(x, F, F') = 1$, then $|F_i| - |F'_i| = 1$, $|F_{i+1}| - |F'_{i+1}| = 0$ and $|F_{i-1}| - |F'_{i-1}| = 0$. The equality (29) holds in this subcase. For subcase (33), it reads $a_{i,j} = -1$ and $a_{i,\theta(j)} = 2$. If $i = \theta(j) = r$, $j = r+1$ and $\mathbf{e}^{\mathfrak{c}}_j(x, F, F') = 1$, then $|F_i| - |F'_i| = 1$, $|F_{i+1}| - |F'_{i+1}| = -1$ and $|F_{i-1}| - |F'_{i-1}| = 0$. If $i = r+1$, $j = r$ and $\mathbf{e}^{\mathfrak{c}}_j(x, F, F') = 1$, then $|F_i| - |F'_i| = 1$, $|F_{i+1}| - |F'_{i+1}| = 0$ and $|F_{i-1}| - |F'_{i-1}| = -1$. Thus the quality (29) holds in this subcase. For subcase (34), there is $(a_{ij}, a_{i,\theta(j)}) = (-1, 0)$ or $(a_{ij}, a_{i,\theta(j)}) = (0, -1)$. If $(a_{ij}, a_{i,\theta(j)}) = (-1, 0)$ and $\mathbf{e}^{\mathfrak{c}}_j(x, F, F') = 1$, then $|F_i| - |F'_i| = 0$, $|F_{i+1}| - |F'_{i+1}| = -1$ $|F_{i-1}| - |F'_{i-1}| = 0$ or $|F_i| - |F'_i| = 0$, $|F_{i+1}| - |F'_{i+1}| = 0$, $|F_{i-1}| - |F'_{i-1}| = -1$. If $(a_{ij}, a_{i,\theta(j)}) = (0, -1)$ and $\mathbf{e}^{\mathfrak{c}}_j(x, F, F') = 1$, then $|F_i| - |F'_i| = 0$, $|F_{i+1}| - |F'_{i+1}| = 1$, $|F_{i-1}| - |F'_{i-1}| = 0$ or $|F_i| - |F'_i| = 0$, $|F_{i+1}| - |F'_{i+1}| = 0$, $|F_{i-1}| - |F'_{i-1}| = 1$. The quality (29) holds in this case as well. Finally for subcase (35), we have $a_{ij} = a_{i,\theta(j)} = 0$ and $|F_i| - |F'_i| = |F_{i+1}| - |F'_{i+1}| = |F_{i-1}| - |F'_{i-1}| = 0$, and hence we have (29). This finishes the proof of the lemma for $\varepsilon = -1$.

The case of $\varepsilon = 1$ can be proved in a similar way, and is left to the reader. □

LEMMA 4.3.3. *If* $a_{ij} = 0$, *then* $[\mathbf{e}^{\mathfrak{c}}_i, \mathbf{e}^{\mathfrak{c}}_j] = \delta_{i,\theta(j)} \mathbf{h}^{\mathfrak{c}}_i$.

PROOF. Let $\varepsilon = -1$. Assume that $i \neq \theta(j)$. By definition, the value of $[\mathbf{e}^{\mathfrak{c}}_i, \mathbf{e}^{\mathfrak{c}}_j]$ at a point $(x, F, F') \in Z_{n,v,\varepsilon}$ is zero unless the point satisfies the following condition.

(36) $$F_k = F'_k, \forall k \neq i, j, \theta(i), \theta(j), F_i \subset F'_i, |F'_i/F_i| = 1, F_j \subset F'_j \text{ and } |F'_j/F_j| = 1.$$

Under this condition, the value of $\mathbf{e}^{\mathfrak{c}}_i \mathbf{e}^{\mathfrak{c}}_j$ at (x, F, F') is equal to $\chi(X)$ where

$$X = \{\tilde{F} \in \mathcal{F}^{\mathfrak{c}}_{n,d} | (x, \tilde{F}) \in T^* \mathcal{F}^{\mathfrak{c}}_{n,d}, F_k = \tilde{F}_k, \forall k \neq i, j, \theta(i), \theta(j), \tilde{F}_i = F'_i, \tilde{F}_j = F_j\}.$$

Observe that X consists of exactly one point, hence the value of $\mathbf{e}^{\mathfrak{c}}_i \mathbf{e}^{\mathfrak{c}}_j$ is 1, provided that (36) holds. A similar argument shows that $\mathbf{e}^{\mathfrak{c}}_j \mathbf{e}^{\mathfrak{c}}_i$ is equal to 1, under the assumption (36). Hence we have $[\mathbf{e}^{\mathfrak{c}}_i, \mathbf{e}^{\mathfrak{c}}_j] = 0$ if $a_{ij} = 0$ and $i \neq \theta(j)$.

Next, we assume that $i = \theta(j)$. Since $a_{ij} = 0$, we have $i \neq r, r+1$. If the point $(x, F, F') \in Z_{n,v,\varepsilon}$ satisfies that $F \neq F'$. Then the evaluation of the $\mathbf{e}^{\mathfrak{c}}_i \mathbf{e}^{\mathfrak{c}}_j$ at the point $(x, F, F') \in Z_{n,v,\varepsilon}$ is zero unless

(37) $$F_k = F'_k, \forall k \neq i, \theta(i), |F_i + F'_i/F_i| = 1 \text{ and } |F_i + F'_i/F'_i| = 1.$$

In this case, it is equal to 1. On the other hand, the evaluation of the $\mathbf{e}^{\mathfrak{c}}_j\mathbf{e}^{\mathfrak{c}}_i$ at the point $(x, F, F') \in Z_{n,v,\varepsilon}$ with $F \neq F'$ is zero unless

$$F_k = F'_k, \forall k \neq i, \theta(i), |F_i/F_i \cap F'_i| = 1, \text{ and } |F'_i/F_i \cap F'_i| = 1. \tag{38}$$

In this case $\mathbf{e}^{\mathfrak{c}}_j\mathbf{e}^{\mathfrak{c}}_i(x, F, F') = 1$. It is well-known that the conditions (37) and (38) are equivalent. Hence we have $[\mathbf{e}^{\mathfrak{c}}_i, \mathbf{e}^{\mathfrak{c}}_j](x, F, F') = 0 = \mathbf{h}^{\mathfrak{c}}_i(x, F, F')$ if $F \neq F'$.

If $(x, F, F') \in Z_{n,v,\varepsilon}$ satisfies $F = F'$, then the value of $\mathbf{e}^{\mathfrak{c}}_i\mathbf{e}^{\mathfrak{c}}_j$ at (x, F, F') is equal to $\chi(X)$ where

$$X = \{\tilde{F}_i | F_i \subset \tilde{F}_i \subseteq F_{i+1} \cap x^{-1}(F_{i-1}), |\tilde{F}_i/F_i| = 1\}, \tag{39}$$

while the value of $\mathbf{e}^{\mathfrak{c}}_j\mathbf{e}^{\mathfrak{c}}_i$ at the same point is equal to $\chi(Y)$ where

$$Y = \{\hat{F}_i | F_{i-1} + x(F_{i+1}) \subseteq \hat{F}_i \subset F_i, |F_i/\hat{F}_i| = 1\}. \tag{40}$$

So the value of the bracket $[\mathbf{e}^{\mathfrak{c}}_i, \mathbf{e}^{\mathfrak{c}}_j]$ at this point is equal to

$$\begin{aligned}\chi(X) - \chi(Y) &= \left|\frac{F_{i+1} \cap x^{-1}(F_{i-1})}{F_i}\right| - \left|\frac{F_i}{F_{i-1} + x(F_{i+1})}\right| \\ &= -2|F_i| + |F_{i+1} \cap x^{-1}(F_{i-1})| + |F_{i-1} + x(F_{i+1})|.\end{aligned} \tag{41}$$

Now the endomorphism x induces an endomorphism, say x_{i+1}, on F_{i+1} with the property that $x_{i+1}(F_{i-1}) \subseteq F_{i-1}$. Hence by passage to quotient, we obtain an endomorphism, say $\bar{x}$, on F_{i+1}/F_{i-1}. By definition, one has $F_{i-1} + x(F_{i+1})/F_{i-1} = \mathrm{im}\bar{x}$ and $F_{i-1} \cap x^{-1}(F_{i-1})/F_{i-1} = \ker \bar{x}$. Hence, there is an equality:

(42)

$$\begin{aligned}|F_{i+1} \cap x^{-1}(F_{i-1})| + |F_{i-1} + x(F_{i+1})| &= |\mathrm{im}\,\bar{x}| + |\ker \bar{x}| + 2|F_{i-1}| \\ &= |F_{i+1}/F_{i-1}| + 2|F_{i-1}| = |F_{i+1}| + |F_{i-1}|.\end{aligned}$$

So one has

$$\chi(X) - \chi(Y) = |F_{i+1}| - 2|F_i| + |F_{i-1}|,$$

which is exactly the value of $\mathbf{h}^{\mathfrak{c}}_i$ at (x, F, F'). Hence we have $[\mathbf{e}^{\mathfrak{c}}_i, \mathbf{e}^{\mathfrak{c}}_j] = \mathbf{h}^{\mathfrak{c}}_i$ if $i = \theta(j)$. The lemma is thus followed for $\varepsilon = -1$. The case $\varepsilon = 1$ can be proved in a similar way. □

LEMMA 4.3.4. *If $a_{ij} = -1$ and $i \neq \theta(i), \theta(j)$, then* $\mathbf{e}^{\mathfrak{c}}_i\mathbf{e}^{\mathfrak{c}}_i\mathbf{e}^{\mathfrak{c}}_j - 2\mathbf{e}^{\mathfrak{c}}_i\mathbf{e}^{\mathfrak{c}}_j\mathbf{e}^{\mathfrak{c}}_i + \mathbf{e}^{\mathfrak{c}}_j\mathbf{e}^{\mathfrak{c}}_i\mathbf{e}^{\mathfrak{c}}_i = 0$.

PROOF. We show the lemma in the case $\varepsilon = -1$. The case $\varepsilon = 1$ can be shown in a similar way, and will be omitted.

By assumption $j = i \pm 1$. We treat the case $j = i+1$. The value of the left-hand side, denoted by S_{ij}, at a point $(x, F, F') \in Z_{n,v,\varepsilon}$ is zero unless it satisfies that

$$F_k = F'_k, \forall k \neq i, j, \theta(i), \theta(j), F_i \subset F'_i, |F'_i/F_i| = 2, F_j \subset F'_j, |F'_j/F_j| = 1. \tag{43}$$

In this case, the value of S_{ij} at (x, F, F') is equal to $\chi(X_1) - 2\chi(X_2) + \chi(X_3)$ where

$$\begin{aligned}X_1 &= \{\tilde{F}_i | F_i \subset \tilde{F}_i \subset F'_i, |\tilde{F}_i/F_i| = 1\} \cong \mathrm{Gr}(1, 2), \\ X_2 &= \{\tilde{F}_i | F_i + x(F'_{i+1}) \subseteq \tilde{F}_i \subset F'_i, |F'_i/\tilde{F}_i| = 1\}, \\ X_3 &= \{\tilde{F}_i | F_i \subset \tilde{F}_i \subset F'_i, |\tilde{F}_i/F_i| = 1, x(F'_{i+1}) \subseteq F_i\}.\end{aligned}$$

(Here $\tilde{F}_i$ runs over all subspaces in V.) If $x(F'_{i+1}) \subseteq F_i$, then $X_2 \cong \mathrm{Gr}(1, 2) \cong X_3$ and so

$$S_{ij}(x, F, F') = 2 - 2.2 + 2 = 0.$$

If $x(F'_{i+1}) \not\subseteq F_i$, then $X_3 = \emptyset$ and $F_i \subsetneq F_i + x(F'_{i+1}) \subseteq F'_i$. The fact $|F'_{i+1}/F_i| = 1$ implies that $|F_i + x(F'_{i+1})| = |F_i| + 1$. So X_2 consists of one point, i.e., $\tilde{F}_i = F_i + x(F'_{i+1})$. In this case we have

$$S_{ij}(x, F, F') = 2 - 2.1 + 0 = 0.$$

If $j = i - 1$, then the evaluation of S_{ij} at a point (x, F, F') is still zero unless it satisfies (43). Under the condition (43), it is equal to $\chi(Y_1) - 2\chi(Y_2) + \chi(Y_3)$ where

$$\begin{aligned} Y_1 &= \{\tilde{F}_i | F_i \subset \tilde{F}_i \subset F'_i, |\tilde{F}_i/F_i| = 1, x(F'_i) \subseteq F_{i-1}\}, \\ Y_2 &= \{\tilde{F}_i | F_i \subset \tilde{F}_i \subseteq F'_i \cap x^{-1}(F_{i-1}), |\tilde{F}_i/F_i| = 1\}, \\ Y_3 &\cong \mathrm{Gr}(1, 2). \end{aligned}$$

If $x(F'_i) \subseteq F_{i-1}$, then we have $Y_1 \cong \mathrm{Gr}(1, 2) \cong Y_2$ and so $S_{ij}(x, F, F') = 2-2.2+2 = 0$. Otherwise $Y_1 = \emptyset$ and Y_2 consists of a point, and so $S_{ij}(x, F, F') = 0-2.1+2 = 0$. This proves the lemma. □

LEMMA 4.3.5. *If $a_{ij} = -1$ and $i = \theta(j)$, then* $\mathbf{e}^{\mathfrak{c}}_i \mathbf{e}^{\mathfrak{c}}_i \mathbf{e}^{\mathfrak{c}}_j - 2\mathbf{e}^{\mathfrak{c}}_i \mathbf{e}^{\mathfrak{c}}_j \mathbf{e}^{\mathfrak{c}}_i + \mathbf{e}^{\mathfrak{c}}_j \mathbf{e}^{\mathfrak{c}}_i \mathbf{e}^{\mathfrak{c}}_i = -4\mathbf{e}^{\mathfrak{c}}_i$.

We summarize the above results in the following proposition, before proving Lemma 4.3.5.

PROPOSITION 4.3.6. *When n is odd, the relations in (15) hold in $\mathbf{S}_{n,v,\varepsilon}$.*

The remaining part of this section is devoted to the proof of Lemma 4.3.5. We assume that $\varepsilon = -1$. The case $a_{ij} = -1$ and $i = \theta(j)$ happens only when n is odd and in this case (i, j) is either $(r, r + 1)$ or $(r + 1, r)$.

Assume that $(i, j) = (r, r + 1)$. We denote the left-hand side of the quality in the lemma by S_{ij} as usual. The value of S_{ij} at a point $(x, F, F') \in Z_{n,v,\varepsilon}$ is zero unless $F_k = F'_k$ for all $k \neq i, \theta(i)$ and $|F_i| + 1 = |F'_{i+1}|$. In this case, we get

(44) $S_{ij}(x, F, F') = \chi(X) - 2\chi(Y) + \chi(Z)$, where

$$\begin{aligned} X &= \{(\tilde{F}_i, \hat{F}_i) | F_i \overset{1}{\subset} \tilde{F}_i \overset{1}{\subset} \hat{F}_i \overset{1}{\supset} F'_i, x(\hat{F}_i) \subseteq F_{i-1}, \hat{F}_i \text{ is isotropic}\}, \\ Y &= \{(\tilde{F}_i, \hat{F}_i) | F_i \overset{1}{\subset} \tilde{F}_i \overset{1}{\supset} \hat{F}_i \overset{1}{\subset} F'_i, x(\tilde{F}_i) \subseteq F_{i-1}, \tilde{F}_i \text{ is isotropic}, x(\hat{F}_i^{\perp}) \subseteq \hat{F}_i\}, \\ Z &= \{(\tilde{F}_i, \hat{F}_i) | F_i \overset{1}{\supset} \tilde{F}_i \overset{1}{\subset} \hat{F}_i \overset{1}{\subset} F'_i, F_{i-1} \subseteq \tilde{F}_i, x(\tilde{F}_i^{\perp}) \subseteq \tilde{F}_i\}. \end{aligned}$$

(Here again $\tilde{F}_i, \hat{F}_i$ are vector subspaces in V.) The analysis is broken into whether F_i is contained in F'_i or not.

We treat the case when $F_i \not\subset F'_i$, which consists of five subcases. We set $\check{G}_i = F_i \cap F'_i$ and $\hat{G}_i = F_i + F'_i$.

<u>Case 1</u>. If $F_i \not\subset F'_i$ and $|\check{G}_i| < |F_i| - 1$, then the sets X, Y and Z are empty. So the value of S_{ij} at (x, F, F') is zero.

<u>Case 2</u>: Assume that $F_i \not\subset F'_i$, $|\check{G}_i| = |F_i| - 1$, $\hat{G}_i$ is isotropic and $x(\check{G}_i^{\perp}) \subseteq \check{G}_i$. Then we have $X \cong Y \cong Z$ so that the S_{ij} in (44) is equal to 0. (Indeed in this case X, Y and Z are isomorphic to $\mathrm{Gr}(1, 2)$.) We now provide explicit isomorphisms of X, Y and Z. The explicit isomorphism from X to Y is defined by

$$(\tilde{F}_i, \hat{F}_i) \mapsto (\tilde{F}_i, \tilde{F}_i \cap F'_i), \quad \forall (\tilde{F}_i, \hat{F}_i) \in X.$$

The inverse of the above map is given by $(\tilde{F}_i, \hat{F}_i) \mapsto (\tilde{F}_i, \tilde{F}_i + F'_i)$ for all $(\tilde{F}_i, \hat{F}_i) \in Y$.

In the same spirit, the map $(\tilde{F}_i, \hat{F}_i) \mapsto (F_i \cap \hat{F}_i, \hat{F}_i)$ defines an isomorphism from Y to Z, so we have the desired isomorphisms $X \cong Y \cong Z$.

Case 3: Assume that $F_i \not\subset F'_i$, $|\check{G}_i| = |F_i| - 1$, $\hat{G}_i$ is not isotropic and $x(\check{G}_i^{\perp}) \subseteq \check{G}_i$. Then we have $X = \emptyset$ and $Z \cong \mathrm{Gr}(1,2)$. By assumption, there exist $u^* \in F_i$ and $v_1^*, v_2^* \in F'_i$ such that $F_i = \check{G}_i \oplus \langle u^* \rangle$, $F'_i = \check{G}_i \oplus \langle v_1^*, v_2^* \rangle$ such that $\langle u^*, v_1^* \rangle = 1$ and $\langle u^*, v_2^* \rangle = 0$. (Here $\langle A \rangle$ is the subspace spanned by the vectors in A.) Consider $\mathring{F}_i = \check{G}_i + \langle u^*, v_2^* \rangle$. This is the only isotropic subspace contained in $\hat{G}_i$ and containing $\check{G}_i$. From this we see that $Y = \{(\mathring{F}_i, \mathring{F}_i \cap F'_i)\}$, and hence $\chi(Y) = 1$. This implies that S_{ij} is equal to zero in this case.

Case 4: Assume that $F_i \not\subset F'_i$, $|F_i \cap F'_i| = |F_i| - 1$, $\hat{G}_i$ is isotropic and $x(\check{G}_i^{\perp}) \not\subseteq \check{G}_i$. The latter condition implies that Z is empty, which we still have $X \cong \mathrm{Gr}(1,2)$ and in particular $\chi(X) = 2$ and $\chi(Z) = 0$. Hence it remains to show that $\chi(Y) = 1$ for S_{ij} to be zero at the point (x, F, F') in this case. The pair (F_i, F'_i) admits the following decomposition.

$$F_i = \check{G}_i \oplus \langle u^* \rangle, \qquad F'_i = \check{G}_i \oplus \langle v_1^*, v_2^* \rangle,$$
$$F_i^{\perp} = \hat{G}_i^{\perp} \oplus \langle v_1, v_2 \rangle, \qquad F_i'^{\perp} = \hat{G}_i^{\perp} \oplus \langle u \rangle,$$

such that $(u, u^*) = (v_1, v_1^*) = (v_2, v_2^*) = 1$, $(u, v_1) = (u, v_2) = (v_1, v_2) = 0$. The condition $x(\check{G}_i^{\perp}) \not\subseteq \check{G}_i$ means that the pair (F_i, F'_i) satisfies one of the following three conditions.

$$x(F_i^{\perp}) \subseteq \check{G}_i, x(F_i'^{\perp}) \not\subseteq F_i,$$
$$x(F_i^{\perp}) \not\subseteq F'_i, x(F_i'^{\perp}) \not\subseteq F_i,$$
$$x(F_i^{\perp}) \not\subseteq F'_i, x(F_i'^{\perp}) \subseteq \check{G}_i.$$

If the pair (F_i, F'_i) satisfies the first condition, then we can further impose the following condition on the above decomposition of (F_i, F'_i):

$$x(u) = v_1^*.$$

Set $\mathring{F}_i = \check{F}_i + \langle v_1^* \rangle = \check{F}_i + x(F'^{\perp})$ so that $\mathring{F}_i^{\perp} = \hat{F}_i^{\perp} \oplus \langle u, v_2 \rangle$. By definition, we have $x(\mathring{F}_i^{\perp}) \subseteq \mathring{F}_i$. This implies that $(F \cap \mathring{F}_i, \mathring{F}_i) \in Y$, and Y contains only this point. Hence $\chi(Y) = 1$.

If the pair (F_i, F'_i) satisfies the second condition, then the argument for the above case still works once we show that $x(v_2) = 0 \bmod \check{F}_i$. Otherwise $x(v_2) = u^* + y$ with $y \in \check{F}_i$ and then the contradiction:

$$1 = (u, u^*) = (u, x(v_2) - y) = (u, x(v_2)) = -(x(u), v_2) = -(v_1^*, v_2) = 0.$$

So we have $\chi(Y) = 1$ in this case as well.

If the pair (F_i, F'_i) satisfies the third condition, then we can impose the condition $x(v_1) = 0 \bmod \check{F}_i$ and $x(v_2) = u^* \bmod \check{F}_i$ on the decomposition of the pair (F_i, F'_i). Set $\mathring{F}_i = \check{F}_i + \langle v_2^* \rangle$ so that $\mathring{F}_i^{\perp} = \hat{F}_i^{\perp} + \langle u, v_1 \rangle$. It is now straightforward to see that $(F \cap \mathring{F}_i, \mathring{F}_i)$ is the unique point in Y and hence $\chi(Y) = 1$. This finishes the analysis of the case $F_i \not\subseteq F'_i$.

Case 5: Assume that $F_i \not\subset F'_i$, $|F_i \cap F'_i| = |F_i| - 1$, $\hat{G}_i$ is not isotropic and $x(\check{G}_i^{\perp}) \not\subseteq \check{G}_i$. Then the sets X and Z are empty. We shall show that Y is empty as well. By assumption, the pair (F_i, F'_i) admits the following decomposition. There exist vectors $u^* \in F_i$, $v_1^*, v_2^* \in F_i'$ and $v_2 \in F_i^\perp$ such that $(u^*, v_1^*) = 1$, $(u^*, v_2^*) = 0$, $(v_2, v_2^*) = 1$, $(v_2, v_1^*) = (v_2, u^*) = 0$ and

$$F_i = \check{G}_i \oplus \langle u^* \rangle, \qquad F_i' = \check{G}_i \oplus \langle v_1^*, v_2^* \rangle,$$
$$F_i^\perp = \hat{G}_i^\perp \oplus \langle u^*, v_2 \rangle, \qquad F_i'^\perp = \hat{G}_i^\perp \oplus \langle v_1^* \rangle.$$

The condition $x(\check{G}_i^\perp) \not\subseteq \check{G}_i$ is equivalent to $x(v_2) = av_1^* + bv_2^* \mod \check{G}_i$ for some $(a, b) \in \mathbb{C}^2 \backslash \{(0, 0)\}$. If $b \neq 0$, then

$$1 = (v_2, v_2^*) = (v_2, b^{-1}x(v_2) + b^{-1}av_1^*) = (v_2, x(v_2)) \in (F_i^\perp, F_i) = \{0\},$$

a contradiction. Hence $b = 0$ so $x(v_2) = av_1^*$. But

$$1 = (u^*, v_1^*) = (u^*, ax(v_2)) \in (F_i, F_i) = \{0\},$$

a contradiction. Hence $a = 0$. This implies that $x(\check{G}_i^\perp) \subseteq \check{G}_i$, yet another contradiction. This shows that Case 5 is vacuous. This finishes the analysis of the case $F_i \not\subset F_i'$, i.e.,

$$S_{ij}(x, F, F') = 0, \quad \text{if } F_i \not\subset F_i'. \tag{45}$$

The remaining case is when $F_i \subset F_i'$. Without loss of generality, we can assume that $n = 3$. Then

$$X \cong \mathrm{Gr}(1, 2) \times A_{x,F'}, Z \cong \mathrm{Gr}(1, 2) \times B_{x,F},$$

where $A_{x,F'}$ and $B_{x,F}$ are from Section 4.1. We have

$$Y = Y_1 \sqcup Y_2, \ Y_1 = \{F = \hat{F}\}, Y_2 = Y - Y_1.$$

It is clear that

$$Y_1 = A_{x,F}, Y_2 = B_{x,F'} - \{F\}.$$

By Lemmas 4.1.1 and 4.1.5, we have

$$\begin{aligned} S_{ij}(F, F') &= \chi(X) - 2\chi(Y) + \chi(Z) \\ &= 2(\chi(A_{x,F'}) - \chi(A_{x,F})) - 2(\chi(B_{x,F'}) - \chi(B_{x,F})) + 2 \\ &= -4 + (-1)^{r_{x,F}} - (-1)^{r_{x,F'}} \\ &= -4, \end{aligned} \tag{46}$$

where the last equality is due to Lemma 4.1.3 if $x(F^\perp) \neq x(F'^\perp)$.

The lemma follows from (45) and (46) in the case $(i, j) = (r, r + 1)$.

The case when $(i, j) = (r+1, r)$ follows from the previous case, once we observe that the pullback of the isomorphism $(x, F, F') \mapsto (x, F', F)$ defines an isomorphism on $\mathbf{S}_{n,v,\varepsilon}$, which swaps $\mathbf{e}_i^\mathfrak{c}$ and $\mathbf{e}_j^\mathfrak{c}$. This finishes the proof for the case $\varepsilon = -1$.

For $\varepsilon = 1$, the proof is the same as the case $\varepsilon = -1$, except that in the proof of (46). But by Lemmas 4.2.2 and 4.2.3, we have

$$\begin{aligned} S_{ij}(F, F') &= 2(\chi(A^\varnothing_{x,F'}) - \chi(A^\varnothing_{x,F})) - 2(\chi(B^\varnothing_{F'}) - \chi(B^\varnothing_F)) + 2 \\ &= -4 + (-1)^{d_{x,F'}} - (-1)^{d_{x,F}} \\ &= -4, \end{aligned}$$

where the last equality is due to Lemma 4.2.1 that $d_{x,F}$ and $d_{x,F'}$ have the same parity. This finishes the proof for the $\varepsilon = 1$ case. The proof of Lemma 4.3.5 is complete.

4.4. Defining relations for n even. In this section, we assume that n is even. We can repeat the analysis in previous section on this case, with some modifications. In particular, we have $F_r^{\perp} = F_r$ for any flag in $\mathcal{F}_{n,v,\varepsilon}$ if $(v,\varepsilon) \neq (\mathrm{odd}, 1)$, and F_r is maximal isotropic if $(v,\varepsilon) = (\mathrm{odd}, 1)$.

Recall $\mathbf{S}_{n,v,\varepsilon}$ is the algebra generated by $\mathbf{e}_i^{\mathfrak{c}}$ and $\mathbf{h}_i^{\mathfrak{c}}$ for $1 \leq i \leq n$, which are the respective restrictions of $\mathbf{e}_{i,\varepsilon}$ and $\mathbf{h}_{i,\varepsilon}$ in $\mathrm{U}_{n,\varepsilon}$.

We now give a conceptual explanation of the generator $\mathbf{e}_r^{\mathfrak{c}}$. There is a natural embedding

$$Z_{n,v,\varepsilon} \to Z_{n+1,v,\varepsilon},$$

which is obvious by definition for $(v,\varepsilon) = (\mathrm{odd}, 1)$ and defined by inserting an extra copy of the r-step in the flags involved if $(v,\varepsilon) \neq (\mathrm{odd}, 1)$. Specifically, for $(v,\varepsilon) \neq (\mathrm{odd}, 1)$, any triple (x, F, F') gets sent to the triple $(x, \breve{F}, \breve{F}')$ where $\breve{F}_i = F_i$ for all $1 \leq i \leq r$, $\breve{F}_i = F_{i-1}$ for all $r+1 \leq i \leq n+1$ and $\breve{F}'$ is defined in a similar manner. The embedding induces an imbedding of algebras:

$$\mathscr{C}_{\mathbf{G}_{v,\varepsilon}}(Z_{n,v,\varepsilon}) \to \mathscr{C}_{\mathbf{G}_{v,\varepsilon}}(Z_{n+1,v,\varepsilon}).$$

Since n is even here, the subalgebra $\mathbf{S}_{n+1,v,\varepsilon}$ of $\mathscr{C}_{\mathbf{G}^{\mathfrak{c}}}(Z_{n+1,v,\varepsilon})$ is the one in Section 4.3. To avoid confusion, let us put a superscript $\breve{}$ on the generators of $\mathbf{S}_{n+1,v,\varepsilon}$, i.e., $\breve{\mathbf{e}}_i$ and $\breve{\mathbf{h}}_i^{\mathfrak{c}}$.

Recall the idempotent element $\breve{\mathbf{j}}_{r,\varepsilon}$ in $\mathbf{S}_{n+1,v,\varepsilon}$ from (25). There is an imbedding of algebras

$$\mathbf{S}_{n,v,\varepsilon} \to \mathbf{S}_{n+1,v,\varepsilon}, \tag{47}$$

such that (26) holds.

The following lemma provides a conceptual explanation of the generator $\mathbf{e}_r^{\mathfrak{c}}$.

LEMMA 4.4.1. *Via the identification* $\mathscr{C}_{\mathbf{G}_{v,\varepsilon}}(Z_{n,v,\varepsilon}) \cong \breve{\mathbf{j}}_{r,\varepsilon}\mathscr{C}_{\mathbf{G}_{v,\varepsilon}}(Z_{n+1,v,\varepsilon})\breve{\mathbf{j}}_{r,\varepsilon}$, *we have*

$$\mathbf{e}_r^{\mathfrak{c}} = (\breve{\mathbf{e}}_{r+1}^{\mathfrak{c}}\breve{\mathbf{e}}_r^{\mathfrak{c}} + \breve{\mathbf{h}}_r^{\mathfrak{c}})\breve{\mathbf{j}}_{r,\varepsilon}.$$

PROOF. Let $\varepsilon = -1$. Without loss of generality, we can assume that $n = 2$. Let $\breve{Z}_{n,v,\varepsilon}$ be the image of the imbedding $Z_{n,v,\varepsilon} \to Z_{n+1,v,\varepsilon}$. It is easy to see that $\breve{\mathbf{h}}_r^{\mathfrak{c}}$ is supported on $\breve{Z}_{n,v,\varepsilon}$ and moreover, $\breve{\mathbf{h}}_r^{\mathfrak{c}}(x, F, F') = -d\delta_{F,F'}$ for all $(x, F, F') \in \breve{Z}_{n,v,\varepsilon}$. On the other hand, it is also true that the function $\breve{\mathbf{e}}_{r+1}^{\mathfrak{c}}\breve{\mathbf{e}}_r^{\mathfrak{c}}\breve{\mathbf{j}}_{r,\varepsilon}$ is supported on $\breve{Z}_{n,v,\varepsilon}$ and moreover for any $(x, F, F') \in \breve{Z}_{n,v,\varepsilon}$ such that $F \neq F'$, the value $\breve{\mathbf{e}}_{r+1}^{\mathfrak{c}}\breve{\mathbf{e}}_r^{\mathfrak{c}}\breve{\mathbf{j}}_{r,\varepsilon}(x, F, F')$ is equal to zero unless $|F_r \cap F_r'| = |F_r| - 1$. In the latter case, it is equal to 1. It remains to show that for all triples (x, F, F') in $\breve{Z}_{n,v,\varepsilon}$,

$$\breve{\mathbf{e}}_{r+1}^{\mathfrak{c}}\breve{\mathbf{e}}_r^{\mathfrak{c}}\breve{\mathbf{j}}_{r,\varepsilon}(x, F, F) = d + \frac{(-1)^{\mathrm{rank}(x)} - 1}{2}. \tag{48}$$

By definition, the left-hand side of the equality (48) is equal to $\chi(S)$ where S is the variety of all subspaces $\hat{F} \subseteq V$ such that $F_r \overset{1}{\subset} \hat{F}$ and $x(\hat{F}) \subseteq \hat{F}^{\perp}$. Applying Lemma 4.1.5, we have

$$\chi(S) = |F| + \frac{(-1)^{\mathrm{rank}(x)} - 1}{2}, \tag{49}$$

where the last equality is by (1). The equality (48) follows immediately from (49). The lemma is thus proved for the case $\varepsilon = -1$.

The proof of the $\varepsilon = 1$ case is similar and omitted. This finishes the proof. □

With Lemma 4.4.1 in hand, the following lemma can be shown by the same argument as the Proof of [**Li21**, Lemma 4.4.2] by using Lemma 4.3.5.

LEMMA 4.4.2. *If $a_{ij} = -1$ and $i = \theta(i)$, then we have*

$$\mathbf{e}^{\mathfrak{c}}_j \mathbf{e}^{\mathfrak{c}}_j \mathbf{e}^{\mathfrak{c}}_i - 2\mathbf{e}^{\mathfrak{c}}_j \mathbf{e}^{\mathfrak{c}}_i \mathbf{e}^{\mathfrak{c}}_j + \mathbf{e}^{\mathfrak{c}}_i \mathbf{e}^{\mathfrak{c}}_j \mathbf{e}^{\mathfrak{c}}_j = 0. \tag{50}$$

$$\mathbf{e}^{\mathfrak{c}}_i \mathbf{e}^{\mathfrak{c}}_i \mathbf{e}^{\mathfrak{c}}_j - 2\mathbf{e}^{\mathfrak{c}}_i \mathbf{e}^{\mathfrak{c}}_j \mathbf{e}^{\mathfrak{c}}_i + \mathbf{e}^{\mathfrak{c}}_j \mathbf{e}^{\mathfrak{c}}_i \mathbf{e}^{\mathfrak{c}}_i = \mathbf{e}^{\mathfrak{c}}_j. \tag{51}$$

PROPOSITION 4.4.3. *When n is even, the relations in (16) hold in $\mathbf{S}_{n,v,\varepsilon}$.*

PROOF. The first equality is clear. The proof of the second equality is similar to its counter part in Proposition 4.3.6. Specifically, it breaks into the following 6 subcases, which can be verified in a similar manner as above. The details are left to the reader.

$$\begin{aligned}
&i = j \neq r,\\
&i = j = r,\\
&i = \theta(j) \neq r,\\
&i \neq j, \theta(j), i, j \neq r,\\
&i = j \pm 1 \text{ or } \theta(j) \pm 1,\\
&i \neq j, \theta(j), j \pm 1, \theta(j) \pm 1.
\end{aligned}$$

The third equality is proved in a way similar to its counterpart in Proposition 4.3.6. The last two is due to Lemma 4.4.2. The proposition is thus proved. □

REMARK 4.4.4. After the work was done, the author received a preprint of [**X22**] on the action of $\mathfrak{sl}^\theta_n$ on the space of constructible functions on the central fiber of the σ-quiver varieties in [**Li19**]. In light of Propositions 4.3.6 and 4.4.3, one can recover the main result in [**X22**] by a standard argument.

We end this section with an example of the semicanonical basis in $\mathbf{S}_{n,v,\varepsilon}$.

EXAMPLE 4.4.5. Let $\varepsilon = -1$. For $n = 2$, $A = \begin{bmatrix} d-1 & 1 \\ 1 & d-1 \end{bmatrix}$, we have

$$s_{A,\varepsilon} = \begin{cases} (\mathbf{e}^{\mathfrak{c}}_r + 1)\mathbf{1}_{\mathrm{ro}(A),\varepsilon}, & \text{if } d \text{ is odd,} \\ \mathbf{e}^{\mathfrak{c}}_r \mathbf{1}_{\mathrm{ro}(A),\varepsilon}, & \text{if } d \text{ is even.} \end{cases}$$

References

[BKLW] Huanchen Bao, Jonathan Kujawa, Yiqiang Li, and Weiqiang Wang, *Geometric Schur duality of classical type*, Transform. Groups **23** (2018), no. 2, 329–389, DOI 10.1007/s00031-017-9447-4. MR3805209

[BW18a] Huanchen Bao and Weiqiang Wang, *A new approach to Kazhdan-Lusztig theory of type B via quantum symmetric pairs* (English, with English and French summaries), Astérisque **402** (2018), vii+134. MR3864017

[BW18b] Huanchen Bao and Weiqiang Wang, *Canonical bases arising from quantum symmetric pairs*, Invent. Math. **213** (2018), no. 3, 1099–1177, DOI 10.1007/s00222-018-0801-5. MR3842062

[BBD82] A. A. Beĭlinson, J. Bernstein, and P. Deligne, *Faisceaux pervers* (French), Analysis and topology on singular spaces, I (Luminy, 1981), Astérisque, vol. 100, Soc. Math. France, Paris, 1982, pp. 5–171. MR751966

[BG99] Alexander Braverman and Dennis Gaitsgory, *On Ginzburg's Lagrangian construction of representations of* GL(n), Math. Res. Lett. **6** (1999), no. 2, 195–201, DOI 10.4310/MRL.1999.v6.n2.a7. MR1689209

[BLM90] A. A. Beilinson, G. Lusztig, and R. MacPherson, *A geometric setting for the quantum deformation of* GL_n, Duke Math. J. **61** (1990), no. 2, 655–677, DOI 10.1215/S0012-7094-90-06124-1. MR1074310

[CG97] Neil Chriss and Victor Ginzburg, *Representation theory and complex geometry*, Birkhäuser Boston, Inc., Boston, MA, 1997. MR1433132

[FLLLW] Zhaobing Fan, Chun-Ju Lai, Yiqiang Li, Li Luo, and Weiqiang Wang, *Affine flag varieties and quantum symmetric pairs*, Mem. Amer. Math. Soc. **265** (2020), no. 1285, v+123, DOI 10.1090/memo/1285. MR4080913

[FL15] Zhaobing Fan and Yiqiang Li, *Positivity of canonical bases under comultiplication*, Int. Math. Res. Not. IMRN **9** (2021), 6871–6931, DOI 10.1093/imrn/rnz047. MR4251292

[G91] Victor Ginzburg, *Lagrangian construction of the enveloping algebra* $U(\mathrm{sl}_n)$ (English, with French summary), C. R. Acad. Sci. Paris Sér. I Math. **312** (1991), no. 12, 907–912. MR1111326

[K14] Stefan Kolb, *Quantum symmetric Kac-Moody pairs*, Adv. Math. **267** (2014), 395–469, DOI 10.1016/j.aim.2014.08.010. MR3269184

[KP82] Hanspeter Kraft and Claudio Procesi, *On the geometry of conjugacy classes in classical groups*, Comment. Math. Helv. **57** (1982), no. 4, 539–602, DOI 10.1007/BF02565876. MR694606

[K02] Shrawan Kumar, *Kac-Moody groups, their flag varieties and representation theory*, Progress in Mathematics, vol. 204, Birkhäuser Boston, Inc., Boston, MA, 2002, DOI 10.1007/978-1-4612-0105-2. MR1923198

[Li10] Yiqiang Li, *Semicanonical bases for Schur algebras*, J. Algebra **324** (2010), no. 3, 347–369, DOI 10.1016/j.jalgebra.2010.03.019. MR2651340

[Li19] Yiqiang Li, *Quiver varieties and symmetric pairs*, Represent. Theory **23** (2019), 1–56, DOI 10.1090/ert/522. MR3900699

[Li21] Yiqiang Li, *Quasi-split symmetric pairs of* $\mathrm{U}(\mathfrak{sl}_n)$ *and Steinberg varieties of classical type*, Represent. Theory **25** (2021), 903–934, DOI 10.1090/ert/570. MR4329193

[LW18] Yiqiang Li and Weiqiang Wang, *Positivity vs negativity of canonical bases*, Bull. Inst. Math. Acad. Sin. (N.S.) **13** (2018), no. 2, 143–198. MR3792711

[LZ] Yiqiang Li and Jieru Zhu, *Quasi-split symmetric pairs of* $U(\mathfrak{gl}_N)$ *and their Schur algebras*, Nagoya Math. J. **245** (2022), 1–27, DOI 10.1017/nmj.2020.16. MR4413360

[L03] G. Lusztig, *Constructible functions on varieties attached to quivers*, Studies in memory of Issai Schur (Chevaleret/Rehovot, 2000), Progr. Math., vol. 210, Birkhäuser Boston, Boston, MA, 2003, pp. 177–223. MR1985727

[N94] Hiraku Nakajima, *Instantons on ALE spaces, quiver varieties, and Kac-Moody algebras*, Duke Math. J. **76** (1994), no. 2, 365–416, DOI 10.1215/S0012-7094-94-07613-8. MR1302318

[N98] Hiraku Nakajima, *Quiver varieties and Kac-Moody algebras*, Duke Math. J. **91** (1998), no. 3, 515–560, DOI 10.1215/S0012-7094-98-09120-7. MR1604167

[Sp87] T. A. Springer, *The classification of involutions of simple algebraic groups*, J. Fac. Sci. Univ. Tokyo Sect. IA Math. **34** (1987), no. 3, 655–670. MR927604

[St74] Robert Steinberg, *Conjugacy classes in algebraic groups*, Lecture Notes in Mathematics, Vol. 366, Springer-Verlag, Berlin-New York, 1974. Notes by Vinay V. Deodhar. MR352279

[X22] R. Xiong, *A geometric realization of symmetric pairs of type AIII*, arXiv:2202.05058, 2022.

Department of Mathematics, University at Buffalo, the State University of New York, Buffalo, New York 14260

Email address: yiqiang@buffalo.edu

SELECTED PUBLISHED TITLES IN THIS SERIES

837 **Yiqiang Li and Changlong Zhong, Editors,** Representation Theory and Flag Varieties, 2026

831 **Irena Swanson and Kei-ichi Watanabe, Editors,** Lectures on Tight Closure and Its Applications, 2025

830 **O. Bezushchak, Yu. Drozd, P. Etingof, V. Futorny, R. Grigorchuk, and E. Zelmanov, Editors,** Modern Algebra Volume 2, 2025

829 **O. Bezushchak, Yu. Drozd, P. Etingof, V. Futorny, R. Grigorchuk, and E. Zelmanov, Editors,** Modern Algebra Volume 1, 2025

828 **Hafedh Herichi, Maria Rosaria Lancia, Therese Basa Landry, Anna Rozanova-Pierrat, and Steffen Winter, Editors,** Fractal Geometry in Pure and Applied Mathematics, 2025

827 **Carmen Caprau, J. Scott Carter, Neslihan Gügümcü, and Sam Nelson, Editors,** Algebraic Structures in Knot Theory, 2025

826 **Ahmed Laghribi and André Leroy, Editors,** Algebraic Structures and Applications, 2025

825 **Saurabh Verma, María A. Navascués, and Amit Priyadarshi, Editors,** Advances in Dimension Theory, Fractal Functions and Measures, 2025

824 **Hanna E. Makaruk and Robert M. Owczarek, Editors,** Inverse Problems: In Memory of Professor Zbigniew Oziewicz, 2025

823 **Mikhail Bershtein, Anton Dzhamay, and Andrei Okounkov, Editors,** The Versatility of Integrability, 2025

822 **Ahmad Barhoumi, Roozbeh Gharakhloo, and Andrei Martinez-Finkelshtein, Editors,** Recent Developments in Orthogonal Polynomials, 2025

821 **Bogdan D. Suceavă, Adara M. Blaga, Marian Ioan Munteanu, Cezar Oniciuc, Shoo Seto, Gabriel-Eduard Vîlcu, and Lihan Wang, Editors,** Recent Advances in Differential Geometry and Related Areas, 2025

820 **Sudeb Mitra, Rakesh Kumar Parmar, Saminathan Ponnusamy, and Toshiyuki Sugawa, Editors,** Geometric Function Theory and Related Topics, 2025

819 **Howard S. Cohl, Roberto S. Costas-Santos, and Robert S. Maier, Editors,** Applications and q-Extensions of Hypergeometric Functions, 2025

818 **Howard S. Cohl, Roberto S. Costas-Santos, and Robert S. Maier, Editors,** Classical Hypergeometric Functions and Generalizations, 2025

817 **Mikhail Khovanov, Joshua Sussan, and Anton Zeitlin, Editors,** From Representation Theory to Mathematical Physics and Back, 2025

816 **Luca F. Di Cerbo and Laurenţiu G. Maxim, Editors,** Geometry and Topology of Aspherical Manifolds, 2025

815 **Daniel Orr and Joshua Jeishing Wen, Editors,** Macdonald Theory and Beyond, 2025

814 **Susan Montgomery and Siu-Hung Ng, Editors,** Quantum Groups, Hopf Algebras, and Applications, 2025

813 **Thomas Creutzig, Julia Plavnik, and David Ridout, Editors,** Quantum Symmetries, 2025

812 **Ross Geoghegan, Craig R. Guilbault, and Kim Ruane, Editors,** Topology at Infinity of Discrete Groups, 2025

811 **Krishnendu Gongopadhyay and Shashikant A. Katre, Editors,** Geometry, Groups and Mathematical Philosophy, 2025

810 **Robert J. Berman and Yanir A. Rubinstein, Editors,** Convex and Complex: Perspectives on Positivity in Geometry, 2025

809 **Fabrice Baudoin and Luca Rizzi, Editors,** New Trends in Sub-Riemannian Geometry, 2025

For a complete list of titles in this series, visit the AMS Bookstore at **www.ams.org/bookstore/conmseries/**.

FSC
www.fsc.org
MIX
Papier | Fördert gute Waldnutzung
FSC® C083411

Kolibri 360 GmbH
Ferdinand-Jühlke-Straße 7
99095 Erfurt, Deutschland
produktsicherheit@kolibri360.de